Earthquake Engineering for Structural Design

Earthquake Engineering
for
Structural Design

Victor Gioncu and Federico M. Mazzolani

Routledge
Taylor & Francis Group

LONDON AND NEW YORK

First published 2011 by Spon Press

2 Park Square, Milton Park, Abingdon, Oxon OX14 4RN
711 Third Avenue, New York, NY 10017, USA

Routledge is an imprint of the Taylor & Francis Group, an informa business

First issued in paperback 2017

British Library Cataloguing in Publication Data
A catalogue record for this book is available from the British Library

Library of Congress Cataloging in Publication Data
Gioncu, Victor.
Earthquake engineering for structural design / Victor Gioncu and Federico Mazzolani.
p. cm.
Includes bibliographical references and index.
I. Mazzolani, Federico M. II. Title.
Earthquakeresistant design.
624.1'762--dc22   2009052791

ISBN13: 978-0-415-46533-5 (hbk)
ISBN13: 978-1-138-11624-5 (pbk)

Contents

Contents   V

Preface   IX

Chapter 1 New Challenges in Seismic Design   1

　1 1 After the last strong earthquakes....................................... 1
　1.2 Global Urbanization and Impact of Earthquakes.................... 4
　1.3 Topics Involved in Seismic Design................................. 11
　1.4 Seismology versus Engineering Seismology........................ 12
　1.5 Earthquake Engineering versus Structural Engineering........... 15
　1.6 Issues for Seismic Design... 26
　1.8 International Activity... 27
　1.9 References... 29

Chapter 2 Living with Earthquakes   32

　2.1 When Earth Shakes 32
　2.2 Historical Earthquakes... 35
　2.3 Great Earthquakes of 20th Century............................ 38
　2.4 Great Earthquakes of the Beginning of 21st Century.......... 59
　2.5 References... 67

Chapter 3 Learning from Earthquakes   71

　3.1 Main Lessons after the Strong Earthquakes....................... 71
　3.2 Lessons for Earthquake Hazard.................................... 75
　3.3 Lessons for Construction Vulnerability........................... 82
　3.4 Lessons for Mitigation of Seismic Risk........................... 89
　3.5 Lessons for Seismic Design... 95
　3.6 References... 99

Chapter 4 Advances in Conception about Earthquakes   102

　4.1 From Engelados to GPS System.................................... 102
　4.2 Towards a New View in Seismology............................... 114
　4.3 Block Models and Seismicity...................................... 120
　4.4 Chaos Theory and Seismology 124
　4.5 Earthquake: Predicting the Unpredictable........................ 132
　4.6 Statistical Seismology and the Theory of
　　　　Multi-Source Data Fusion.. 136
　4.7 References.. 138

Chapter 5 Tectonic Plates and Faults   143

5.1 Plate Tectonics.. 143
5.2 Plate Boundary Types.. 144
5.3 Diffuse Zones and Fault Types.............................. 155
5.4 World's Main Plate Boundaries............................. 171
5.5 References.. 176

Chapter 6 Faults and Earthquakes   179

6.1 Anatomy of an Earthquake................................... 179
6.2 Crustal Interplate Earthquake............................... 183
6.3 Crustal Intraplate Earthquakes............................. 190
6.4 Intraslab (inslab) Earthquakes.............................. 205
6.5 World's Seismic Zones.. 209
6.6 References.. 212

Chapter 7 Earthquakes and Ground Motions   217

7.1 Ground Motions Evaluation and Earthquake Engineering.. 217
7.2 Source Characteristics... 220
7.3 Path Effects.. 227
7.4 Ground Response to Earthquakes........................... 236
7.5 Peculiar Features of Near-source Ground Motions.......... 249
7.6 Ground Motions Peculiarities of Crustal Intraplate
       Earthquakes... 275
7.7 Ground Motions Peculiarities of Intraslab Sources......... 283
7.8 References 285

Chapter 8 Ground Motions and Structures   295

8.1 Structure Influence on Ground Motions .... .295
8.2 Foundation Responses... 298
8.3 Structure Responses... 306
8.4 Developments of Materials for seismic-resistant structures 334
8.5 Developments of Steel Structural Systems 341
8.6 Soil-foundation-structure Interaction........................ 359
8.7 Seismic Vulnerability Factors................................ 366
8.8 References 377

Chapter 9 Advances in Seismic Design Methodologies   386

9.1 Challenges in Seismic Design................................ 386
9.2 Performance-based Seismic Design: Implication of Owners,
       Users and Society... 390

9.3 Development of Multi-level Base Seismic Design................ 392
9.4 Response Spectra as Representation of Ground Motions.......... 404
9.5 Seismic Analysis Procedures.. 418
9.6 Behavior of Non-structural Components during Earthquake...... 441
9.7 References.. ..446

Chapter 10 Challenges for Next Code Generation of Seismic Codes   451

10.1 Developments of Seismic Design Codes......................... 451
10.2 Progresses in Seismic Design Codes..................................... 456
10.3 Challenges for New Design Approaches......................... 466
10.4 Characteristics of Earthquakes in Function of Source............. 471
10.5 Structural Response to Interplate Crustal Earthquakes:
     Near-source Ground Motions.................................... 476
10.6 Structural Response to Intraplate Crustal Earthquakes:
     Low-to Moderate Ground Motions for Crustal Fractures......... 494
10.7 Structural Response to Intraslab Deep Earthquakes:
     Long Duration Earthquakes................................. 512
10.8 Recommendations for Developing Simple but Reliable New
     Code Provisions.. 519
10.9 References... 528

Appendix : Glossary   533
A.1 Engineering Seismology... 533
A.2 Earthquake Engineering... 543
A.3 References... 552

Index   553

Preface

Earthquakes were the cause of more than 1.5 million deaths worldwide during the 20th Century. During the beginning of the 21st Century the number of deaths was about half a million. This is an unacceptable finding, because earthquakes can no longer be regarded as natural disasters, since the main cause of this huge number of casualties is the inadequate seismic resistance of the building stock, lifelines and industry, which could be avoided. Earthquakes do not kill people, but the building collapse can do it. It is an unbelievable situation that, after a century of research works, each strong earthquake brings new surprises and creates the situation that new lessons have to be learnt. After a series of devastating earthquakes during the last years of the past century (1994 Northridge, 1995 Kobe, 1999 Kocaeli and Taiwan earthquakes), it has been recognized by society that both seismic hazard and risk have to be reassessed.

Important progress was made in the last period, but many problems remain unsatisfactorily solved. Therefore, now is the right moment to analyze the level of current knowledge and to identify the challenges for future research works and for the next code generation. This is the main intention of this book. The progress in understanding and controlling the complex phenomena of the earthquake production can be analyzed both from scientific and practical points of view.

From the scientific point of view, the main effort must be directed towards the inner understanding of the complex phenomenon of an earthquake. Some new fundamental disciplines, developed in the last decades, must be deeply studied. The most important among them is the *Earth Science*, which explores the different processes and transformations produced in the evolution of the Earth and looks at how it is likely to evolve in the future. The Earth Science, more than any other science, is the embodiment of the emerging new sciences, where the focus is on *Complex Non-linear Systems*: Evolutionary and Self-Organized Dissipative Systems, Bifurcation, Catastrophes and Chaos Theories. The Earth surface can be considered as a complex of self-organized systems, consisting in tectonic plates and mantle, the equilibrium being assured by the mantle convection. The earthquakes represent a sudden destruction of this equilibrium. Considering the *Chaos Theory*, an essential discipline for Complex Non-linear Systems, which underlines the importance of small perturbation in the initial conditions (*Butterfly effect*), it is possible to understand the great difficulties in the prediction of the characteristics of the next earthquakes. These aspects show the limits of the probabilistic analysis in case of earthquake. Due to small differences in the initial conditions, for the same source, each new event is very different from the already occurred one, breaking all probabilistic rules.

From the practical point of view, the main problem is to determine the progress in research works and to transfer it into practice. Today a ***Structural Design*** has grown within the new multi-disciplinary development fields of ***Engineering Seismology*** and ***Earthquake Engineering.*** Engineering Seismology is a branch of Seismology, having the purpose to transfer the seismological new knowledge to Structural Engineers, in such a way that it should be used in practice. Earthquake Engineering, having the task to solve the analysis of structure under seismic actions, is a branch of the more general field of Structural Engineering Science. It must develop specific methodologies for analyzing the effects of seismic actions on structures, very different from those used in case of other actions, like dead, live, wind, snow, etc. loads. ***Seismic Design***, collecting the data given by Engineering Seismology in terms of seismic actions and using the methodologies proposed by Earthquake Engineering, performs a complex examination of the constructions, including numerical analyses, design aspects and solutions for details.

The basic concepts of today's Earthquake Engineering were born almost 70 years ago, when the knowledge about the seismic actions and structural response were rather poor. Many initial concepts were changed due to the progress in research works, but additional improvements still remain to be concretized for reaching a satisfactory level of seismic design. Today, the transfer of the new knowledge to practice is not completely done and, therefore, many new important developments are predictable in the near future.

The challenge for a proper seismic design is to solve the balance between earthquake demand and structural capacity. The earthquake demand corresponds to the effects of earthquake on the structure and depends on the ground motion modelling. Structural capacity is the structural ability to resist these effects without failure. Looking to the development of Engineering Seismology and Earthquake Engineering, it can be clearly observed that the major effort of researchers was directed towards the structural response analysis. Therefore, the structural response can be predicted fairly confidently, but these achievements remain without real effects, if the evaluation of seismic actions is not accurate and rather doubtful. In fact, the prediction of ground motions is still far from a satisfactory level, due to both the complexity of the seismic phenomena and the communication lack between seismologists and engineers. So, the reduction of uncertainties in the ground motion modelling is now the main challenge in seismic design.

This target is possible only if the impressive progress in Seismology will be transferred by Engineering Seismology into Seismic Design. It is very clear that any progress is impossible without considering the new amount of knowledge, recently cumulated in Seismology about the different source types: interplate, intraplate and intraslab. There are big differences between the ground motions generated by these sources so that ignorance of these aspects can be considered a shortcoming in seismic design. In this context, the Engineering Seismology is now paying more concern to establish the differences in the main characteristics of the sources. At the same time, the task of Earthquake Engineering is to take more care about the structural response for these different ground motions. A very important conclusion for the structural designer must be the following: the seismic design of structures must be done in function of the ground motions types, including all essential features corresponding to each type (near-source effects, pulse periods, duration, ductility demands, etc). The design based only on the earthquake magnitude and the corresponding spectrum does not assure a proper and safe seismic design.

The preparation of this book was a quite difficult activity for the authors, due to the necessity to approach the domain of Seismology, which is not very familiar to structural engineers. But they were fully aware that, without this effort, the existing gap between Engineering Seismology and Earthquake Engineering should be very far to be filled.

The book is divided in to three distinct parts. In the first (Chapters 1 to 4), general aspects are considered: new challenges in seismic design, great historical and recent earthquakes, main lessons after these earthquakes and the advances in conceptions about earthquakes. The second part, corresponding to Engineering Seismology (Chapters 5 to 7), is devoted to identify the cause producing earthquakes: tectonic plates, faults, earthquakes and ground motions. The last part, corresponding to Earthquake Engineering (Chapters 8 to 10), presents the problems of Seismic Design: ground motions and structures, advances in seismic design methodologies, progress in codification and challenges for next code generation.

After the completion of the manuscript of this book, a very important earthquake occurred on 6 April 2009 in the Abruzzo region of central Italy, where more than 1500 buildings collapsed and the balance of victims was 297. The city of L'Aquila was the most damaged, being situated in the epicentral area. This city is very rich in many cultural treasures, being known as the city of 99 churches, 99 squares and 99 fountains. The magnitude of the earthquake was M 5.8, framing in the category of moderate earthquakes, but due to the near-source effects, soil conditions, as well as the presence of a lot of historical buildings without any upgrading against earthquakes, the damage was very huge. It is interesting to notice that this earthquake, because of its characteristics, frames exactly in the typologies, which are forecast in this book. In fact, the earthquakes that occur in central Italy belong to the category of interplate collision type, being the results of the collision between the Adria and the Tyrrhenian micro-plates. A result of this collision was the formation of the Apennine Mountains, which are located along the collision fault. The sources of these earthquakes are situated in the superficial crustal zone, the depth of source being at around 10 km. The earthquake is produced by a normal fault, but very important ground motions amplification is produced in the L'Aquila zone by the presence of a sedimentary basin, filled by lacustrine sediments, with a maximum depth of about 250 m in the center of L'Aquila. All the characteristics of this earthquake belong to the category of near-source effects: pulse, amplification in the field of short periods affecting mainly rigid structures (the masonry structures were more damaged than the reinforced concrete frames), importance of vertical component (higher than horizontal component) in the zone near to the epicenter, influence of soil conditions, short significant duration, many aftershocks until the fault equilibrium is restored, etc. The recorded accelerations were about 0.35g for the stations in the vicinity of L'Aquila city, while in the central area, these figures were double, even more than

0.70g. This earthquake is an example of the way in which the effects of an earthquake must be examined by structural engineers, to analyze the damage to the building stock, accompanied by a very careful examination of the local seismological conditions.

The effective approach presented in the book was further confirmed at the beginning of 2010 when there was string of significant earthquakes.

The Haiti earthquake was a strong magnitude M 7.0, with the epicenter near Port-au-Prince, Haiti's capital. The earthquake occurred on 12 January 2010. The Haitian Government reported that an estimated 230,000 people died, 300,000 injured and 1,000,000 made homeless. They also estimated that 250,000 residences and 30,000 commercial buildings had collapsed or were severely damaged. Important notable buildings were significantly damaged or destroyed, including the Presidential Palace, the National Assembly building, and the Port-au-Prince Cathedral. Why did this earthquake transform into such catastrophic event? The answer relates to two geological and social conditions. The island of Hispaniola is seismically very active and has a rich history of destructive earthquakes due to the fact that it is situated on the boundary between Carribean and North American tectonic plates. In the zone of Haiti, the fault is characterized by a crustal strike-slip system, which produces very destructive earthquakes, following a long duration of ground motions. The shallow depth of source (13 km) in the vicinity of Port-au-Prince characterized the earthquake as near-source, where the ground motions were considerably amplified. In addition, these effects were amplified by poor building conditions, mainly due to the lack of seismic codes and control of executions. So, the majority of old and new buildings fit into the category of non-engineering structures, which, in addition to the near-source effects, explains the devastating characteristics of this earthquake.

The Chilean earthquake occurred off the pacific coast of Chile on February 27, rating a magnitude of M 8.8 (the 7th strong earthquake recorded in the world). The epicenter of the earthquake was offshore at a 35 km depth, 115 km from Chile's second largest city, Conception, and 335 km from Santiago, the capital. A moderate tsunami also occurred. The Chilean officials confirmed the number of death, 279. At least 500,000 homes were estimated to be damaged and some buildings collapsed in Conception and Santiago. It is interesting to explain why, in spite of very high intensity of this earthquake, the damage was not as large as in Haiti. The earthquake took place along the boundary between the Nazca and South American tectonic plates, a thrust-faulting mechanism caused by the subduction of the Nazca plate beneath the South American plate. The Chilian coast has suffered many megathrust earthquakes along this plate boundary, including the strongest earthquake ever measured of M 9.5. The epicenter was sufficiently far enough from the main cities, that the near-source effects were not so strong as in the Haiti earthquake. In addition, the Chilean seismic code is one of the best in existence and the control of execution is very well organized.

The Yushu-China (Qinghai province) earthquake struck on 14 April 2010, and registered a magnitude of M 6.9. According to official sources, 2200 people died, and 1434 were severely injured. Qinghai lies in the Northeastern part of the Tibetan Plateau, which was formed due to the collision of Indian with the Eurasian plates, along of one the most active fault lines. Extensive structural damage was reported, mainly due to the non-engineering structures. The importance of this earthquake is related to the relevance of the earthquake potential of Tibetan Plateau, where many other strong earthquakes have occurred (see 2008 Sichuan earthquake).

All figures relating to the aforementioned earthquakes are derived from contemporary press reports via Wikipedia.

The book is especially designed for the specialists who are mainly interested in understanding the complex phenomena characterizing the earthquake formation as well as the consequences of different earthquake types on the structural behaviors. Today, they must be fully aware that any progress in Structural Seismic Design cannot develop without considering the important progress in Seismology. This is the challenge for the next code generation.

The book is based on a selection of the results of many papers, reports, conference proceedings and also internet information. At the same time, the authors' experience in earthquake research and seismic design of structures has played an important part in the book elaboration. Being aware that this is an attempt, maybe the first, to present the problems of both Engineering Seismology and Earthquake Engineering in a unitary way, for trying to provide the bases of a proper Structural Design, this book is predisposed to receive criticisms from the specialists belonging to both fields. The authors are open to accept all constructive critics.

Victor Gioncu   Federico M. Mazzolani

May, 2010

Chapter 1

New Challenges in Seismic Design

1.1 AFTER THE LAST STRONG EARTHQUAKES

1.1.1 Impact of Northridge and Kobe Earthquakes on Seismic Design

Earthquakes represent the largest potential source of casualties and damage for inhabited areas due to natural hazard. Although the location varies, the pattern is the same: an earthquake strikes without warning, leaving cities in rubble and killing tens to hundreds of thousands of people. Worldwide during the 20th Century, there were ten earthquakes killing more than 50,000 people and over 100 earthquakes killing more than 1000 people (FEMA 383, 2003). Every year, something like five thousand to ten thousand people die during earthquakes worldwide. The 1976 Tangshan-China (magnitude M 8.0), the worst earthquake in recent times, killed over 600,000. Among these terrifying data, the moderate 1994 Northridge in Los Angeles (magnitude M 6.7), which killed 60 people, and 1995 Kobe in Japan (magnitude M 6.9), which killed 5600 people, seemed to be relatively insignificant. Nevertheless, these two earthquakes have changed the direction of earthquake engineering research throughout the World (Blakeborough, 2002). Two main reasons produced this crucial change.

The first reason lies not in the number of dead, but in their economic costs. Each event was a direct hit by a moderate earthquake on a dense built-up area. In Northridge, around 15,000 buildings had to be demolished, resulting in a total loss ranging from $15bn to $40bn. In Kobe, 180,000 buildings were destroyed or seriously damaged, the repair costs being estimated in the range of $90bn to $150bn. Each earthquake set a record loss for natural disasters both for the USA and Japan, respectively. Following these earthquakes, it was immediately apparent that the old principles for seismic design had to change. Whereas the previous principles had been primarily oriented to safeguard buildings against collapse, the new and more refined rules are devoted to reduce the damage costs, by keeping the non-structural elements and the structures in an acceptable damage level. So, the principles of *Performance Based Seismic Design* were set up.

The second reason lies in the fact that both earthquakes struck very densely populated areas. Only looking to these experiences, it is sufficient to appreciate the catastrophic potential of even a moderate earthquake when its epicenter is situated in an urban area. The differences in damage costs can be justified in the fact that the epicenter of the Kobe earthquake was located beneath a highly urbanized region, while the one of the Northridge earthquake was positioned beneath the Northern edge of Los Angeles. But this enormous damage can be considered minor in comparison with the potential losses in big cities situated in seismic areas, like Mexico City, San Francisco, Tokyo, etc. A recent calculation model predicted losses of $100bn to $200bn for a Californian earthquake and over $1000bn for a Tokyo earthquake, if the 1923 earthquake were to happen today with the same magnitude (Gioncu and Mazzolani, 2002). The growing world urban population increases the number of cities located in seismic areas over some earthquake sources. The Northridge and Kobe earthquakes generate a new research direction, the effects of *near-source earthquakes* on the structures situated in epicentral areas.

Therefore, the structural response can be predicted fairly confidently, but these achievements remain without real effects if the accurate determining of the seismic actions is doubtful (Gioncu and Mazzolani, 2006). The *elimination of the gap* between the advances in Engineering Seismology and Earthquake Engineering is a major challenge of the new approaches in Seismic Design.

1.1.2 Ways to Develop These New Challenges

One of the main goals in Seismic Design is improving the understanding of earthquakes and their effects. This activity comprises the range of disciplines from Seismology and Engineering Seismology to Earthquake Engineering. The following aspects must be considered to achieve this target (FEMA 383, 2003):

Improve earthquake monitoring. Seismic hazard identification and risk assessment are critical components of earthquake mitigation strategy. Under this goal, a monitoring system, based on the regional networks of instrumented stations, on the use of satellite-based observations (GPS monitoring stations) and associated data centers, has been developed (FEMA 383, 2003). The most useful data for seismic design are obtained from the seismic stations, by means of recorded accelerations, velocities or displacements. The realization of an efficient network of stations is a very difficult task. The station site and spacing requirements for seismological researches are very different from those for earthquake engineering purposes. Seismologists, interested in the study of the Earth structure, want their stations to be located at quiet sites, as far away from any human activity as possible. On the other hand, earthquake engineers want instruments in the built environment of urban areas. Strong motion recordings useful for engineering purposes must be within 20 to 50 km from the earthquake-rupture (depending on site soil characteristics). A regional seismic network with stations spaced by several hundreds of kilometers does not yield the information about the near-source strong ground motions, required for earthquake engineering purposes. Studies indicate that a station spacing of about 1 km or less is necessary in order to reduce the observed variances in strong motions. One of the main problems of strong earthquake monitoring refers to the uninhibited or lightly populates areas, where the implantation of a dense station network is impossible. Therefore, the problem of seismic monitoring is not technical, but a political and financial problem (Lee, 2002).

Improve understanding of earthquake occurrence. In the last decades, Seismology has made significant progresses in understanding the basic physics of earthquakes. Together with these progresses, modern technologies such as Global Positioning System (GPS) allow seismologists to forecast the overall long-term seismic activity. Yet, earthquakes continue to be a major threat to our society, as we have witnessed during the seismic events of the last century and the beginning of the new one. A major difficulty is due to the fact that an earthquake involves a large number of elementary processes, so that, even if we understand the physics governing each elementary process, the complex interaction between them makes accurate forecasts of earthquakes very difficult (Kanamori, 2001). The Plate Tectonics and Continental Drift Theories are a starting point for understanding the forces within the Earth causing earthquakes. Three major types of plate boundaries are recognized: divergent, convergent or transformed, depending on whether the plates move away, toward, or laterally passing one to another, respectively. Ninety percent of the world's earthquakes occur along plate boundaries. The earthquake types depend on the boundary types. For instance, *subduction* occurs where one plate converges toward another plate, moves beneath it, and plunges as much as several hundred kilometers into the Earth's interior. In function of depth, two different types of earthquake occur: interplate crustal and intraslab deep earthquake, with very different physical characteristics. The remaining 10 percent occurs in areas away from present plate boundaries, being the results of ancient plate boundary configurations. The seismic movements of these earthquakes are very different from the ones produced at present boundaries. All these aspects must be considered for the understanding of earthquake occurrence.

Improve fundamental knowledge of earthquake effects. Among the most important contributions to reducing earthquake losses, there are both the improving of understanding and the modelling earthquake effects, including the source properties, the wave propagation from the source to site and the local conditions characterizing the site. This task implies the development of methods to generate synthetic seismograms for the expected future earthquakes, incorporating improved understanding of the rupture process and information about the fault type and the properties of the surrounding earth's crust. At the same time, the identification of the parameters of ground motions causing soil liquefaction, land sliding and damage of structures (such as peak acceleration, ground velocity and displacement, shaking duration, spectral content, etc.). The effects of near-source are of primary importance for urbanized areas. These seismograms must accurately simulate these parameters used by structural engineers in the seismic design process.

Improve the seismic design of structures. A new facet of Earthquake Engineering research concerning the seismic structural response is based on the reliance of integrated experimentation, theory, databases and model-based computer simulations. Under this objective the priorities refer to improving the understanding of behaviour and collapse mechanisms of various classes of structures under different earthquake types, in order to establish new methodologies for performance-based earthquake engineering. These new methodologies must consider different design philosophies for structures situated in low to moderate and strong seismic areas. The objective implies also developing new materials, new technologies and new structural systems for earthquake resistant structures.

Start development of next generation performance-based codes. The goal of these activities assures the ability to reduce seismic vulnerability of structural systems, learning from the lessons given by the last strong earthquakes and from the remarkable knowledge development during the last decades in the frame of Seismology. The main considered task is to transfer the accumulated results from the academic research works to the design practice, filling the existing gap between these to activities, disseminating upgrade guidelines and new codes, cooperating with professional associations, promoting education for practicing professionals.

1.2 GLOBAL URBANIZATION AND IMPACT OF EARTHQUAKES

1.2.1 Urban Revolution

The urban revolution is the process which produces the transformation of villages or small cities into large, socially complex, civilized urban centers, as a result of the population increasing. The total urbanization of the society is an inevitable process.

The world population has exploded, and continues to explode, both in number and global distribution during the past few centuries. Between 100 BC and about 1600 AD global population doubled from perhaps 300 million to more than 600 million. In the following 200 years, improvements in medicine, living conditions and a dramatic reduction of mortality rates, resulted in a second doubling in the world population to 1200 million. Between 1800 and 1950 the global population doubled for the third time, reaching 2500 million. The fourth doubling occurred in less than 40 years bringing global population in 1990 to more than 5000 million. In 2000 the world population reached 6100 million. Although the rate of population increase seems now to be reduced, a doubling of population in the next 50 years is predicted (Bilham, 1995, 2000).

But the most important observation is the global urbanization produced in the last decades. If in 1800 only 2% of the world's population was urbanized, in 1950, the percentage increased to 30%, while in 2000, this value reached 47%. In 2008 it is expected that more than half the population will be living in urban areas, and by 2030, more than 60% will live in these areas (UN-Habitat, 2005). The increase is not uniform in the world: the population in urban areas in developing countries will grow from 0.5 billion in 1960 to 4.1 billion in 2020, while in developed countries, the increase will be lower, from 0.55 billion in 1960 to 1.2 billion in 2020 (Fig. 1.1).

If in 1950 there was only one city with a population over 10 million inhabitants (New York City), in 2005 there are 24 mega cities. Five of these cities exceed 20 million inhabitants (Brinkhoff, 2005). The most dynamic urbanization occurs in developing countries of Asia, Latin America and Africa. By 2030 these continents will have a higher number of urban areas than any other major area in the world. In the developed countries, with few exceptions, urban populations have grown just a little.

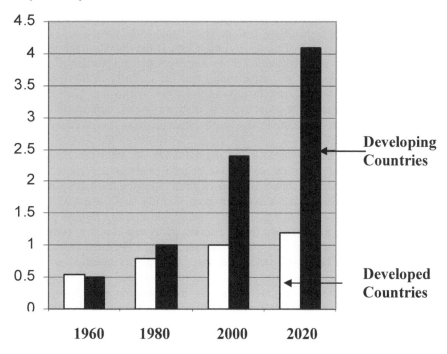

Figure1.1 Global urbanization: Increase of world's population in developed and developing countries (modified after Bilham, 2000)

Table 1.1 Principal Agglomerations in the World (at the level of 2005)

Definition Cities	Population million	Total number	Situated in seismic areas	Percentage %
Super mega	> 20	5	4	80
Mega	> 10	19	15	79
Super	> 5	35	19	54
Big	> 2	128	78	61
Important	> 1	241	150	62
Total		428	266	62

1.2.2 Earthquakes and Urbanization

From Table 1.1 it is very important to notice that most population growth has occurred in the cities situated in seismic areas. So, from the five super-mega cities, Tokyo, Mexico City, Seoul, New York and Sao Paolo, only the last one is situated in a non-seismic area and of the mega cities 79% is developed in seismic areas. Fig. 1.2 shows the world distribution of seismic areas and the position of the most important cities. One can see, astonishing, that the majority of these cities are situated in seismic-prone areas, in this respect Asia, Africa and Latin America being at the first place. The great majority of these seismic areas belong to the developing countries where the level of earthquake preparedness programs is very low.

Earthquakes do not kill people, but the buildings do. With the exception of tsunamis and landslides, most earthquake related fatalities are caused by the collapse of people's homes upon them, or by fires following the earthquakes. The worst earthquake in recorded history occurred in 1556 in Shansi, China, whose magnitude measured about M 8.0. This earthquake, situated in a densely populated area, killed 830,000 people, most living in caves excavated in poorly consolidated loess (wind deposit silt and clay). Earthquakes located in isolated areas far from human population rarely cause any deaths. So, major earthquakes are not damaging, if they occur in uninhabited or lightly populated areas. However, an earthquake does not need to be very strong in magnitude for causing serious damage when it occurs in a heavily populated area. For examples, the five largest losses were: over $100 billion for the Kobe earthquake, M 6.9; $40 billion for the Northridge earthquake, M 6.7; $14 billion for the Chi-Chi earthquake, M 7.6; $14 billion for the Armenia earthquake, M 6.7; and $12 billion for the Izmit earthquake, M 7.6. Note that three of these five earthquakes are not called major earthquakes, because their magnitudes are below M 7.0 (Lee, 2002).

As it is shown, a doubling in world population is expected in the next 50 years, and this would increase the number of important cities. Now 62% of the main world's cities are located in seismic-prone areas and this percentage must increase in the future. A simple calculation shows that 1 billion new housing starts are expected in the next few decades. These are the houses that will pose a future threat to the next generation of urban dwellers and structural engineering. Comprehensive studies elaborated by reinsurance companies, specialized in disaster business, estimate that the global direct costs of natural disasters will top $ 300 bn annually by 2050, the majority being produced by earthquakes.

Therefore, now is the time to prepare the urban habitants against the future earthquakes. A new great problem is identified for the next period: the mitigation of disaster risk for seismic urbanized areas, in order to reduce fatalities and economic losses.

"More effective prevention strategies would save not only tens of billions of dollars, but save tens of thousands of lives. Funds currently spent on intervention and relief could be devoted to enhancing equitable and sustainable development instead, which would further reduce the risk for war and disaster. Building a culture of prevention is not easy. While the costs of preventions have to be paid in

present, its benefits lie in a distant future. Moreover, the benefits are not tangible: they are the disasters that did NOT happen" (Annan, 1999).

1.2.3   Seismic Problems of Developed and Developing Countries

Gross Domestic Product (GDP) represents the marked value of goods and services produced by labor and property in a country (BEA, 2005). The GDP per capita is the ratio:

GDP per capita = GDP/total population

being used as an indicator giving an idea of goods and services available to the residents of a country, and representing the material well-being of the population (WDI, 2005).

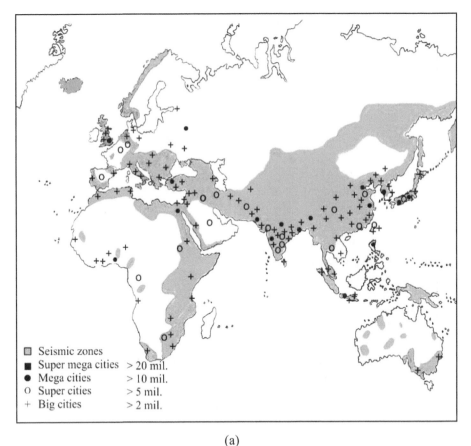

(a)

Figure 1.2 (continued)

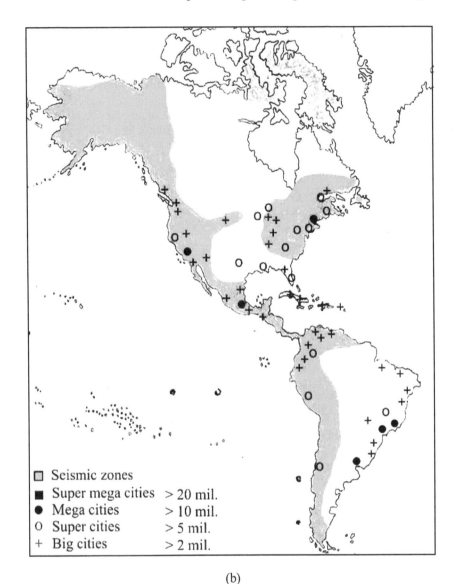

(b)

Figure 1.2 Global urbanization and seismic areas. (a) Africa, Eurasia and Australia; (b) North, Central and South America

This indicator allows subdividing the world countries into the three categories (Wikipedia, nd):

-   *Developed countries* (or industrial, more economically strong countries), for which GDP per capita is greater than $20,000 US;
-   *Transition countries* for which GDP per capita is less than $20,000 US but greater than $5000;

- *Developing countries* (less developed, economically weak, under-developed countries), for which GDP per capita is less than $5000 US.

This framing is also related to the specific seismic problems of a country belonging to the above classification.

Ninety percent of the world's major disasters in during 1990-1999 were in developing countries. However, these countries have made fewer efforts than developed countries to adapt their physical environments to mitigate the impact of disasters or to insure themselves against disaster risk (Freeman et al, 2003).

In the recent decades, the seismic disaster risk in urban centers in both developed and developing countries is increased. The seismic problems in developed countries are related to the increasing of urban system and population, whereas in developing countries the main source of increasing seismic risk can be attributed to the overcrowding of cities with non-engineered buildings, inadequate infrastructures and services, together with environmental degradation (Erdik, 1996). It is, therefore, clear that the tasks for the mitigation of seismic risk are different according to the two situations (Fig. 1.3).

For *Developed Countries,* in spite of the improvement of code provisions, the costs of losses due to earthquakes continue to increase, also under the condition that the average magnitude of earthquakes remains approximately constant. The maximum economical losses are produced when strong earthquakes happen near big and modern cities. The number of fatalities is nevertheless reduced, because new buildings are erected respecting the anti-seismic rules, having as a main purpose the protection of life. Therefore, the main problems of developed countries are economical.

Today there is an increasing number of people and buildings in earthquake-prone areas, meaning that earthquakes affect more and more buildings, facilities, roads, bridges, dams, etc., each year. Over half of the world population is concentrated in urban areas covering just 4 percent of the world surface. Rapid urbanization is a distinctive feature of world development and the number of mega-cities will increase in future. As shown in the previous section, many of these cities are situated in seismically active areas where the concentration of people continues to grow and the damage produced during previous earthquakes is expected to be magnified during the next seismic events. Therefore, one of the most important lessons after an earthquake is the awareness *to improve the seismic design conception,* in order to reduce the economical consequences of future important earthquakes. The revisions of existing code provisions towards performance-based design needs are suggested to be a proper way.

On the other hand, in the *Developing Countries,* the situation is different. Even light or moderate earthquakes are acting in the areas of the old cities where the number of non-engineered buildings (buildings erected ignoring any anti-seismic rules) is dominant. Although is very important to mention that during 2003 Iran-Bam earthquake, the great majority of damage was not produced in historical buildings but in the new buildings, erected without any anti-seismic measures (Zahrai and Heidarzadeh, 2004). In these cases the main disaster is not only economical, as in the first case, but it derives from the human fatalities due to the collapse of old and non-engineered buildings. For these countries the most

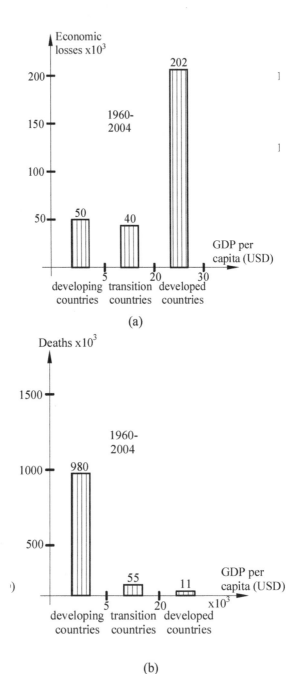

(a)

(b)
Figure 1.3 Seismic losses versus GDP per capita (in US $). (a) Economic losses;
(b) Deaths

important aspect for the mitigation of seismic risk is *the improvement of the seismic resistance of both existing and new buildings*, in order to reduce the human losses. The urgent need for earthquake strengthening of the existing buildings designed and erected before the existence of codified rules, as well as the need to increase the site control of application of the code provisions for new buildings, is very important issues for the next period.

1.3 TOPICS INVOLVED IN SEISMIC DESIGN

In view of the mitigation of seismic risk some distinct topics have been developed (Fig. 1.4) (Gioncu and Mazzolani, 2003):

Engineering Seismology, developed to solve the problems of the Earthquake hazard, is a branch of *Seismology*, having the purpose to use the seismological knowledge for the seismic design of buildings, by proposing the seismic actions function of the source and site characteristics.

Earthquake Engineering, with the task to solve the problems of construction vulnerability, is a branch of more general field, the *Structural Engineering* Science, having the purpose to develop specific methodologies for analyzing the effects of seismic actions on constructions, very different from that used in case of other actions like dead, live, wind, snow, etc., loads.

Seismic Design collects the data given by Engineering Seismology referring to seismic loads and using the methodologies proposed by the Earthquake Engineering and performs a complex examination of structures, including numerical analysis, structural conformation, solutions for details, and eventually an engineering overview on the designed structure. The main scope of Seismic Design is to obtain the economical victory over a strong earthquake by reducing structural damage controlled by the designer.

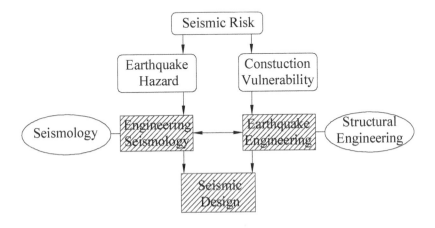

Figure 1.4 Topics involved in seismic design

1.4 SEISMOLOGY VERSUS ENGINEERING SEISMOLOGY

1.4.1 Seismologists and Engineers

The seismological and engineering approaches are typical for the difference between science and application (Smith, 2001).

The science of *Seismology* is involved with the study of tectonic plate movements associated to important earthquakes. The *seismologists*, as Earth scientists, are specialized in geophysics and they are devoted to analyze the genesis and propagation of seismic waves in geological materials. Some of them study the relation between faults, stress and seismicity, others interpret the mechanisms of rupture from seismic wave data, others integrate geoscientific information in order to define zones of seismicity, and finally others collaborate with engineers trying to minimize the damage caused to construction. But the physicists working in this field use conceptions and methodologies different from that used by engineers. They have also the privilege of proposing, testing and discarding erroneous hypotheses. The explanation of earthquake generation is an example of this approach. From chemical reactions to the continental drift or Earth expansion, all types of hypotheses were used without any consequence for these incorrect assertions. Now, the causes of earthquakes being clarified and the Theory of Plate Tectonics is accepted by all scientists, the interest of seismologists is mainly concentrated in understanding the faults, positions and movements, remaining in the field of qualitative description of phenomena.

Structural Engineering is a field of engineering which deals with the design of structural systems with the purpose of supporting and resisting various loading conditions. A *Structural Engineer* is most commonly involved in the design of buildings and other civil engineering structures. Contrary to the seismologist, the structural engineers' activity must rely on proven principles and data, as wrong hypotheses produce impermissible life and economic losses. For an engineer, the approaches for the structural design must be quantitative. Forces caused by earthquakes must be evaluated and the structure members are proportioned to have a resistance greater than these forces. The main purpose of structural design is to produce a suitable structure, even in the case of strong earthquakes. Reducing future earthquake losses depends on the understanding and quantification of the damaging effects of earthquakes. Using data from regional seismic networks, the research in this area is devoted to discover how the characteristics of the earthquake source, wave propagation effects and near-surface geological deposits could control the strong shaking. Specific studies are also investigating the factors which govern the susceptibility to ground failure from land sliding and liquefaction. All these aspects are framed in the so-called earthquake hazard.

It can happen that the interests of seismologists and structural engineers frequently overlap, but in different ways. Both are interested in knowing the positions of active faults. The seismologists want to know how these faults move, in function of their typology and characteristic, while the engineers are concerned with the potential for the future earthquakes on these faults and their implication for building behavior in order to use such information for the structural designs. This aspect is a matter of common conflict between science and practice. The continuous transferring of earthquake hazard information and research findings into the domain of practicing engineers should be an activity of primary importance.

In this context, *Engineering Seismology* has the specific task to transform the qualitative knowledge of Seismology in quantitative information, which shall be suitable to be used in the frame of *Earthquake Engineering*, intending to fill the gap between scientific knowledge and practical questions of structural design. The engineers should work more closely to seismologists to solve many problems in both fields of Engineering Seismology and Earthquake Engineering. Many problems can be solved with their further joint efforts, purposely oriented to gain faster achievements (Hu, 2001).

1.4.2 Tasks of Seismology

Seismology (from the Greek words *Seismos* meaning earthquake and *Logos* meaning science) is the science of earthquakes, being a branch of a more general science of *Geophysics*, which refers to the Earth structure. Its objective is the study and elaboration of theories concerning the generation of an earthquake and the propagation of seismic waves. At the same time Seismology is involved with record and interpretation of recorded seismograms.

Astronomy can predict planetary orbits accurately, because it can rely on the well-established theories on gravity and attraction forces. Unfortunately, Seismology does not possess such clear theories, being forced to adopt only some empirical approaches. One knows more about the structure of the Universe than the interior of our Earth.

One of main aspects of *Seismology* is source characterization and this involves:
- Establishing the main tectonic plates and their boundaries.
- Use of GPS stations to determine the movements of tectonic plates and to establish the stressed zones.
- Better understanding of the rock friction and rupture physics in the fault zones.
- Establishing the main fault types.
- Discovery and documentation of large pre-historical and historical earthquakes.
- Use seismic stations to record ground motions and to process these data.
- Quantification of seismic hazard, determining magnitude and recurrence of earthquakes.
- Dissemination of collected data for all interested fields.
- Elaboration of seismic maps, containing the known faults, seismic areas and their characteristics.
- Last but not least, the future earthquakes.

1.4.3 Tasks of Engineering Seismology

While Seismology is involved with the phenomena produced in the Earth interior, *Engineering Seismology* relates the phenomena produced at the Earth surface, directly affecting the buildings situated in seismic areas.
- Determination of influence of traveled path and attenuation curves.
- Effects of site condition types.
- Description of spatial movements, directivity, incoherence, etc.
- Modification of ground motions due to topographic irregularities and alluvial valleys.
- Effects of soil liquefaction.
- Establishing the main ground motion types.
- Digitalization of recorded accelerations, velocities and displacements.
- Establishing the dominant natural period of ground motions.
- Determination of ground motion duration.
- Determination of number of significant pulses.
- Characteristics of near-field and far-field earthquakes

A basic aspect is related to the *general information concerning the seismic region,* which is developed by means of the following activities:
- Preparing data concerning these historical earthquakes in regions where earthquakes occurred in the past.
- Examining the movements of tectonic plates in regions with very intense seismic activity.
- Evaluating the potential of strong ground motions occuring in regions with moderate seismic activity.
- Controlling quiescent regions, containing hidden inactive faults, which are determined by using modern prospecting, but which can be transformed in to active faults.

A second aspect is related to the *information concerning source characteristics,* which is devoted to evaluate:
- Focal mechanism type, considering the presence of the tectonic plate boundaries, in case of inter-plate earthquakes, and the existing faults in a plate, in case of intra-plate earthquakes.
- Focal depth, knowing the great differences between surface and deep sources.
- Directivity of the fault ruptures along an existing fault, knowing the tendencies of plate movements.
- Spatial characteristics of ground motions.

A very important aspect is related to this *information concerning the geotechnical local conditions,* which is faced to establish:
- Stratification in horizontal and vertical planes, soil layer thickness.
- Dynamic properties of these soil layers.
- Hydrological regime of the regions, related to the influence on amplification of ground motions.

- Existence of relief and basin irregularities, knowing that they can produce important amplification of ground motions.
- Possibility of occurrence of landslides or ground liquefaction.
- Determination of influence of traveled path and attenuation curves.
- Effects of site condition types.
- Description of spatial movements, directivity, incoherence, etc.
- Modification of ground motions due to topographic irregularities and alluvial valleys.
- Effects of soil liquefaction.
- Establishing the main ground motion types.
- Digitalization of recorded accelerations, velocities and displacements.
- Establishing the dominant natural period of ground motions.
- Determination of ground motion duration.
- Determination of number of significant pulses.
- Characteristics of near-field and far-field earthquakes

Finally, it is very important to obtain the *characteristics of design earthquake* for the analyzed site, by means of:

- Presentation of possible accelerogram sets.
- Maximum possible earthquake magnitude.
- Magnitude recurrence.
- Dominate natural period.
- Earthquake duration.
- Attenuation curves in function of the distance from epicenter.

One must have in mind that all information given by Engineering Seismology has a great character of incertitude, due to the difficulties to know in detail the crust and mantle characteristics and to transform these characteristics in quantitative values for design purposes. The limited number of records, obtained in the same place during different earthquakes, originating more or less from the same source, impedes the use of statistical procedures to determine the values of these characteristics.

1.5 EARTHQUAKE ENGINEERING VERSUS STRUCTURAL ENGINEERING

1.5.1 Earthquake Loads versus Static Loads

In order to understand the difficulties in defining the earthquake loads, it can be useful to compare them with conventional loads, for instance the wind loads, both being horizontal but physically different (Murty, 2003):

- According to the common design philosophy the basis are: the pressure on the exposed surface areas in case of wind and the inertia forces, resulting from the random motions of the ground at the base of structure, in case of earthquakes (Fig. 1.5a).
- Wind forces on the structure have a non-zero component superposed with a relatively small oscillating component. Thus under wind forces, the structure may experience small fluctuations in the stress field, but reversal

of stresses occurs only when the direction of wind reverses, which happens only over a large duration of time measured at the level of hours.

- On the other hand, the motion of ground during the earthquake is cyclic around the neutral position of the structure. Thus, the stress state in the structure due to seismic actions undergoes many complete reversals, which happens during the small duration of an earthquake, at the level of seconds (Fig.1.5b).

- Maximum wind forces are reached in average intervals measured in the days, while for the maximum earthquake forces the intervals are measured in dozens of years (Fig. 1.5c), which can be greater than the existence of the building.

- The risk of exceeding the maximum wind forces is very high, while it is reduced for the earthquake maximum forces. During the life of a structure, the maximum wind forces can be reached many times, while the maximum earthquake forces never are reached or reached one or maximum two times.

Considering these differences, the seismic design philosophy is very different in comparison with the design philosophy for wind actions.

1.5.2 Bases of Seismic Design Philosophy

The main actions that a structure is asked to carry (dead, live, snow, etc.) are essentially static loads. Their maximum values can be easily determined using probabilistic approaches. The earthquake does not belong to such kind of actions. The input is the dynamic motion of the ground and the dynamic characteristics of structure determine how much that motion is amplified or de-amplified in terms of structural response. Because the input is given under form of dynamic displacements, the requested structural resistance can be interpreted as an energy dissipation capacity rather than a strength capacity. This energy can be interpreted as the product of forces and displacements. Therefore, the structure performance against earthquake can be improved either by increasing its strength or by increasing its deformability, or by increasing both (Harris, 1989).

The design of structures under static loads is commonly based upon the limit states criteria. The response of structures to this kind of action, determined as the maximum possible values using a probabilistic approach, is analyzed with a linear elastic model (only in very special cases a non-linear model is used). The load effects in the members are then compared with their resistance capacity, thus giving rules for sizing members so that the probability of exceeding various limit states (serviceability or ultimate) is sufficiently low. So, no damage for these static loads is accepted.

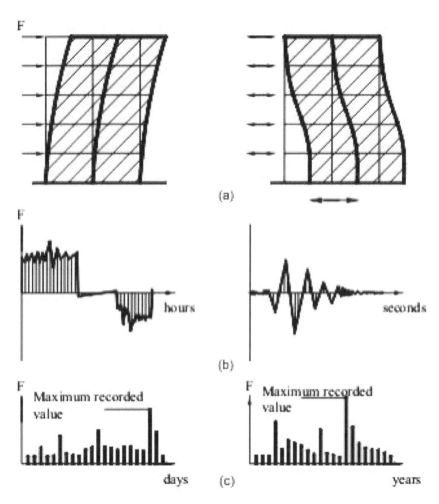

Figure 1.5 Wind loads versus earthquake loads: (a) Actions; (b) Variation in time; (c) Maximum recorded values

The seismic design against earthquake loads is based on a completely different philosophy, due to some very justifiable reasons. First, the design for the largest credible seismic load, resulting from the strongest expected earthquake in the structure site, is unreasonable and economically unacceptable. While the maximum values for static loads can arise frequently, these maximum seismic loads are very rare during the structure life. Therefore, the design earthquakes are selected at a given level, which is smaller than the one associated to the largest possible load. Secondly, the determination of the proper seismic loads is a very difficult task, due to the complexity of seismic phenomena. Thus, one must accept the fact that, in some cases, the earthquake actions could exceed the design values. As a consequence, the structures can occasionally fail to exhibit their expected performance under these events, which exceed the design values, and, consequently, they may suffer local damage, due to the loss of resistance in some members. But a properly designed structure must preserve its general integrity, which consists of the quality of being able to avoid local damage, the structure remaining stable as a whole. This goal can be achieved by arranging structural elements, which gives stability to the entire structural system, and by assuring a sufficient ductility to the members and connections (Gioncu and Mazzolani, 2002).

The above criteria are assumed as a basis of the modern design philosophy, according to which the design lateral force is obtained by dividing the maximum force that a structure will experience by a *reduction factor q* (Fig. 1.6a), which is specified by seismic design codes. The use of this factor is possible provided that the structure can stably withstand a structural damage in the range of plastic deformations without collapse due to loss of strength. This property is called *ductility,* which represents the ability of the structure to undergo plastic deformations without any significant reduction of strength. There are three ductility types: good (high), medium and poor (low) ductility (Fig. 1.6b).

But it must be recognized that, beyond these considerations, there is a very important incertitude in determining the seismic actions. This incertitude is due to uncontrolled and unpredictable aspects, many of them being not well understood from the physical standpoint. Reliable statistical analyses of earthquakes are rather difficult since existing recorded data cover only too short time intervals as compared with the geological period responsible for the seismic processes.

Finally, an important aspect of seismic design is the realization of the important role of the quality of conception and construction. Earthquake does not respect theories and calculations in case of a poor quality project and execution (ESDEP, 2005).

Therefore, it becomes clear that the strategies for insuring reliability against such events could be based on modeling these phenomena in the most simple and conservative way to compensate the lack of actual information. There are some fundamental theories, developed only in the last period, which can be used to understand the phenomenon of earthquake generation and to use the actual knowledge for a proper seismic structure design. These new branches of the earthquake science will be presented in the following chapters.

1.5.3 Tasks of Earthquake Engineering

The general field of *Structural Engineering* deals with the investigation of factors influencing structural behavior and in the analysis of structures under these factors. A complete structural analysis requires the evaluation of the interaction of the structure with soil, foundations and non-structural elements. The spatiality of the structural schemes, the material properties and the action characteristics must be taken into account. The complexity of the analysis implies many difficulties, which today are largely overcome by using computer aids.

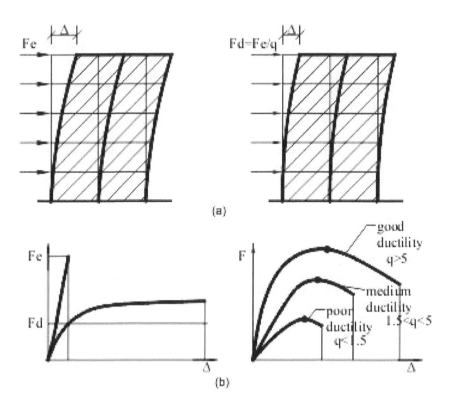

Figure 1.6 Design philosophy for earthquake loads: (a) Definition of reduction
factor q; (b) Good, medium and poor ductility

The *Seismic Design* introduces additional difficulties with respect to the above
ones. The most important aspect is the request that the structure is able to resist
severe earthquakes with acceptable damage but without collapse, by exploiting its
ductility properties. This modern design philosophy requires special design
methodologies, very different from the conventional ones. This is the basic reason
why *Earthquake Engineering* can be considered as a new branch of Structural
Engineering, its main tasks being the evaluation of the construction vulnerability
against seismic actions.

A fundamental aspect is the *consideration of the earthquake type in the
analysis,* which must involve the following problems:

-   Development of specific methodologies for interplate and intraplate
    earthquake types, by using different ground motions records and response
    spectra.
-   Identification of the differences between the near-field and far-field
    earthquakes, the pulse characteristics for the former being very important
    and the number of important cycles for the second.

- Examination of the influence of local soil conditions as a function of stratification and dynamic properties of soil layers.

The determination of the *structural response* must be setup by considering some specific aspects of seismic design:

- Soil-foundation-structure-non-structural elements interaction, considering the actual properties of each element and the modification of these properties in function of the earthquake level.
- Spatiality of response, considering the horizontal and vertical components of seismic actions.
- Effects of structural and accidental torsions, produced by the structural asymmetry and incoherency of seismic actions.
- Influence of horizontal and vertical irregularities on the overall structural response.
- Pounding of two neighboring buildings when the seismic separation is not sufficient.

The need of developing some *specific design methodologies* comes out from the following requirements:

- Improvement of the equivalent lateral force analysis, by proposing a set of response spectra as a function of ground motion characteristics. Different spectra for near-field, far-field, interplate, intraplate earthquakes are required.
- Improvement of the static elastic-plastic incremental pushover analysis, by the determination of a proper loading distribution along the structure height taking into account the vibration mode influence.
- Development of the incremental methodology for time-history analysis, in order to evaluate the collapse condition of structures. Considering that the natural period of severe earthquakes is longer than the one for moderate earthquakes, the methodology must consider the contemporary modification of accelerations and natural periods.

1.6 ISSUES FOR SEISMIC DESIGN

1.6.1 Design Team

The main purpose of this topic is to filter the complex information obtained from Engineering Seismology and Earthquake Engineering, by using the engineering judgments.

A design earthquake related to a design action, which ensures to the structure an adequate seismic safety, is proposed instead of a realistic earthquake. The definition of such a design earthquake requires a series of engineering judgments on seismological and structural matters, together with a close collaboration among seismologists, geotechnists and structural engineers. The philosophy of engineering seismic design states that the design should accomplish the objectives of multi-level approaches (different behaviors for minor, moderate or severe earthquakes), and the seismic design requiring the corresponding design loads. All these issues are intended to prepare code provisions in order to protect the public from life losses and buildings structural collapse.

A complete building design requires the combined effort of professionals in several different disciplines (Fig. 1.7). First of all, the architect and the structural engineer, who are directly involved in the building design. The design process starts from the decision of the owner (who obtains the financial support from a lender) to build a construction, based on a socio-economic study of feasibility. Once the decision to build is made, the next step consists of making an architectural pre-design of the building, considering all economical and functional aspects. A multi-disciplinary team can be involved in this project, for collaborating with the architect as coordinator of this team. Since the beginning, the architect should be especially mindful of the restrictions imposed by the facilities and the equipments required for the building functions, and in addition, as a principal restriction, the need to have a structure creating spaces for operations and supporting in a safe and economical way the different actions (first of them in importance being the seismic ones) (De Buen, 1996). Not only the structural engineer, but the architect as well, should be familiar with the structural requirements in seismic areas, in order to conceive a good building configuration. So, as a new recent discipline, *Earthquake Architecture* (Arnold, 1996, Charleson and Taylor, 2000, 2004, Giuliani, 1992, 2000, Giuliani et al, 1996) is born, with the scope to describe a degree of architectural expression related to the structural aspects in connection with the earthquake actions. This topic is involved with regional, urban and building seismic design problems.

The structural engineer becomes the key player to satisfy the requirements of the architectural conceptions. The main objective of an engineer involved in the design process of a specific seismic-resistant building, is to produce a structure which satisfies the functional requirements both aesthetically and economically and which can safely perform its intended use under all potential loads and environmental actions. He or she must be in contact with the builder, the material and equipment suppliers and building officials. But the main co-workers are the *geotechical engineer* and the *seismologist*, who have the responsibility to determine the characteristics of site environment given by the earthquake source, propagation path and local soil conditions. The complex activity to design the fondation and the superstructure, in the conditions required by the architectural and site demands, is framed in the field of a new topic called *Seismic Design*. The main purpose of this topic is the effort to put all the aspects of this very complex problem in a practical format to be correctly implemented in the modern codes.

An important aspect is the *establishment of strategies for seismic design* as a function of the *earthquake type* being based on:

- Specific methodologies for active zones, especially for important urbanized areas where the near-field earthquakes can produce important damage.

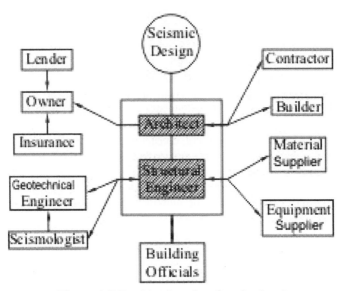

Figure 1.7 Combined works of professionals

-   Specific methodologies for moderate seismic areas, having less complex problems than the ones for strong seismic areas.

Some *collateral problems* must be considered, such as:

-   Wind design methodology versus seismic design. For establishing the seismic design philosophy when the wind load is more severe than the seismic load.
-   Effects of after-earthquake fires, on buildings already damaged during the earthquake, for evaluating the residual structural resistance to survive.

Finally, the most important activity of seismic design is the *codification:*

-   Elaboration of codes for macro and micro-zonation, indicating the presence of different fault types and difficult soil conditions.
-   Elaboration of up-to-date codes for structural design, considering also the most recent knowledge.
-   Elaboration of provisions for the design of non-structural elements, including partitions, claddings and equipments.

1.6.2 New Challenge: Earthquake Architecture

Earthquake Architecture is intended to have the meaning of an approach to architectural design which is based upon earthquake engineering issues as its primary source of inspiration. The requirement of integrating architecture with structural design is a complex problem. When the structure is designed, it is expected to look elegant, appealing, and above all, statically correct. Harmony between structure and architectural form is the key for a successful expression.

Ideally, the structure should visually clarify and enrich the form (Ali, 1990, Sev, 2001, Charleson and Taylor, 2004).

Some examples in which the structural conception gave important contribution to the architectural aesthetic are presented in Figure 1.8. In seismic design a key solution is to use bracing systems. If the architects have the ability to use these façade bracings in a creative modality, some very successful architectural expressions can be obtained.

The new aseismic systems used for energy dissipation have also an important influence on architectural configurations (Mezzi et al, 2004). Figure 1.9 shows some buildings where new bracing systems equipped with dissipative devices are located in the façades, enriching the architectural expression.

A very clear domain is where the bracing systems, applied to the inexpressive existing façades of buildings (Charleson et al, 2002, Reitherman, 2005), make a real contribution to the improvement of the architectural value through the seismic strengthening of existing buildings (Fig. 1.10).

(a)   (b)

Figure 1.8 (continues)

(c)   (d)

Figure 1.8 Bracing systems improving the architectural features of the building facades: a) ALCOA Building, San Francisco; b) John Handcock Building, Chicago; c) Hotel de Las Artes, Barcelona; d) Hearst Tower, New York

(a)

Figure 1.9 (continues)

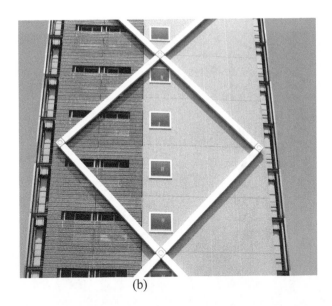
(b)

Figure 1.9 Architectural features of buildings with diagonal dampers on the facade:a) The building of the Tokyo Institute of Technology, Japan; b) The building of the Yokohama Institute of Technology, Japan

(a)

Figure 1.10 (continues)

(b)

Figure 1.10 Architectural expression of seismic strengthening: (a) Apartment building in Berkeley, California: (b) The University Hall in Berkeley, California

1.7 SOME ETHICAL DILEMMA IN SEISMIC DESIGN

The Interests of economy, safety function and aesthetics rarely, if never, work together. More typically they run in opposite directions. In many situations the owners, – structural engineers and architects play a different role with different opinions in relation to the problem of reducing the construction costs. This conflict of interest must be regulated by the *professional conscience.* For protecting the designer against this kind of pressures some professional codes have been elaborated by the State administrations. Structural engineers and architects, in their roles as building design professionals, enter into a special agreement with the general public: the State will protect their monopoly on certain segments the construction industry by allowing only registered engineers and architects to design buildings; the design professionals will have, in return, the duty to protect the public welfare in the built environment, by respecting the code rules. Therefore, the main purpose of a seismic code, as any other building code, is to protect life, health and public welfare. In ethical terms, such a code regulates the given actions on the basis of their consequences for the public (Spector, 1997).

The code mission is quite clear. The application, however, requires judgments and compromises due to the fact that a big gap still exists between the progressive knowledge and codification. Academics and researchers are devoted to research activity, but structural engineers are professionals and not researchers. So, they are basically conservative and resist the new concepts. A new knowledge is very difficult to implement in codes. Therefore, the code provisions must be, always, a compromise between new and old knowledge and procedures; otherwise the new methodology will be rejected by the designers.

In this situation, a progressive designer, who knows more about the actual structural behavior during an earthquake than the codified one, has to face an ethical problem, whether he has to consider or not in practice his advanced knowledge. He has three choices:

-   To accept the code provisions without any reserve, ignoring his knowledge;
-   To reject the code of practice as inadequate, by substituting it with some personal provisions;
-   To try to interpret the code provisions in the light of the new knowledge.

This attitude towards the use of the code is an ethical dilemma. The first position can be rejected because he renounces his ability as progressive designer. The second one is very dangerous because his personal judgment can lapse into arrogance. Therefore, the only way to solve this dilemma is the last attitude, to respect the spirit of the code, but introducing in design the new knowledge. The balance between old and new provisions is the dilemma, which must be solved by the progressive designers. Unfortunately, ethical theory offers little help in this dilemma (Spector, 1997).

1.8 INTERNATIONAL ACTIVITY

1.8.1   Seismology and Engineering Seismology

Seismology and Engineering Seismology are, in the truest sense, the global sciences, not limited by political boundaries. Thus, they can only be effectively practiced through international cooperation. When it was evident that seismic signals propagate through the Earth over large distances, the international organization for coordinating, collecting and distributing the data as well as promoting earthquake research became a pressing need. During two "International Conferences on Seimology" held in 1901 and 1903 in Strasbourg, the establishment of an "International Association of Seimology" (IAS) was proposed in 1904. This can be considered as the root of what finally became the *International Association of Seimology and Physics of the Earth's Interior* (IASPEI) founded in 1951 (Agnew, 2002, Adams, 2002, Berckhemer, 2002). IASPEI promotes the study of earthquakes and other seismic sources, the propagation of seismic waves, and the Earth's internal structure, properties and processes. Scientists participating to IASPEI initiated to coordinate research and scientific exchanges which requires the cooperation in the earthquake prone-countries. IASPEI scientific activities are categorized by the following themes: Earth structure and geodynamics earthquake sources – prediction and modeling, tectonophysics, earthquake hazard, risk and strong ground motions, seismological observation, interpretation of data, and education.

Work on specific topics is carried out through commissions, sub-commissions, committees and working groups formed to meet specific needs of new, exciting problems as they emerge.   One of these commissions is the *European Seismological Commission* (ECS), having the aim to promote seismological studies and projects in Europe, countries bordering the Mediterranean Sea and immediate neighbors: the area from the Mid-Atlantic Ridge to the Urals and the Artic Ocean to Northern Africa. The activity of ECS is mainly focused in organizing every two years the General Assembly. The following objectives are of particular interest: to facilitate exchange of ideas and to organize personal meetings among scientists, to encourage the cooperation between individuals and organizations as well as between other European and non-European scientific and engineering communities by establishing Working Groups on seismological problems, to promote inter-disciplinary studies involving Seimology, and to organize training courses for young scientists.

Engineering Seismology scientific activity (Bolt, 2004) is framed in the general activity of IASPEI, focusing the seismological research topics applicable to the engineering communities as probabilistic seismic hazard assessment, their implications versus building codes for a seismic resistant design, and the earthquake risk assessment as a basis for risk management. Furthermore, engineering seismology basic research is performed to supply suitable datasets such as regional earthquake catalogues, numerical models and investigations on the earthquakes generating potential geological fault zones, and attenuation relations for potentially damaging ground motion parameters.

1.8.2   Earthquake Engineering

Earthquake Engineering is a 20th Century development. In 1963 the *International Association for Earthquake Engineering* (IAEE) was founded, aiming at organizing World Conferences on Earthquake Engineering (WCEE). Since the IAEE foundation, these Conferences have been held every four years. The last ones, presenting very important progresses in seismic concepts were held in Madrid in 1992, Acapulco in 1996, Auckland in 2000, Vancouver in 2004, and Beijing 2008. These World Conferences represent a wide arena giving the opportunity to scientists, engineers, industrial professionals and government officials to present their scientific and engineering works, to exchange ideas and knowledge for the mitigation of seismic risk. They also provide a common platform for delegates for all over the world to initiate new cooperations. For instance, about 2500 delegates from about 73 countries were present during the 13th Vancouver Conference and 2300 technical papers were included in the Conference Proceedings (available as a CD-ROM containing approximately 27,000 pages).

In 1964 the *European Association for Earthquake Engineering* (EASS) (initially called European Commission for Earthquake Engineering) was founded during the Skopje Conference, having the same tasks as the IAEE organization. EASS organizes European Conferences on Earthquake Engineering (ECEE) every four years (in alternation with the WCEE Conferences). The last important ones were in Vienna in 1994, in Paris in 1998, in London in 2002, in Geneva in 2006, where for the first time a joint Conference of ECEE and ESC (European Seismological Commission) take place as the First European Conference on Earthquake Engineering and Seismology (ECEES). This was an excellent forum for idea exchanges between seismologists and earthquake engineers.

Similar chain of conferences is organized by *US Earthquake Engineering Research Institute* (EERI) as National Conference on Earthquake Engineering (NCEE), the last ones being organized in Seattle in 1998, in Boston in 2002 and San Francisco in 2006 (to commemorate 100 years from the 1906 San Francisco earthquake). The EERI, founded in 1949 in Oakland (California), is a national nonprofit technical society of engineers, geoscientists, architects, planners, public officials and social scientists, having the main objective to reduce earthquake risk by advancing both the science and practice of Earthquake Engineering.

1.9   REFERENCES

nd – no date: the web site is periodically modified

Adams, R.D. (2002): International seismology. In International Handbook of Earthquake & Engineering Seismology (eds. W.H.K. Lee et al), Academic Press, Amsterdam, 29-37

Agnew, D.C. (2002): History of Seismology. In International Handbook of Earthquake & Engineering Seismology (eds. W.H.K. Lee et al), Academic Press, Amsterdam, 3-11

Ali, M.M. (1990): Integration of structural form and aesthetics in tall building design. Tall Building Design (eds. L.S. Beedle and D.B. Rice), Van Nostrand Reinhold Company, New York, 3-12

Annan, K. (1999): Introduction to UN Secretary-General's annual report on the work of the Organization of United Nations.

Arnold, Ch. (1996): Architectural aspects of seismic resistant design. In 11th World Conference on Earthquake Engineering, Acapulco, 23-28 June, 1996, CD ROM 2003

BEA-Bureau of Economic Analysis (2005): Glossary Index
http:// www.bea.doc.gov/bea/glossary.htm

Berckhemer, H. (2002): Foreword. In International Handbook of Earthquake & Engineering Seismology (eds. W.H.K. Lee et al) Academic Press, Amsterdam, XVII-XVIII

Bilham, R. (1995): Global fatalities from earthquakes in the past 2000 years: prognosis for the next 30. In Reduction and Predictability of Natural Disasters. (eds. J. Rundle, F. Klein , D. Turcotte), Santa Fe Institute, Studies in the Science Complexity, Vol. XXV, 19-31

Bilham, R. (2000): Urban earthquake fatalities – a safer world or worse to come?
http://ciresColorado.edu/bilham/urbanSRL.pdf

Blakeborough, T. (2002): Northridge and Kobe- that are revolutionizing engineering practice. SOUE xzNews.
http://www.soue.org.uk/souenews/issue2/earthquakes.html

Bolt, B.A. (2004): Engineering Seismology. In Earthquake Engineering: From Engineering Seismology to Performance-Based Engineering (eds. Y. Bozorgnia and V.V. Bertero), CRC Press, Boca raton, 2.1- 2.35

Brinkhoff, Th. (2005): The principal agglomerations in the world.
http://www.citypopulation.de

Charleson, A., Taylor, M. (2000): Towards earthquake architecture. In 12th World Conference on Earthquake Engineering, Auckland, 30 January-4 February 2000, CD ROM 858

Charleson, A., Taylor, M. (2004): Earthquake architecture explorations. 13th World Conference on Earthquake Engineering, Vancouver, 1-6 August, 2004, Paper 596

Charleson, A, Preston, J., Taylor M. (2002): Architectural expression of seismic strengthening, Earthquake Spectra, Vol.17, No.3, 417-426

De Buen, O. (1996): Earthquake resistant design: A view from the practice. In 11th World Conference on Earthquake Engineering, Acapulco, 23-28 June 1996, CD ROM 2003

Erdik, M.(1996): Seismic risk analysis for urban systems. In 11th World Conference on Earthquake Engineering, Acapulco, 23-28 June 1996, CDROM 2017

ESDEP WG 14 (2005): Seismic design. Lecture 17.1: An overall of the seismic behaviour of structural systems
http:/www.kuleuven.ac.b.

FEMA 383 (2003): Expanding and Using Knowledge to Reduce Earthquake Losses. NEHRP Program, Strategic Plan 2001-2005

Freeman, P.K., Keen, M., Mani, M. (2003): Being prepared. Natural disasters are becoming more frequent, more destructive, and deadlier, and poor countries are being hit the hardest. Finance & Development, September, 42-45

Gioncu, V., Mazzolani, F.M. (2002): Ductility of Seismic Resistant Steel Structures. Spon Press, London

Gioncu, V., Mazzolani, F.M. (2003): Challenges in design of steel structures subjected to exceptional earthquakes. In Behaviour of Steel Structures in Seismic Areas, STESSA 2003 (ed. F.M. Mazzolani), Naples, 9-12 June 2003, Balkema, 89-95

Gioncu, V., Mazzolani, F.M. (2006): Influence of earthquake types on the design of seismic-resistant steel structures. Part 1: Challenge for the new design approaches. Part 2: Structural responses for different earthquake types. In Behaviour of Steel Structures in Seismic Areas, STESSA 2006 (eds. F.M. Mazzolani and A. Wada), Yokohama, 14-17 August 2006, Balkema, 113-120, 121-127

Giuliani, H. (1992): A new approach for the integral solution of building design. In 10th World Conference on Earthquake Engineering, Madrid, 19-24 July 1992, Balkema, Rotterdam, 3619-3622

Giuliani, H. (2000): Seismic resistant architecture: A theory for the architectural design of buildings in seismic zones. In 12th World Conference on Earthquake Engineering, Auckland, 30 January-4 February 2000, CD ROM 2456

Giuliani, H., De Acosta, R., Yacante, M.I., Camora, A.M., Giuliani, H.L. (1996): Seismic resistant architecture on building scale: A morphological answer. In 11th World Conference on Earthquake Engineering. Acapulco, 23-28 June 1996, CD ROM 1067

Harris, J.R. (1989): The need for system reliability studies in earthquake engineering. In New Directions in Structural System Reliability (ed. D.M. Frangopol), Boulder, 12-14 September 1988, University of Colorado, 177-179

Hu, Y. (2001): Strengthen the link between engineering and seismology. In China-US Millenium Symposium of Earthquake Engineering: Earthquake Engineering Frontiers in the New Millenium (eds. B.F. Spencer and Y.X. Hu), Beijing, 8-11 November 2001

Kanamori, H (2001): Future directions in seismology for earthquake damage mitigation. Linbeck Distinguished Lecture Series in Earthquake Engineering: Challenges for the New Millenium.
http://www.nd.edu/~linbeck/

Lee, W.H.K. (2002): Challenges on observational seismology. In International Handbook of Earthquake & Engineering Seismology, Part A (eds. W.H.K. Lee et all), Academic Press, Amsterdam, 269-281

Mezzi, M., Parducci, A., Verducci, P. (2004): Architectural and structural configurations of buildings with innovative aseismic systems. 13th World Conference on Earthquake Engineering, Vancouver, 1-6 August 2004, Paper No. 1318

Murty, C.V.R. (2003): Seismic design provisions for earthen structures in Indian seismic code. CPTFTEGE, 23-24 September 2003

Reitherman, R. (2005): The expression of seismic design. Consortium of University for Research in Earthquake Engineering, CUREE Report

Sev, A. (2001): Integrating architecture and structural form in tall steel building design. CTBUH Review, Vol. 1, No. 2, 1-8

Smith, K.G. (2001): Innovation in earthquake resistant concrete structure design philosophies: A century progress since Hunnebique's patent. Engineering Structures, Vol. 23, 72-81

Spector, T. (1997): Ethical dilemmas and seismic design. Earthquake Spectra, Vol. 13, No. 3, 489-504

UN-Habitat (2005): United Nations human settlements program.
http://www. unhabitat.org

WDI (2005): World development indicators: GDP per capita
http://www.cgdev.org

Wikipedia (nd): Developed country, developing country
http://en.wikipedia.org

Zahrai, S.M., Heidarzadeh. M. (2004): Seismic performances of existing buildings during the 2003 Bam earthquake. In 13th World Conference on Earthquake Engineering, Vancouver, 1-6 August 2004, Paper No. 1715

Chapter 2

Living with Earthquakes

2.1 WHEN THE EARTH SHAKES

2.1.1 Earth as Living Body in Permanent Motion

A severe earthquake is a terrifying experience. Relatives, homes and goods can be lost in only a few minutes. Compared to any other natural event, the earthquake is the most frightful one, because it undermines the basic stability of human existence and the confidence that this stability is under control. When this stability is destroyed by the violence of nature, fright and panic reach an intolerable level.

As natural as wind and snow are in the atmosphere, the seismic movements develop inside of the Earth. The entire lifetime of the Earth is a continuous sequence of underground movements with more than a billion quakes per year, an average of one quake every 30 seconds. Therefore, the Earth is like a living body in permanent motion. This movement is produced by the convection currents developed in the viscous mantle due to the prevailing high temperature and pressure gradients between the crust and the core. These convection currents result in a circulation of the Earth masses: hot molten lava comes out and the cold rock mass goes down into the Earth (Fig. 2.1).

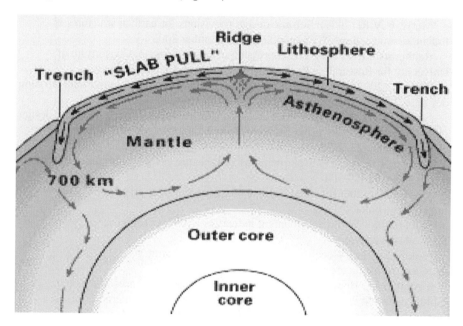

Figure 2.1 Convection currents into the Earth (USGS, nd)

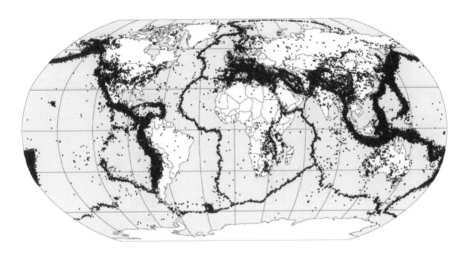

Figure 2.2 Earthquakes and tectonic borders (Wikipedia, NASA, 2007)

Many such local circulations are taking place in different regions underneath the Earth surface, leading some portions of the Earth undergoing movements in different directions along the surface, and producing sliding of crust and some portions of the mantle on the hot molten outer core. These movements of Earth masses produce a division of the crust in some portions called tectonic plates, moving in different directions and with different velocities (Fig. 2.2). Sometimes equilibrium among plates exists due to the shear forces along the plate boundaries, named faults. But when the strain energy is exceeded in some fault portions, a sudden slipping occurs, being the source of the *earthquake*. The comparison between the distribution of the epicenters of strong earthquakes and the tectonic plate borders shows very clearly that the main cause of earthquakes is the relative movements of tectonic plates. The majority of earthquakes in the world occur along these borders, named *interplate earthquakes,* but there are also earthquakes shaking the zones within the plate itself, far from the plate borders, named *intra-plate earthquakes.*

Earthquakes occur everywhere through the world. Any minute of a day, the Internet site for the European-Mediterranean Seismological Centre (EMSC) presents a map with the European earthquakes produced in the last few hours, days or weeks. The US Geological Survey gives the same information about the world earthquakes (Fig. 2.3 shows an example of this, with the strong South Taiwan earthquake).

The continuous movement of the Earth can be noticed by the fact that every day thousands of small ground movements are recorded in the world, every week a moderate earthquake is recorded in some place. At least one significant earthquake causing damage and injuries occurs every month, while every year two or three strong earthquakes produce important economic losses, killing thousands of people. Statistically, the frequency of earthquake occurrence is given in Table 2.1.

Table 2.1 Frequency of Occurrence of Earthquakes Based on Observations since 1900 (Mazzolani, 2002, USGS, nd)

Descriptor	Magnitude	Average annually
Very great	> 9.0	1 at 20 years
Great	8.0 - 8.9	1
Major	7.0 - 7.9	18
Strong	6.0 - 6.9	120
Moderate	5.0 - 5.9	800
Light	4.0 - 4.9	6200
Minor	3.0 – 3.9	49000
Very minor	2.0 – 2.9	about 1000 per day
	1.0 – 1.9	about 8000 per day

One can see that each year about thousand earthquakes exceed the magnitude 5, which is considered as the lower limit for the damaging earthquakes. But even earthquakes considered to be strong do not necessarily create critical situations if the rules for seismic design have been observed. Contrary, even moderate earthquakes acting on old and degraded buildings can produce significant damage.

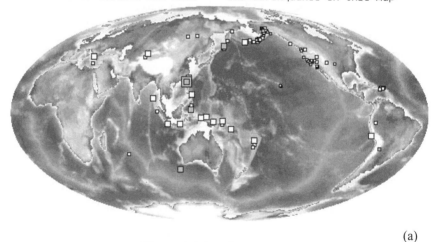

(a)

Figure 2.3 (continues)

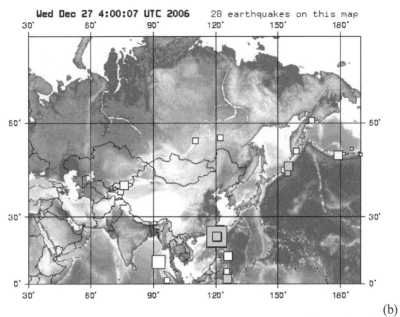

(b)

Figure 2.3 Earthquakes in the world on 27.12.2006: (a) Worldwide earthquakes;
(b) Asia region with strong Taiwan earthquake (M 7.1) (USGS, nd)

The number of worldwide damaging earthquakes from 1990 to 2009 is presented in Table 2.2. One can see that this number does not vary very much, showing some constancy in the annual earthquake number, ranging from 1100 to 1900. Contrary, the number of estimated deaths varies very much if the earthquakes occur or not in urbanized zones containing many non-engineered buildings or producing secondary effects, like tsunamis (e.g. the 2004 Sumatra earthquake).

2.2 HISTORICAL EARTHQUAKES

The historical earthquakes, in the period until 1900, which produced an important number of fatalities and/or a very high magnitude, are presented in Table 2.3. One can see that, even if the density of population in seismic areas was small, the number of deaths was in some cases very high.

The *226 BC Rhodes earthquake* destroyed one of the most important sculptures of Antiquity, the Colossus (Wikipedia, nd).

The *1349 Rome earthquake* caused the collapse of the outer South side wall of the Roman Colosseum (Wikipedia, nd).

The World's deadliest earthquake seems to be the *1556 China-Shaanxsi* earthquake in central China (Wikipedia, nd). It struck a region where most people lived in caves carved from soft rock. These dwellings collapsed during the earthquake, killing an estimated 830,000 people.

Table 2.2 Number of Worldwide Damaging Earthquakes from 1990 to 2009
(USGS, nd)

Year	Magnitude				Total	Estimated deaths
	5.0-5.9	6.0-6.9	7.0-7.9	8.0-9.9		
1990	1635	115	12	0	1762	52000
1991	1469	105	11	0	1585	2300
1992	1541	104	23	0	1668	3800
1993	1449	141	15	1	1606	10000
1994	1542	161	13	2	1718	1050
1995	1327	185	22	3	1537	8000
1996	1223	160	21	1	1405	500
1997	1118	125	20	0	1263	3000
1998	979	113	14	2	1108	9000
1999	1106	123	23	0	1252	23000
2000	1345	158	14	1	1518	200
2001	1243	126	15	1	1385	21500
2002	1086	132	13	0	1231	1700
2003	1203	140	14	1	1358	33819
2004	1118	139	13	2	1272	284000
2005	1700	144	10	1	1855	89354
2006	1427	132	10	1	1570	6666
2007	1696	167	14	4	1881	789
2008	1768	168	12	0	1948	88011
2009	1465	125	15	1	1606	1748

Table 2.3 Historical Earthquakes

Year and date		Location	Magnitude	Deaths
464 BC		Sparta		
226 BC		Greece, Rhodes		
365		Crete, Knossos		50000
526	20.05	Syria, Antiochia		250000
844		Syria, Damascus		70000
856		Iran, Damghan		200000
856		Greece, Corinth		45000
893		India, Daipur		180000
893	23.03	Iran, Ardabil		150000
1138	09.08	Syria, Aleppo		230000
1268		Turkey, Anatolia		60000
1290		China, Chihli		100000
1349		Italy, Rome		
1456		Napoli		80000
1556	21.01	China, Shaanxi	8.0	830000

Year and date		Location	Magnitude	Deaths
1662		China, Anhwei		300000
1667		Caucasia, Shemakha		80000
1693	11.01	Italy, Sicilia		100000
1700	26.01	Canada, Cascadia	9.0	-
1727	08.11	Iran, Tabriz		77000
1730	30.12	Japan, Hokkaido Island		137000
1731		China, Beijing		100000
1737	11.10	India, Calcutta		300000
1755	01.11	Portugal, Lisbon	8.7	70000
1780		Iran, Tabriz		100000
1783	04. 02	Italy, Calabria		50000
1811	-	USA, New Madrid	8.6	-
1856	12.10	Greece, Creta	7.8	-
1883	26.08	Java		100000
1897		India, Gujarat	8.0	1600

One of the best known historical events is the *1755 Lisbon Portugal earthquake* (Fig. 2.4), when a magnitude 8.7 ground motion, produced on the Atlantic Sea, struck the Iberian Peninsula (Wikipedia, nd). The tsunami was measured at 6m near Lisbon and 20m at Cadiz in Spain, killing thousands along the shores. Fires burned for 6 days after the initial shock and half the houses in Lisbon were ruined.

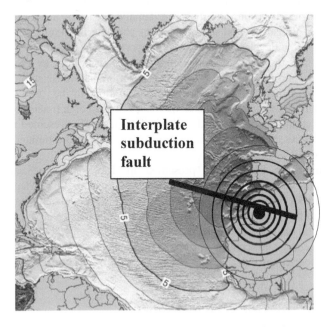

Interplate subduction fault

Figure 2.4 1755 Lisbon earthquake and the great tsunami (after NGDC, nd)

Because of the marked loss of life and mass destruction, this earthquake remains one of the most devastating tragedies from Europe's history. This earthquake is considered as the birthday for the science of Seismology, due to the first studies concerning the wave propagation from the source to the site (Agnew, 2002).

A series of three earthquakes centered near *New Madrid, Missouri, US*, in 1811 and 1812 marked the strongest intraplate earthquakes to occur in North America (Wikipedia, nd). Although the magnitude estimates for the pre-instrumentally recorded earthquakes are uncertain, there is no question that these events were unique for the Eastern North America and rare elsewhere. Similar-sized earthquakes are known to have occurred only in China and India. These earthquakes greatly affected the topography of the North America continent. Lake St. Francis, 64 km long and 1 km wide, formed when a piece of sunken ground flooded with groundwater, and additional shifts in surface structure caused the Mississippi to change its course.

2.3 GREAT EARTHQUAKES OF 20th CENTURY

The 20th Century can be divided into three very distinct periods. The first one (until 1950) is characterized by very timid attempts to develop a consistent explanation of the nature of earthquakes. During this period no records are available (with the exception of the historical record of 1940 El Centro earthquake, the first recorded one). Due to the absence of anti-seismic rules for the protection of buildings situated in seismic areas, the number of deaths was very large. The second period (1950-1980), is characterized by a coherent theory development, but due to a reduced number of instrumented seismic stations, limited information concerning the characteristics of ground motions and damage exists for these earthquakes. A dense network of instrumental seismic stations characterizes the last period of 20th Century, where the developed theory was supported by a large number of records. At the same time, the tremendous development of Seismology, as a Science of the Earth behavior, and Engineering Seismology, as a branch of Seismology, devoted to supply the structural designer with information to be used in practical application, have had a great influence in the developing of anti-seismic concepts.

The main earthquakes produced in the 20th Century are presented in Table 2.4, where date, places, magnitude, deaths, earthquake type and estimated damage (in US $) are given.

Examining the earthquakes presented in Table 2.4, one can see that, fortunately, the very strong events (Chile, 1960, Alaska, 1964, Aleutian Islands, 1957, Kamchatka, 1923 and 1957, etc.) occurred in less populated areas. Also, it is very clear that the great number of deaths occur in developing countries, while the great damage is recorded in the developed ones. This means that the sensitivity for seismic protection and consequently the problems of seismic design are very different in these countries.

Table 2.4 Earthquakes of 20th Century producing great economical or cultural damage, important number of deaths , >10000, or having great magnitude, M >7

Year	Date	Location	Magnitude	Deaths	Damage (billion $)
1905	04.04	India, Kangra	8.6	20000	
1906	18.04	USA , San Francisco	7.8	1000	0.5
1906	17.08	Chile,Valparaiso	8.2	20000	
1906	31.01	Ecuador	8.8		
1908	28.12	Italy, Messina	7.5	83000	
1920	16.12	China , Gansu	8.6	220000	
1923	03.02	Kamchatka	8.5		
1923	01.09	Japan , Kanto	8.3	143000	
1927	22.05	China , Xining	8.3	200000	
1932	25.12	China, Gansu	7.6	70000	
1935	30.05	Pakistan, Quelta	7.5	60000	
1938	01.02	Indonesia	8.5		
1939	25.12	Chile, Conception	8.3	25000	
1939	26.12	Turkey, Erzincan	7.9	25000	
1940	10.11	Romania,Vrancea	7.4		
1940	18.05	USA, El Centro	7.1		
1948	05.10	Turkmenistan, Ashgabat	7.3	110000	
1950	15.08	India, China border	8.6		
1952	04.11	Kamchatka	9.0		
1957	09.03	Aleutian Islands	9.1		
1960	24.04	Morocco, Agadir	5.9	12000	
1960	22.05	Chile, Valdivia	9.5	6000	0.5
1963	13.10	Kuril Islands	8.5		
1964	28.03	USA, Alaska, Anchorage	9.2	116	0.4
1970	31.05	Peru, Ancash	8.1	66000	0.5
1976	27.07	China, Tangshan	8.0	250000 (660000)	2
1977	04.04	Romania, Vrancea	7.2	1600	2
1980	10.10	Algeria, El Asnam	7.3	9700	2
1985	19.09	Mexico, Mexico City	8.1	20000	5
1988	07.12	Armenia, Spitak	7.1	25000	16
1989	17.10	USA, Loma Prieta	7.1	70	8
1990	21. 07	Iran, Manjil	7.7	40000	
1993	29.09	India, Killari	6.3	23000	
1994	17. 01	USA , Northridge	6.7	63	40
1995	17. 01	Japan, Kobe	6.9	5600	140
1997	26.09	Italia, Umbria	5.9	11 (great cultural losses)	
1997	10. 05	Iran, Ardebil	7.1	1600	
1998	30. 05	Afghanistan	7.1	5000	
1999	17. 08	Turkey, Izmit	7.4	20000	30
1999	21. 09	Taiwan, Chi- Chi	7.3	2500	1

In the following, the main earthquakes that occurred in the last century are commented on in order to identify their influence on the main advances in seismic design.

The first very important seismic event produced in the 20th Century is the *1906 US San Francisco earthquake* (Fig. 2.5) (USGS, nd, Wikipedia, nd). A break 430 km long along the San Andreas Fault initiated an earthquake of 7.8 magnitude and sparked widespread fire in the San Francisco area. The earthquake damaged large portions of the city, especially the older buildings, which were not structurally sound, and caused large sections of the city which had been built upon filled land to resettle and sink. A horizontal displacement of 6.4 m was measured and in many places broke fence lines and split roadways. Since that day, this earthquake remains one of the most devastating to hit the San Francisco area. The main shock epicenter occurred about 3 km from the city, being the first case when an earthquake occurred close to a very urbanized area. At the same time, the resulting fire after the earthquake shows the damaging effects of the out of control subsequent fires. It has been estimated that about 90% of the total destruction was a consequence of the subsequent fires. The destruction that resulted from this earthquake opened the eyes of many people to the hazards of living in an earthquake-prone zone. At the same time, the importance of this earthquake for seismic design was the establishment of the relation between faults and earthquakes, being the start of a scientific conception, for structural design. After this earthquake, the Seismological Society of America was established, being considered like the birth date of Earthquake Engineering in US.

The second very important event was the Messina earthquake. On December 28, 1908 a strong earthquake (M 7.5) hit Messina in Sicily, as well as Reggio Calabria (Wikipedia, nd). The earthquake arose in the Messina Strait, which separates Sicily from Calabria. After the earthquake, a tsunami was formed striking many coastal habitats with 15m waves and causing even more destruction than the earthquake itself. This is one of the largest tsunami ever to have occurred. The cause of this earthquake was due to the existing fault system, resulting from the subduction of the African tectonic plate under the Eurasian one. With 83,000 deaths, the Messina earthquake is, until today, the most murderous among the European earthquakes. According to Housner (2002), this earthquake can be considered the origin in Italy of practical earthquake engineering, because the Commission report appears to be the first engineering recommendation for the design of earthquake-resistant structures, by using the equivalent static method of analysis.

The third very important earthquake of the beginning of the 20th Century was the *1923 Japan Kanto (Tokyo) earthquake* (Fig. 2.6), (Wikipedia, nd), produced in the area around Sagami Bay with a magnitude of 8.3. The origin of this earthquake was the presence of a fault system associated to the existence of four tectonic plates (Pacific, Philippine, North American and Eurasian plates) near the Japan Isles. In the frame of this fault system, the Pacific plate subducted under the other

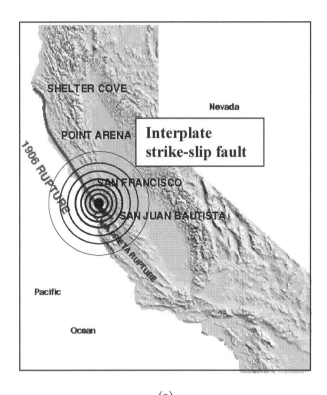

(a)

(b)

Figure 2.5 1906 San Francisco earthquake. (a) Epicenter location;
(b) Collapse of City Hall (after USGS, nd)

plates. The fault displacement under the bay was measured to be nearly 240 m. The large movements of ground in the Tokyo-Yokohama area had catastrophic consequences; more than half of the building stock in the two cities was severely damaged or ruined completely. However, most of the damage was due to the devastating fires, which followed after the earthquake. The earthquake generated a tsunami measuring 12 m high at Atami. After this earthquake, the Earthquake Research Institute of the Tokyo University was the first research group formed for studying both seismology and earthquake engineering. These researches significantly contributed to the progress of the earthquake knowledge in Japan.

The *1940 US El-Centro earthquake* is very important in the history of the development of seismic design, being the first time when an earthquake was recorded in digital form which then could be used in structural analysis. But the instrument, which recorded this accelerogram, was attached to the concrete floor of the El Centro Terminal Substation Building and not in a free-field location, which can give some doubts about the actual earthquake characteristics. The record may have under-represented the high frequency motions of the ground, because of the soil interaction of the massive foundation with the surrounding soft soil (Vibrationdata, 2007). Unfortunately, due to the fact that the knowledge about the actual characteristics of different types of earthquakes was limited for a long time, this record was used for many years in areas and for structures where such type of earthquake acceleration can never occur.

The world's largest ground motion was the *1960 Chile earthquake,* which occurred off the coast of central Chile (Fig. 2.7). The quakes, whose magnitude of 9.5 has never been reached up to now by other recorded ones, resulted from the rupture along the boundary of the subduction between the Nazca and the South American Plates, which was 1000 km long and 200 km wide. Its tremors caused widespread destruction, changes in land formations and one of the largest tsunami recorded in the Pacific region (Wikipedia, nd). The coastal mountains uplifted about 2 m, the hills of the Andes about 0.5 m and some offshore islands about 6 m. The cities of Valdivia, Conception and Puerto Mont suffered the heaviest destruction, while many smaller villages were completely destroyed. The effects of the tsunami produced important damage in Chile (the 25m high sea waves washed along the coasts of Hawaii, Japan and Philippines). Although the earthquake magnitude was very high, there was not an enormous number of victims because the population was alerted that something was going to happen by previous shakes and underground noise.

The great *1964 Great Alaskan earthquake* (Fig. 2.8) (also called Good Friday earthquake) (Wikipedia, nd), with magnitude 9.2, was the largest ground motion

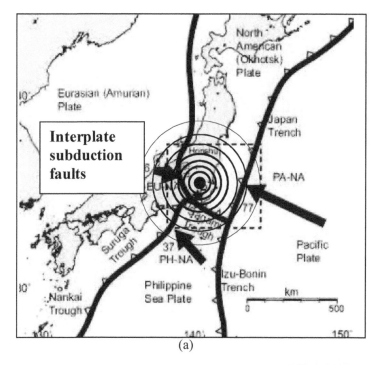

(a)

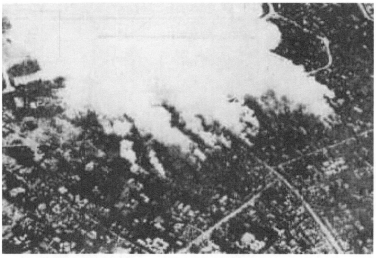

(b)

Figure 2.6 1923 Kanto (Tokyo) earthquake. (a) Tectonic plates in Japan (after Nyst et al, 2006); (b) Devastating fire (USGS, nd)

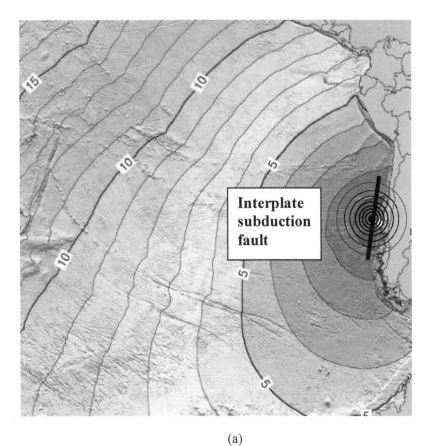

(a)

(b)

Figure 2.7 1960 Chile earthquake: (a) Location and tsunami;
(b) Damage of buildings (after NGDC, nd)

in North America and the second largest ever recorded (after the Chile earthquake). The rupture was about 800 km long, producing massive avalanches and landslides. Important areas were uplifted by nearly 11.5 m and generated a devastating tsunami, with a wave of 67 m measured at Valdez Inlet. However, this earthquake caused only 115 deaths, 9 due to ground motions and 106 due to the tsunami, and low monetary losses, considerably small for a quake of this size. This was primarily due to the low population density and the type of houses, mainly made of wood.

The 4th most deadly earthquake of the world in the 20th Century was the undersea *1970 Peru Ancash earthquake* (Wikipedia nd), with a magnitude of 8.1. This earthquake was the result of the subduction of the Nazca tectonic plate under the South American tectonic plate. Combined with a resultant landslide, it was the worst catastrophic natural disaster ever recorded in the history of Peru. Many coastal cities were significantly damaged, but the Andean valley, a steep valley paralleling the coast, suffered the worst disaster. The ground motions destabilized

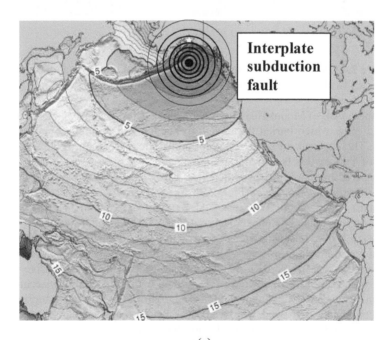

(a)
Figure 2.8 (continues)

(b)

Figure 2.8 1964 Alaskan earthquake. (a) Location of epicenter and tsunami;
(b) Bridge damage on the Cooper River (NGDC, nd, USGS, nd)

the Northern wall of Mount Huascaran, inducing a catastrophic avalanche of rocks, mud and snow which buried the towns of Yungay and Ranrahirca in only three minutes. The government of Peru has forbidden the excavation of the area where the city of Yungay was buried, declaring it a national cemetery.

Considered by many to be the deadliest earthquake of 20th Century, the *1976 China Tangshan earthquake* (Fig. 2.9) (also known as the Great Tangshan earthquake) (RMS, 2006, Wikipedia, nd) brought devastation to many areas in North-western China. The quake occurred when a 150 km portion of the Tan-Lu fault system broke, releasing enough energy to trigger a magnitude 8.0. It is very interesting to notice that this very catastrophic earthquake occurred in an area far from the tectonic plate borders. The area experienced 10km of extensive surface faulting which ran through downtown Tangshan, with horizontal displacements up to 1.5 m. The densely populated, industrial coal-mining city of Tangshan was almost completely destroyed. The event also caused major damage in the city of Tianjin located 100 km Southwest of Tangshan and moderate damage in Beijing, located approximately 140 km to the West. It is believed to be the largest earthquake of the 20th Century for the number of deaths. Due to the earthquake's occurrence in the middle of the night, the damage caused a very high number of fatalities among the inhabitants of Tangshan city and surrounding communities. The first earthquake was followed by a major 7.8 magnitude aftershock some 16 hours later, increasing the total of deaths. The official death count was approximately 250,000, but many specialists consider that the real number exceeded 600,000 deaths.

The *1977 Romania Vrancea earthquake* (Fig. 2.10) (Balan et al, 1982) is considered as one of the most damaging earthquakes in Europe, because a very densely populated area was affected. Most of the damage was concentrated in Romania's capital Bucharest, at a distance of 160 km from the epicenter, where about 33 large buildings collapsed. Due to the deep depth of the source (94 km), the affected area was very large. In Bulgaria, many buildings were damaged. Three apartment buildings in the town of Svishtov collapsed, killing more than 100 people.

A 50km slip off the Pacific coast resulted in one of the most damaging quakes in the history of the Americas, the *1985 Mexico City earthquake* (Fig. 2.11) (Wikipedia, nd, Gioncu and Mazzolani, 2002). The epicenter was off the Pacific coast of the Mexican state of Michoacan (therefore in many studies this event is called the Michoacan earthquake), where the Cocos tectonic plate subducts the North American tectonic plate. Mexico City, more than 300 km from the epicenter, suffered major building damage, because a large part of the city was built on the filled land of an ancient lake bed. Due to these very bad soil conditions, the buildings experienced roughly five times the ground shaking than the outlying areas. Many buildings collapsed because of soil foundation loss, pounding effects or insufficient ductility. It is very important to notice that, during this earthquake, the first (and, fortunately, the only one case) of steel structure global failure occurred.

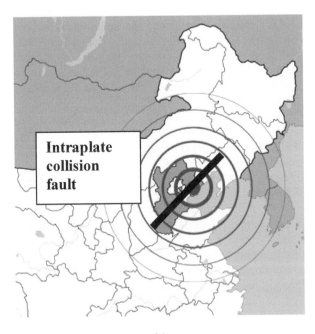

(a)

Figure 2.9 (continues)

(b)

Figure 2.9 1976 China Tangshan earthquake. (a) Location of earthquake;
(b) Air view of the devastation after the earthquake (USGS, nd)

The *1989 Loma Prieta earthquake* (Fig. 2.12) (Wikipedia, nd) occurred when the crustal rocks comprising the Pacific and North American plates abruptly slipped 2 m along the boundary of the San Andreas fault system in the direction of San Francisco city. The epicenter of the quake was an unpopulated area in the Aptos and Cruz Mountains, near the Loma Prieta Peak. The rupture initiated at the depth of 18 km and extended 35 km along the fault system without breaking the surface of the Earth. The number of deaths (only 63) was remarkably low due to the introduction of modern concepts in design. The worst disaster produced by this earthquake with the most number of fatalities was due to the collapse of the Cypress Street Viaduct on the Nimitz Freeway. One 15m section of the San Francisco Oakland Bay Bridge also collapsed.

The *1994 US Northridge earthquake* (Fig. 2.13) (Wikipedia, nd) was unusual because the epicenter was within a very densely populated metropolitan area of Los Angeles. Despite the area's proximity to the San Andreas Fault, the Northridge quake did not occur along this fault, but rather on a previously undiscovered blind fault. The rupture was initiated at 18 km depth and then spread Northwest. Many commercial buildings and bridges were damaged and, in terms of property damage, the earthquake is one of the most costly natural disasters in US history. This was the first time that an earthquake occurred in a very populated zone and the special effects of near-fault ground motions (very high velocities and accelerations and important vertical components of ground motion) were noticed. Many reinforced concrete (RC) buildings and bridges were severely damaged. But, from the structural engineering point of view, one of the great surprises of this earthquake was the discovery of widespread damage to welded connections in modern steel moment frame buildings. After this earthquake, important changes were made in design and construction practice, also due to the revision of the building codes.

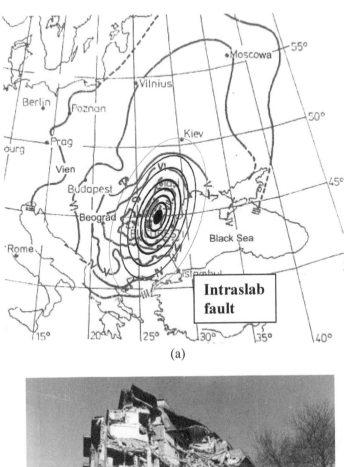

(a)

(b)

Figure 2.10 1977 Vrancea earthquake. (a) Location of earthquake and affected area (after Balan et al, 1982); (b) Failure of an RC building

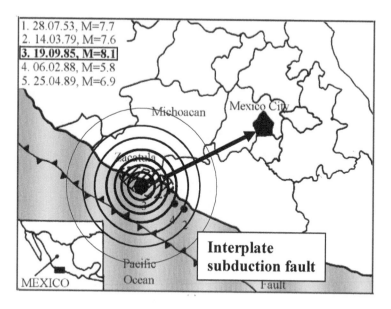

(a)

(b)

Figures 2.11(continues)

(c)

Figure 2.11 1985 Mexico City earthquake. (a) Location of epicenter (after Gioncu and Mazzolani, 2002); (b) RC building collapses (USGS, nd) ; (c) Steel Pino Suarez Building collapses (NGDC, nd)

(a)
Figure 2.12 (continues)

(b)

Figure 2.12 1989 Loma Prieta earthquake. (a) Location on the San Andreas fault;
(b) Damage of San Franciso-Oakland Bay Bridge (USGS, nd)

The Northridge earthquake marked a new direction in structural research, the
effect of near-source earthquakes being now one of the most studied subjects.

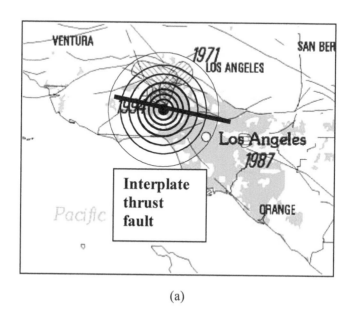

(a)

Figure 2.13 (continues)

(b)

(c)

Figure 2.13 US Northridge earthquake. (a) Fault system and location of earthquake; (b) Collapsed garage; (c) Freeway collapse (USGS, nd)

The *1995 Japan Kobe earthquake* (Fig. 2.14) (Wikipedia, nd) (known also as the Hanshin earthquake) was the most devastating earthquake ever to hit Japan. There are many coincidences with the 1994 Northridge earthquake. Kobe is located near the dangerous intersection of three tectonic plates: the Pacific, Eurasian and Philippine, but the earthquake occurred on an unknown secondary fault with the epicenter on the Northern end of Awaji Island. The proximity of the epicenter (20 km away from the city of Kobe), and the propagation of rupture directly beneath the highly populated region, help to explain the high level of destruction. In addition, the effects of post-earthquake fire and liquefaction strongly contributed to produce a very high level of damage. The same damage to steel connections as in the case of the Northridge earthquake was noticed, indicating that welded connections were one of the weakest locations in steel moment frames.

The damage of highways was the most impressive image of this earthquake: ten spans of the Hanshin Expressway Route 43 in three locations in Kobe and Nishinomiya were knocked down. The Kobe earthquake was listed in the Guinness Book of Records as the *costliest natural disaster to befall any one Country*.

The Northern part of Anatolia was stuck by the *1999 Izmit earthquake* (Fig. 2.15) (Wikipedia, nd) (also called the Kocaeli earthquake), which can be considered among the largest seismic events to have occurred in the Eastern Mediterranean Basin during the last century, due to the number of fatalities and damage. The rupture of the Anatolian fault (with similarities with the San Andreas US fault) has the length of 145 km in a zone with a population of 20 million inhabitants (one third of Turkey's total population) and encompassed nearly half

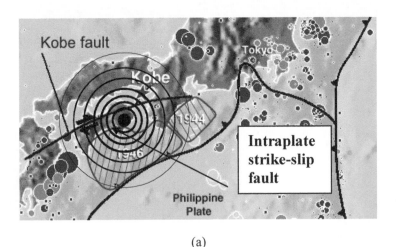

(a)

Figure 2.14 (continues)

(b)

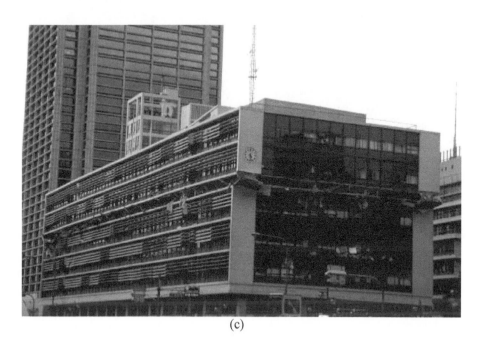

(c)

Figure 2.14 1995 Japan Kobe earthquake. (a) Location of epicenter (USGS, nd); (b) Collapse of concrete building (c) Collapse of City Hall (courtesy from Fischinger et al, 1998)

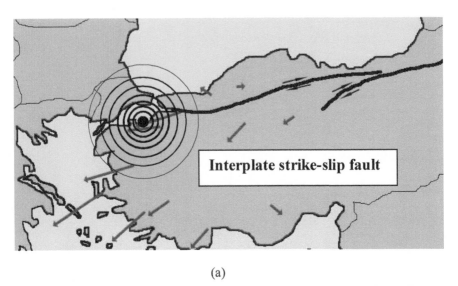

(a)

(b)

(c)

Figure 2.15 1999 Turkey Izmit (Kocaeli) earthquake. (a) Location of earthquake; North Anatolian fault; (b), (c) RC buildings collapse (USGS, nd)

of Turkey's industry. The predominant structural system used for buildings in Turkey consists of reinforced concrete frames, with masonry infills. Unfortunately, poor structural conception (soft stories, strong beams and weak columns) and execution (low material qualities and poor detailing) caused the collapse of a large number of buildings. This earthquake clearly demonstrated that improperly constructed buildings kill people.

The last great earthquake of the 20th Century was the *1999 Chi-Chi earthquake* (Fig. 2.16) (Wikipedia, nd), which devastated the geographical center of the Taiwan island along the Chelungpu fault, being created from the subduction of the Philippine tectonic plate beneath the Eurasia tectonic plate. Being one of the world high earthquake risk regions, the island is heavily implemented with instrumentation to record seismic data, having the possibility to record the near-fault ground motion characteristics. Therefore, during the Chi-Chi earthquake very high accelerations, velocities and vertical movements were recorded, never recorded during other earthquakes. This gives to the specialists the possibility to have a right imagine about the seismic actions striking the structures in the near-fault areas. The powerful forces moved mountains, cut off rivers, caused liquefaction and damaged bridges, highways and concrete dams and overturned many high buildings. As in the case of the Izmit earthquake, the poor quality of control at job sites has been blamed as the primary culprit for most building damage.

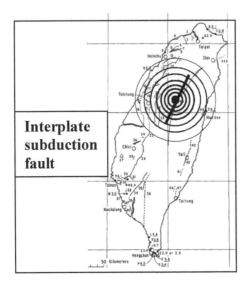

Interplate subduction fault

(a)

(b)

Figure 2.16 1999 Chi-Chi earthquake. (a) Location of earthquake (USGS, nd);
(b) RC buildings collapse (NGDC, nd)

2.4 GREAT EARTHQUAKES OF THE BEGINNING OF 21st CENTURY

The new century began with very large number of casualties following earthquakes, presented in Table 2.5.

Table 2.5 The great earthquakes of period 2000-2006

Year	Date	Location	Magnitude	Deaths	Damage (billion $)
2001	13.01	El Salvador	7.9	1100	1.3
2001	26.01	India, Bhuj	7.9	20000	5.5
2001	23.06	Peru, Atico	8.4	2000	
2003	26.12	Iran, Bam	6.7	35000	large cultural damage
2004	26.12	Indonesia, Sumatra	9.0	283000	13.5
2005	28.03	Indonesia, Sumatra	8.7	1520	
2005	08.10	Pakistan, Kashmir	7.6	79000	3.5
2006	02.01	Sandwich Islands	7.4		
2006	27.01	Indonesia, Banda Sea	7.6		
2006	20.04	Russia, Koryakia	7.6		0.6
2006	03.05	Pacific, Tonga	7.9		
2006	17.07	Indonesia, Java	7.7		
2006	27.12	Taiwan	7.2		
2007	12.01	Kuril Island	8.1		
2007	21.01	Molucca Sea	7.5		
2007	01.04	Solomon Islands	8.1	28	
2007	15. 08	Peru	8.0	650	
2007	13.09	Sumatra	8.4		
2007	14.11	Chile	7.7	2	
2007	29.11	Martinique Islands	7.4		
2007	09.12	Fiji Islands	7.8		
2007	19.12	Aleutian Islands	7.2		
2008	12.05	Sichuan, China	7.9	67180	
2009	06.04	L'Aquila, Italia	6.3	281	cultural damage

The *2001 Gujarat earthquake* (Wikipedia, nd) (also called the Bhuj earthquake) was the most devastating earthquake in India in recent history. It occurred at a great distance from any plate boundary, where the majority of earthquakes of this size happen according to the theory of the Tectonic Plates. Because of this, the area was not completely prepared for an earthquake of such intensity, causing so much devastation. The quake destroyed 90% of the homes in Bhuj and more than 50 multi-story buildings collapsed in Ahmedabad (4.5 million inhabitants).

An important earthquake, the *2001 Peru Atico* earthquake took place on the coast of Peru. Due to the geotechnical feature, some unusual damage occurred in areas situated 300 to 500 km Southeast far from the original epicenter, which was caused by landslides and rock falls. The reduced number of fatalities in relation with the high magnitude is due to the large distance of the epicenter from the major population centers.

The *2003 Iran Bam earthquake* (Fig. 2.17) (Wikipedia, nd) destroyed the ancient citadel of Bam, dating back around 2500 years (during the Persian period), and considered the biggest adobe construction of the world. The city benefited from tourism, with an increasing number of people visiting the ancient citadel. The earthquake destroyed 70% of the city, the Bam citadel was leveled to the ground, producing a very large number of fatalities due to the collapse of old buildings. This earthquake was responsible for the irreparable destruction of one of the main important monumental cities in the world.

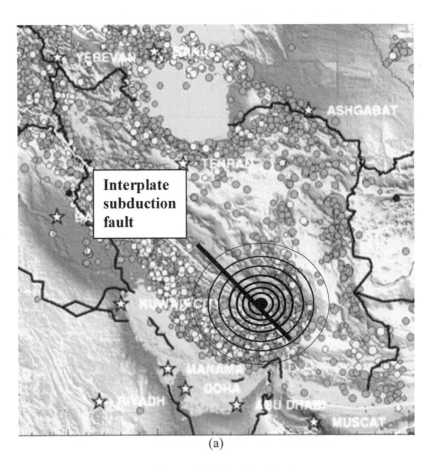

(a)

Figure 2.17 (continues)

(a)

(b)
Figure 2.17 2003 Iran Bam earthquake. (a) Location of Bam epicenter (USGS, nd); (b) Bam citadel before and after earthquake (FEMA, nd)

The World's third strongest of 9.0 M recorded earthquake as magnitude (after Chile and Alaska earthquakes), the *2004 Indian Ocean earthquake* (Wikipedia, nd) (also called Great Sumatra-Andaman earthquake) occurred in one of the most seismically active zones on Earth (Fig. 2.18), the West coast of Sumatra, due to the subduction of the India tectonic plate under the Eurasia tectonic plate. The earthquake caused major damage to buildings and infrastructure, predominantly in North Sumatra. But the great disaster was caused by the tsunami, whose origin was due to the fact that the epicenter was located in the open sea. It catastrophically affected Indonesia, Sri Lanka, India, Thailand and the Maldives. Damage and deaths were also produced on the East African coast (Somalia, Tanzania, and Kenya) and in Bangladesh, Burma, the Seychelles and Malaysia. The great number of fatalities was mainly due to the lack of an alarm system. It is very interesting to notice that the following 2005 great earthquake, which occurred practically in the same place, with a magnitude close to that of the 2004 earthquake (see Table 2.5), did not give any sign of a possible tsunami.

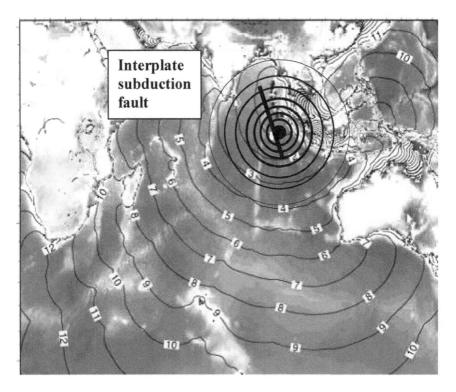

(a)

Figure 2. 18 (continues)

(b)

Figure 2.18 2004 Indian Ocean earthquake. (a) Location of epicenter and tsunami; (after NGDC, nd) (b) Produced tsunami (NOAA, 2004)

The *2005 Kashmir earthquake* (Wikipedia, nd) occurred not far from the border of Afghanistan and India along a fault associated to the Indian and Eurasian tectonic plates. It was the strongest natural disaster in Pakistan's history in terms of victims, destruction of infrastructure and economic assets.

For other earthquakes that occurred in this period, even if their magnitude was high, there were no reports about a high number of deaths and important damage, because they were located in scarcely populated areas.

The *2008 Sichuan (China) earthquake* (known also as Wenchuan earthquake) (Wikipedia, nd), magnitude 8.0 M, occurred on 12 May 2008, with the epicenter in the Wenchuan County, Sichuan Western province of China (Fig. 2.19a). The seismicity of this region is a result of Northward convergence of the Indian Plate (represented by Himalayan Mountains and Tibetan Plateau) against Eurasian Plate (represented by the very rigid Sichuan Plateau of South China Block). As a result of this collision the Logomen Shan Fault formed, a collision thrust fault type. Along this fault a high degree of stress concentration occurs and, finally, this caused a sudden dislocation in fault, leading to the violent Sichuan earthquake. The earthquake lasted about two minutes and released 30 times the energy of the 1995 Kobe (Japan) earthquake. The shallowness of epicenter (about 19 km), the density of population of this region and the presence of many non-resistant structures and school buildings, greatly increase the severity of the earthquake, producing the collapse of many unprepared buildings (Fig. 2.19b,c). The main lesson learned after this earthquake is that even an earthquake measuring magnitude 8.0 M, which is considered a big one, need not necessarily be a calamity, it is the building vulnerability which changes it into a disaster.

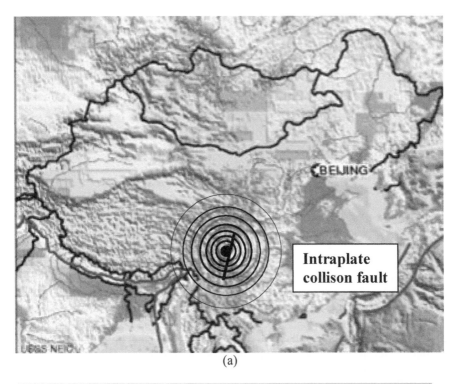

(a)

(b)

Figure 2.19 (continues)

(c)

Figure 2.19 2008 Sichuan (China) Earthquake: (a) Location;
(b), (c) Damaged buildings (USGS, nd)

A very damaging earthquakes occurred on 6 April 2009 in central Italy (Fig. 2.20a), in the Abruzzo region, where the city of L'Aquila (situated in the epicentral area), very rich in cultural treasures, was most damaged (Fig. 2.20b). The earthquake magnitude was not very high, but due to near-source conditions, bad soil conditions (sedimentary basin with maximum depth of about 250 m) and seismic unprepared historical buildings, the damage was very significant.

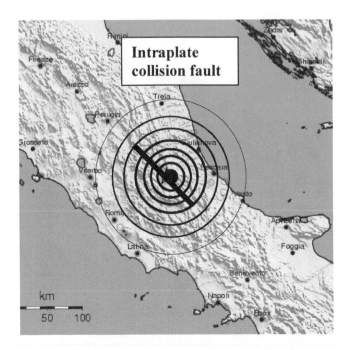

(a)

(b)

Figure 2.20 2009 L'Aquila earthquake: (a) Location;
(b) Church damage

2.5 REFERENCES

nd – no date: the web site is periodically modified

Agnew, D.C. (2002): History of Seismology. In International Handbook of Earthquake & Engineering Seismology (eds. W.H.K. Lee et al), Academic Press, Amsterdam, 3-11

Balan, S., Cristescu, S., Cornea, I. (1982): The Romanian Earthquake of 4 March 1977, Editura Academiei Romane, Bucuresti

Blak, B. (2000): The North Anatolian fault zone. Boundary between the Arabian Plate and the Anatolian Block.
http://geologyindy/byu.edu/faculty/rah/tectonics/students20%Presentation/2000

Earthquake-Wikipedia (2007): Image-Quake epicentres 1963-98.
http:/en.wikipedia.org/wiki/Earthquake

EDM (2000): Report on the Chi-Chi Taiwan earthquake of Sepember 21, 1999. Technical report no.7
http:// www.edm.bosai.go.jp/Taiwan 1999/report/Chapter 2.3.pdf

EDM (2001):Gujarat, India, earthquake of January 26, 2001
http://gees.usc.edu/GEES/RecentEQ/India_Gujarat?photos/photobyrediff/capt9.jpg

EMSC, European-Mediterannean Seismological Centre
http://www.emsc-csem.org/-223k

FEMA (nd): Bam earthquake response
http://www.fema.gov/bamearthquake

FHWA(1994): The Northridge earthquake: progress made, lessons learned in seismic-resistant bridge design.
http://www.tfhrc.gov/pubrds/summer94/p94su26.htm

Fischinger, M., Cerovsek, Tomo, Turk, Z.(1998): EASY. Earthquake engineering slide information system. A hypermedia learning toll. Electronic Journal of Information Technology in Construction, Vol. 3, str. 1-10

Gioncu, V., Mazzolani, F.M (2002): Ductility of Seismic Resistant Steel Structures. Spon Press, London

Housner, G.W. (2002): Historical view of earthquake engineering. In International Handbook of Earthquake & Engineering Seismology, Academic Press, Amsterdam, 13-18

ISDR (2001): Remembering the worst earthquake in Latin America
http://www.crid.or.cr/crid/CD_EIRD_Inform/imagenes/ing/No1_2001/pagina23.htm

JSCE (2001): Provisional report on the June 23, 2001, Atico earthquake, Peru
http:/jsce.or.jp/report/14/01/provisional_n.pdf

Mazzolani, F.M. (2002): Structural integrity under exceptional actions: basic definitions and field of activity, In COST Seminar, Lisbon, 19-20 April 2002, 67-80

NGDC (nd): Geologic Hazards Photos. Earthquake of September 19, 1985, Mexico City
http://www.ngdc.noaa.gov/seg /cdroms/geohazards_v2/documents/647003.htm
NGDC (nd): Tsunami travel time maps for the Atlantic, Indian and Pacific Oceans.
http://www.ngdc.noaa.gov/seg/hazard/tsu_travel_time.shtml
NGDC (nd): The great Hanshi-Awaji (Kobe) Earthquake, Japan, January 17 1995
http://wwwngdc.noaa/hazard/data/geohazard/
NGCD (nd): Natural hazard slide sets. Chi-Chi, Taiwan earthquake
http:// www. ngdc.noaa.gov/nndc/struts
NOAA, Center for Tsunami (2004): December 26, 2004 Indonesian Sumatra earthquake
http:// netr.pmel.noaa.gov/sumatra2004.html.photo
Nyst, M., Nishimura, T., Pollitz, F.F., Thatcher, W. (2006): The 1923 Kanto Earthquake Re-evaluated using a newly augmented geodetic data set.
http://quake.usgs.gov/research/deformation/kanto_1923.pdf
Pararas_Carayannis, G. (2007): The great earthquake of 19 September 1985 and the major earthquake of 21 September 1985 in Mexico
http://www.drgeorgepc.com/Tsunami1985Mexic0.html
RMS (2006): The great Tangshan earthquake. 30-year retrospective
http://www.rms.com/Publications/1976Tangshan.pdf
UNESCO (2004): Bam earthquake, UNESCO's response
http://portal.unesco.org/en/ev.php
USGS (nd): This dynamic earth: The story of plate tectonics.
http://pubs.usgs.gov/publications/text/dynamic.html
USGS (nd): Progress towards a safer future since the 1989 Loma Prieta earthquake.
http://pubs.usgs.gov/fs/1999/fs 151-99/
USGS (nd): Response to an urban earthquake, Northridge 94
http:// pubs.usgs/of/1996/ofr-96-0263/
USGS (nd): Open-file report 99-47
http;// pubs.usgs.gov/1999/of99-447
USGS (nd): Magnitude 6.6 Southeastern Iran, 2003 December 16
http:// neic.usgs.gov/neis/eq_depot/2003/
USGS (nd): Latest Earthquakes in the World. Past 7 Days
http:// earthquake.usgs.gov/recenteqsww/-49k
USGS (nd): The great 1906 San Francisco earthquake
http://earthquake.usgs.gov/regional/nca/1906/18april/index.php
USGS (nd): Earthquake facts and statistics
USGS (nd): Multimedia gallery home/ Photos
http://gallery.usgs.gov/tags/Sichuan
http://www.neic.cr.usgs.gov/neis/eqlists/eqstats.html
USGS (nd): Kanto, Japan, 23 September. Magnitude 7.9
http://earthquake.usgs/gov/world/events/1923_09_01.php
USGS (nd): Prince William Sound, Alaska, magnitude 9.2

http://earthquake.usgs.gov/earthquake/states/events/1964_03_28_pics1.php
    USGS (nd): Tangshan, China, 1976 July 27, 7.5 Magnitude
http|// earthquake.usgs.gov/earthquakes/world/events/ 1976_07_27.php
    USGS (nd): Earthquake Topics in Kobe, Japan
http://earthquake.usgs.gov/ learn/topics/
    USGS (nd): Damage photos from Izmit earthquake
http://earthquake.usgs.gov/research/geology/turkey/8-26-99.php
    Vibrationdata (2007): Welcome to Vibrationdata. El Centro earthquake page
http://www.vibrationdata.com./elcentro.htm
    Wikipedia (nd): 226BC
http://en.wikipedia.org/wiki/226BC
    Wikipedia (nd): Colosseum
http://en.wikipedia.org/wiki/Colosseum
    Wikipedia (nd): 1556 China-Shaanxi earthquake
http://en.wikipedia.org/wiki/1556_Shaanxi_earthquake
    Wikipedia (nd): 1755 Lisbon earthquake.
http://en.wikipedia.org/wiki/1755_Lisbon_earthquake
    Wikipedia (nd): New Madrid earthquake
http://en. Wikipedia.org?wiki/New_Madrid_Earthquake
    Wikipedia (nd):1906 San Francisco earthquake.
http://en.wikipedia.org/wiki/1906_Sam_Francisco_earthquaake
    Wikipedia (nd): 1908 Messina earthquake
http://en.wikipedia.org/Messina_Earthquake
    Wikipedia (nd): 1923 Great Kanto earthquake
http://en.wikipedia.org/wiki/1923_Great_Kanto_earthquake
    Wikipedia (nd): 1960 Valdivia earthquake
http://en.wikipedia.org/wiki/1960_Valdivia_earthquake
    Wikipedia (nd): Good Friday Earthquake
http://en.wikipedia.org/wiki/Goog_Friday-Earthquake
    Wikipedia (nd): 1970 Ancash earthquake
http://en.wikipedia.org/wiki/1970_Ancash_earthquake
    Wikipedia (nd): 1976 Tangshan earthquake
http://en.wikipedia.org/wiki/Tanshan_earthquake
    Wikipedia (nd): Loma Prieta earthquake
http://en.wikipedia.org?wiki/Loma_Prieta_Earthquake
    Wikipedia (nd): Bay Bridge collapse
http://en.wikipedia.org/wiki/Image:Bay_Bridge_collapse_2.jpg
    Wikipedia (nd): Northridge earthquake
http://en.wikipedia.org/wiki/1994_Northridge_Earthquake
    Wikipedia (nd): Great Hanshin earthquake
http://en.wikipedia.org/wiki/ Great_Hanshin_earthquake
    Wikipedia (nd): 1999 Izmit earthquake
http://en.wikipedia.org/wki/1999_%BOsmit_earthquake
    Wikipedia (nd): 1999 Chi-Chi earthquake
http://en.wikipedia.org?wiki/Chi-Chi_earthquake
    Wikipedia (nd): 2001 Gujarat earthquake

http://en.wikipedia.org/wiki/2001_Gujarat_earthquake
    Wikipedia (nd): Bam, Iran
http://en.wikipedia.org/wiki/Bam_Iran
    Wikipedia (nd): 2005 Kashmir earthquake
http://en.wikipedia.org/wiki/2005_Kashmir_earthquake
    WWU, Network Path (2007): Alaska earthquake, Prince William Sound, 1964
http://www.smate.wwu.edu/teched/geology/eq-Alaska64.html

Chapter 3

Learning from Earthquakes

3.1 MAIN LESSONS AFTER THE STRONG EARTHQUAKES

3.1.1 The Paradox of Design: Success through Failure

In spite of the great efforts made in recent years in solving satisfactorily the problem of building design in seismic areas, recent earthquakes are capable of producing more damage today than ever before. Besides, a marked increase of financial losses can be observed (Fig. 3.1), especially during the catastrophic decade 1985-1995, marked by the Mexico City, Loma Prieta, Northridge and Kobe earthquakes. The main reason of this remarkable increase of damage costs in recent years is due to the concentration of population and industrialization in high seismicity regions. The rapid and in many cases uncontrolled urbanization of metropolitan regions contributed to the vulnerability of their communities and infrastructure. More recent earthquakes have occurred in these regions and the events of Izmit and Taiwan produced great disasters due to the proximity of highly urbanized areas. These events clearly shown that the design methodologies and code provisions are not infallible. This is true especially in cases where the earthquake sources are near to a densely populated area.

The paradox of structural engineers is that while engineers can learn what not to do from failures, they do not learn how to do from successes (Petroski, 1985, 2006). The failure of a structure contributes more to the evolution of design concepts than a structure successfully standing without accidents, on the condition that the engineers have the capability to understand what happened (Gioncu and Mazzolani, 2002).

Structural failures occur in part because the design process is subject to flaws and failings of human nature. Failures persist, because the design process is fundamentally carried out by a human mind in a human context. The structural failures cannot be eliminated by the development of refined computational models. A given situation where no structural failures occur, could be produced by an over conservative design, wasting resources which might be better applied elsewhere in the society (Petroski, 2006). Therefore, the failure of structures may be regarded as a factor of progress in design, not a calamity, under the condition to learn the lesson of each catastrophic event.

So, the main purpose of modern seismic design is only to reduce economic losses, and at the same time to save human life, not to completely eliminate them. The learning from disasters is the key to solve this problem. It is very interesting to notice the hurry in which the structural engineers across the world rushed to the sites of some produced quakes to study the damage. Their objectives: to gain a better understanding of the geological faults, to determine why buildings and structures failed, and to share these findings with other earthquake-prone zones (Hays et al, 1999).

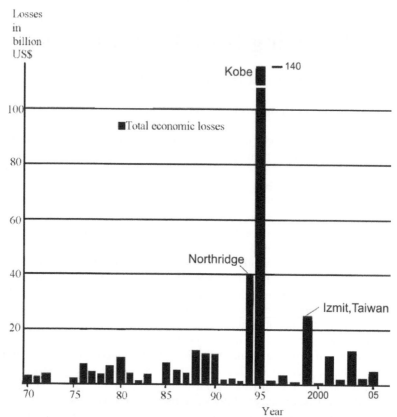

Figure 3.1 Cost of losses due to earthquakes (after Gioncu and Mazzolani, 2002)

3.1.2 Post-earthquake Investigations

The purpose of post-earthquake investigations is to observe and document the earthquake processes and the effects on man-made works. Progress in earthquake engineering has been significantly influenced by the experiences in destructive ground motions. While a great deal can be learned about construction vulnerability through laboratory and analytical studies, the most effective teacher is the examination of an earthquake on full-scale structures. No mathematical model can be accepted unless it correctly explains what happens in nature. No seismic disaster preparedness plan can be confidently implemented unless its principles have been already tested through actual use (McClure, 1989).

The post-earthquake investigations can be subdivided into seismic hazard examination and construction vulnerability studies.

The examination of produced earthquakes can be separated in some important periods, considering the evolution of ideas about the causes of the earthquake generation, the development of network instrumentation all over the world and the achievement of data banks. During the *historical period* (until 1910), the idea about the cause of earthquakes varies between mythology and natural phenomena. In the *early period* (1910-1950), the causes were related to the presence of faults and the theory of tectonic plates of Wegener was elaborated and confirmed by in situ observations. During the *maturated period* (1950-1975) the development of a world network of instrumentations and the scientific examination of produced earthquakes began. The *modern period,* after 1975, when the large network of instrumentations and the use of GPS allow studying the tectonic plate movement and the effect of these movements on the Earth, gave the possibility to understand the rupture mechanism and to study the different earthquake types.

The post-earthquake examination of damaged buildings has succeeded during the same periods, from primitive to modern methodologies. In the last period it is recognized that, although there are tragic aspects, a damaging earthquake provides a unique laboratory for multi-disciplinary studies to document, understand and explain what happened. Such studies are beneficial to minimize the potential impacts of future earthquakes and to provide information, which can be transferred to other communities, in order to mitigate seismic risk (Hays, 1996). This information is derived from seismological, geotechnical, architectural, structural engineering and socio-economical studies. The result of these multi-disciplinary investigations has provided critical data, showing the vulnerability of a community versus an earthquake, depending on physical factors (magnitude, focal depth, proximity to the urban center, directivity, soil properties, etc.), as well as social factors (degree of prevention, mitigation, preparedness measures, etc.). Each damaging earthquake provides unique experiences and data, increasing the basic understanding of the earthquake complex phenomena and the social dimensions of an earthquake disaster.

By post-earthquake investigations of damaging earthquakes produced during both historical and modern periods, some very important lessons can be learned. These lessons refer to the earthquake hazard, construction vulnerability and mitigation of seismic risk.

3.13 Seismic Risk Mitigation

Unfortunately both scientists and engineers have no means to warn of or to accurately forecast an earthquake; the only possibility they have is to try to mitigate its consequences.

The last important earthquakes produced both in old and new urban areas have demonstrated that the social and economic impact of building damages, loss of functions and business interruptions were very huge, stressing the necessity to develop a transparent seismic design methodology to be applied to build new safe constructions, as well as to strengthen existing ones (Fajfar and Krawinkler, 1997).

In the past, for a long time, among the builders there was a defeatist attitude to consider an earthquake as a fatal force, not possible to resist. After the great 1906 San Francisco earthquake, Professor Charles Derleth (University of California) wrote (Roesset, 1992):

"... Many engineers appear to have the idea that earthquake stresses in framed structures can be calculated so that rational design to resist earthquake destruction can be made, just as one may allow for dead and live loads, or wind and impact stresses. An attempt to calculate earthquake stresses is futile. Such calculations could lead to no practical conclusions of values...".

This quotation reflects, in the past time, the negative attitude of engineers concerning seismic design.

Recently, thanks to the important progress in Seismology and Structural Engineering, a contrary attitude has been developed, with the conviction that all phenomena produced by an earthquake, even when very complex, can be controlled by analyses. Indeed, in the last thirty years, the understanding of the nature of earthquakes, of the effects of site soil characteristics and of the structural response of structures subjected to seismic actions has made real advances. The installation of a dense instrumentation in high earthquake risk areas provided a great amount of information about the main characteristics of ground motions. At the same time, due to the development of personal computers and special computer programs, the structural analysis of earthquakes now is no more an unsolvable problem. Therefore, today, the structural engineers are confident to have reached the stage where the actual structural performance during strong ground motions can be satisfactorily predicted, explained and quantified.

But, the recent important earthquakes (as 1989 Loma Prieta, 1994 Northridge, 1995 Kobe, 1999 Kocaeli, 1999 Chi-Chi) have shown that sometimes the reality is different from this optimistic attitude. The lessons learned after these events have shown that, despite the above-mentioned significant progress, there are still many unsolved problems and today it is more appropriate to speak only about *advances in seismic design* rather than about infallible solutions. Unfortunately, for the future, one can expect that earthquakes will continue to produce damage and large loss of lives. The real tragedy is that these human losses are due to the failure of the constructions and not due to the earthquake itself, the builders being the makers of tools for killing people.

Therefore, the *mitigation of seismic risk* is a moral duty of the structural engineers. This mitigation is possible only when the problems connected to the earthquakes are correctly approached. After Bertero (1997), the definition of seismic risk is:

"...the probability that social and/or economic consequence of earthquakes will equal or exceed specified values at a site, at various sites, or in an area, during a specified exposure time...".

After Ambraseys (2002), Grossi (2004), the seismic risk can be defined as:

(Seismic risk) = (Seismic hazard) (Vulnerability) (Losses)

where:

- *Seismic hazard* is the probability of occurrence, at a severity level and at a specific period of time, for a given area, of a potential damaging earthquake, which is beyond human control, but the knowledge of it is possible. The earthquake hazard is typically determined using a combination of seismological, morphological, geological and geotechnical investigations, combined with the history of earthquakes in the region. It depends not only on the seismicity, but also on collateral hazards: liquefaction, landslide, etc.

- *Vulnerability* is the degree of loss resulting from the occurrence of an earthquake of a given magnitude. The vulnerability refers to foundations, structural elements and non-structural elements of all built environment, including buildings and different facilities, roads, bridges, dams, population density, preparedness, building and lifeline damage, fire following, etc.
- *Losses* refer to the direct losses (failure of buildings and lifelines), indirect losses (interruption of activities), usually specified as a monetary value (in US $), and secondary losses (short and long-term impacts on the overall economy and socio-economic conditions).

If the earthquakes in some areas are inevitable, being out of human control, the vulnerability must be supervised and this should be the main objective of *earthquake preparedness programs*. Trends in mitigation of seismic risk refer to improving the *Seismic hazard* knowledge and to using new design methodologies aiming to reduce the vulnerability and losses. These trends differ in function of the degree of the economical development in the examined region.

Development in computer technologies, including Geographic Information System (GIS), Data Base Management System (DBMS), and Knowledge-Based Expert System (KBES), have made it possible to develop innovative methods for earthquake damage and loss estimation.

3.2 LESSONS FOR EARTHQUAKE HAZARD

A representative listing of topics involved in the earthquake hazard examination might consider the seismological, geotechnical and engineering aspects.

A better understanding of the *seismological aspects* of the earthquake phenomena results from the cooperation between geophysicists and seismologists. In spite of the complexity of an earthquake manifestation, real progress was marked in the classification of earthquakes. Nowadays it is very clear that there are some basic earthquake types with very different characteristics (Burkhard, 2001, Gioncu and Mazzolani, 2003) (Fig. 3.2).

- Earthquakes tend to recur where they have occurred in the past, especially along the boundary of tectonic plates (i.e. the Californian earthquakes) but major devastating earthquakes can occur also in areas where no previous events have been recorded (see New-Madrid, USA, or Bluj, India, earthquakes). There are very few areas in the world that are immune from the earthquake danger.

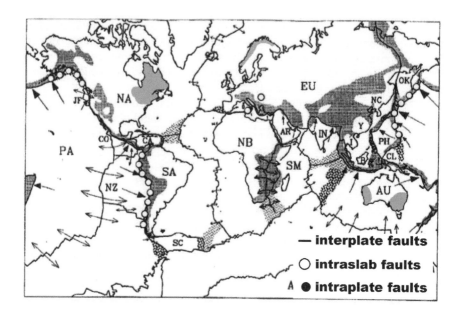

Figure 3.2 Different earthquake types (modified after Burkhard, 2001)

- The fault movements can be characterized by the tendency for the tectonic plates of moving away (creating faults, valleys and gulfs), moving towards each other (creating mountains due to the subduction or by crust folding) or by slipping. Each of these fault movements produces earthquakes with very different characteristics.

- *Interplate crustal earthquakes*, produced in the crust at the contact edges of two major tectonic plates; the Californian, Anatolian, or Iranian earthquakes are the most representative for this earthquake type. One must emphasize that these contact edges represent only the actual configuration of tectonic plate, ignoring the ancient steps in the Earth's history (see the Theory of cycles , Chapter 4).

- *Intraplate crustal earthquakes* produced in the same plate along some existing inland faults, the most representative for this type being the Central European, Chinese, Japanese, Indian and Australian earthquakes. These faults were produced during the formation of the actual Earth's crust, due to the ancient collisions of some tectonic plates, or by rupture of the existing plates due to the internal forces.

- *Intraslab earthquakes* produced in the same subducting slab, under the crust at large depths, the most representative for this type being the Canadian, Japanese and South American earthquakes, as well as the Sicily, Vrancea (Romania) and Southern Greece European earthquakes.

- The positions of active faults producing interplate earthquakes are very well known due to the fact that the tectonic plate boundaries are well defined. There are some main faults along the Circum-Pacific Ring and Alpide-Himalayan belt, which regularly produce great earthquakes.

- The positions of blind faults producing intraplate earthquakes are more difficult to be detected, being inactive for many years, but potentially they occur in near stable continental regions. The presence of a recently discovered, inactive until now, blind fault in the western area of Timisoara City, Romania, (Fig. 3.3) creates many problems to structural designers. The 1976 Tangshan China earthquake (250,000–660,000 fatalities) is an example of the damaging potential of this earthquake type. Unfortunately, the geological basis for intraplate seismogenesis is still poorly understood. A truth is established concerning these earthquake types: the presence of a fault potentially indicates a possible earthquake, but it is not sure that this will occur in the future.

- The interplate earthquakes occur in some very long fault zones, which appear to be segmented. So, the sources move from a segment to another, the case of the San Andreas, Mexican and Anatolian faults being significant for these migratory earthquakes. Figure 3.4 shows the case of the Anatolian fault, where the produced earthquakes advance from East to West, indicating the strike-slip direction of the fault movement.

- Generally, the boundary plate contacts are characterized by the presence of many parallel or diffuse faults, which can produce large earthquakes at the same level as the principal fault. The well-known San Andreas Fault has some parallel faults as Hayward, Rodgers, Calaveras, Greenville, and San Gregory. One can see that the probability now of the occurrence of a strong earthquake (the waiting *Big One*) is higher for Hayward fault than for San Andreas (Fig .3.5).

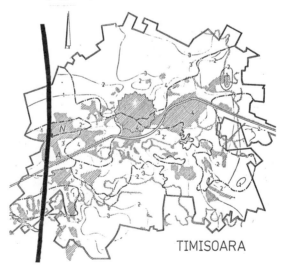

TIMISOARA

Figure 3.3 The map of Timisoara City, Romania, with the inactive, until now, Western fault

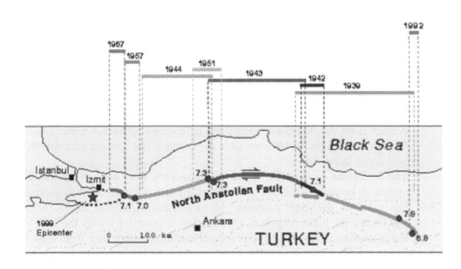

Figure 3.4 Migration of epicenters along the North Anatolian Fault
(USGS, nd)

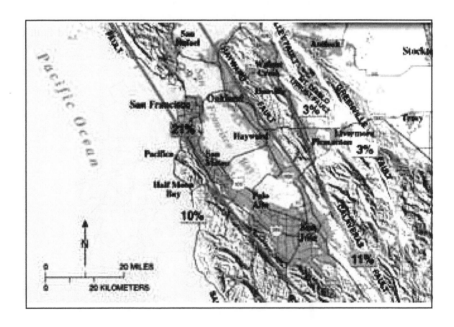

Figure 3.5 San Francisco's fault system (FEMA, 2003)

The *geotechnical aspects of earthquakes* are involved with the lessons considering the influence of travel path and site soil on the ground motions.

-   The influence of travel path and site soil depends on the earthquake type, being very different for interplate or intraplate earthquakes.
-   For interplate crustal sources, the travel path and local soil conditions modify the basic characteristics of the source, acting as the band-pass filters. The bad soil conditions of 1906 and 1989 San Francisco, 1977 Bucharest and 1985 Mexico City have a great influence on site ground motions, overshadowing the source characteristics (see Chapter 2).
-   For intraplate crustal earthquakes, characterized by high-frequency movements, the influence of site conditions is more reduced. Notable exceptions have been observed in case of shallow soil deposits and soil liquefaction, when significant amplification of ground motions can occurs. The soil liquefaction during the 1811 New-Madrid earthquake produced a very strong earthquake (with magnitude over 8) in a region without evident faults. The soil deposits and liquefaction played also a very important role during the Niigata and Kobe earthquakes (Fig. 3.6).
-   Soft soil, valley or relief effects can produce important ground motion amplifications. The Mexico-City is the most demonstrative case about valley and soft soil effects (Fig. 3.7).
-   The presence of buildings and the high density of built-space can be a very important factor, producing an important amplification of ground motions, especially in case of soft soil conditions. It is possible that this effect has

Figure 3.6 Overturning of some buildings during the Niigata earthquake, due to the soil liquefaction (USGS, nd)

Figure 3.7 Amplification of ground motions in Mexico City due to soft soil
and valley effect (Gioncu and Mazzolani, 2002)

given an important contribution on the very high amplification of records
in the Kobe earthquake, where the magnitude of source was moderate, but
the recorded ground motions in densely erected areas were very high.
Many debates arose after the erection of Taipei 101 Building, due to the
increasing trend of the number and the values of magnitude of ground
motions that occurred under the building site.

The *engineering aspects of earthquakes* refer to the lessons intended to present
the seismological observations at the level to be used for building design practice.

- Focal depth has a considerable influence on the earthquake characteristics.
  Surface sources have influence on a reduced area around an epicenter (see
  Californian and Anatolian earthquakes), while the deep sources affect a
  large area (for instance Michoacan and Vrancea earthquakes), producing
  large damage at locations situated far from the epicenter (see Chapter 1).
- Rupture areas, characterized by length and width, directly influence
  earthquake magnitudes and durations.
- Ground motions in near fault zones are characterized by directivity, as a
  function of the direction of the fault rupture (in many cases very important
  surface displacements along the fault are observed, Figure 3.8). It is very

important to note the difference between the characteristics of parallel and normal ground motions. This directivity was very evident during the Loma Prieta, Northridge, Kobe, Kocaeli and Chi-Chi earthquakes.

- For the near-field sites (an area within a distance of 15-20 km from the epicenter) the earthquake characteristics are very different in comparison with the ones recorded at the far-field sites. The three directions recorded of accelerations, velocities and displacements obtained at Northridge, Kobe and Chi-Chi prove this very important aspect.

- For surface sources in the near-field sites, even moderate earthquakes in magnitude can produce high levels of peak accelerations and velocities. The 1999 moderate Athens earthquake produced high accelerations.

- Recurrence periods, characteristic for a given place, are very different for interplate and intraplate earthquakes, because the former move along a very well determined fault, while the latter are characteristic for an area, where the epicenter positions are undetermined and in continuously changing. For interplate earthquakes it is possible to use statistical methodologies, which, in contrast, are useless for the intraplate earthquakes due to the absence of sufficient data on the same site.

Examining all these aspects, it results that there are three main earthquake types: crustal interplate and intraplate earthquakes, and deep intraslab earthquakes. Both earthquake hazard and seismic hazard analyses must be carried out on the bases of this classification of earthquake types.

Figure 3.8 Surface rupture effects during 1906 San Francisco earthquake
(USGS, nd)

3.3 LESSONS FOR CONSTRUCTION VULNERABILITY

During the last important earthquakes, the difference between good and poor building performances can deliver many lessons regarded from different points of view (McClure, 1989, Mazzolani, 2002, Gioncu and Mazzolani, 2002, 2003).

The *analysis aspects* refer to the lessons for improving the analysis methodologies.

- For a proper seismic design the estimation of seismic actions must consider all characteristics of earthquakes: interplate, intraplate, intraslab, near-field or far-field position and site influence, in function of site position related to the source.
- Incorporating in the analysis the checking of rigidity, strength and ductility is a measure, which mostly affects the code attitude to reduce the damage and to prevent collapse.
- Special distinction must be made between required values (resulting from earthquake characteristics) and available capacity (function of the structure features).
- Designing structures using code provisions does not always safeguard against damage in case of exceptional earthquakes, because they refer to normal earthquakes only. Therefore, an improved robustness (overall load bearing capacity which a structure is able to provide) must be provided to the structure when such exceptional earthquakes are foreseeable in the building site.
- Exceptional earthquakes can be defined as the ones not considered in the codes, or considered in a wrong way, but also when the actual intensity of the earthquake is higher than the design one. After this definition, the majority of damaging earthquakes can be framed into this category (Mazzolani, 2002).
- In case of exceptional earthquakes, the design code does not safeguard the structure against excessive plastic deformations. Potential failure modes must be identified and ductility should be provided at all locations where plastic deformations occur. Therefore, detailing for ductility and redundancy provide safety against collapse.
- Stiffness and strength of some secondary structural elements (e.g. staircases) and non-structural elements (e.g. stiff infilled walls), which are not considered as a part of the lateral force-resisting system, may strongly affect the seismic response of buildings, especially in case of uncontrolled distribution of such elements.
- Buildings that experience successive earthquakes may suffer progressive weakening or eventual collapse. From the 33 high buildings collapsed in Bucharest during the 1977 Vrancea earthquake, 31 were partial damaged during the 1940 earthquake.
- Traditional code provisions consider methodologies based on seismic forces, but in the last period the interest seems to move towards methodologies based on seismic displacements.

The *structural conformation aspects* refer to the lessons to improve the global behavior of the structure, using appropriate constructional rules.

- Uncertainness on the seismic input and structural response requires a particular emphasis on conceptual design and detailing aspects rather than on the analysis issue.
- Well-conformed, well-detailed and well-constructed buildings have the chance of good performance during large earthquakes, without collapse or excessive damage, even if the analysis does not consider all the behavioral aspects in a correct way.
- Well-conformed structure means well-coordinated architectural and engineering aspects, which requires that the structural engineer should be involved early in the conceptual design.
- Poor construction practice and lack of quality control can lead to severe damage or collapse (see 1999 Kocaeli earthquake).
- Building horizontal and vertical conformations on seismic performance are recognized to be of primary importance. Buildings having irregular in plane shapes (in L, U or T or corner buildings) and uncontrolled setback elevations or soft levels generally perform in a poor way. These irregularities must be avoided or at least minimized (Fig. 3.9).
- Buildings having no uniform distribution of resistant elements in plan have shown a bad performance, by experiencing severe tensional effects.
- Special attention must be paid to the horizontal diaphragms (roof and floor systems) in order to uniformly transfer the seismic forces to all vertical resistant elements.
- Severe pounding damage can result when adjacent buildings do not have adequate separation, giving enough space to move independently without touching. During the Mexico City earthquake a lot of damage was produced due to these poundings (Fig. 3.10).
- Weak columns and strong beams configuration, resulting for large span multi-story frames, can give rise to column failure and global structural collapse.
- Dual structures (moment-resisting frames combined with braced or infilled frames) generally performed very well during the last earthquakes, provided that bracings do not create some discontinuities. But in some cases, when the infilling is not controlled by the designer, due to the irregularity in distribution, it is preferable to use frames only.

Figure 3.9. Collapse of corner buildings; (a) Bucharest RC building; (b) Kobe steel building (courtesy of Fischinger et al, 1998)

- Discontinuities of mass, stiffness and strength in elevation give rise to local damages very difficult to avoid. Soft stories create hazardous conditions. During the Kocaeli (Turkey) earthquake many buildings having "pilotis structures" (free frames at the first level and rigid walls above) collapsed (Fig. 3.11).

The *conformation of structural element* refers to the lessons to improve the local behavior of the structure.

- The weakest points in building structures are typically the connectionsamong structural elements. Designing adequate connections is normally more difficult than providing adequate strength and ductility to structural members. Many structural failures during 1994 Northridge and 1995 Kobe earthquakes have been caused by inadequate detailing of connections (Fig. 3.12).
- Overall collapse may occur when local strength and ductility of structural elements (columns, beams and joints) are insufficient. For steel structures, the most known case is the Pino Suarez building, collapsed during the 1985 Mexico City earthquake, due to lack of ductility of trusses and columns (Fig. 3.13) (Gioncu and Mazzolani, 2002).
- Structural redundancy providing redistribution of forces can prevent collapse when individual members deteriorate during an earthquake. The complete failure of an element can be produced by local buckling or by fracture (in case of steel elements) or by crushing (in case of reinforced

Figure 3.10 Collapse of Hotel de Carlo due to pounding (NGDC, nd)

Figure 3.11 Collapse of first level for pilotis structure during the Kocaeli earthquake (USGS, nd)

Figure 3.12 Failure of some connections during Kobe earthquake
(Gioncu and Mazzolani, 2002)

(a)
Figure 3. 13 (continues)

(b)

Figure 3.13 Pino Suarez building: (a) Collapse due to lack of ductility
(NGDC, nd); (b) Collapse of column (Gioncu and Mazzolani, 2002)

concrete elements). As the post-buckling behavior can preserve some
carrying capacity, it is recommended to design the elements against the
ultimate local fracture failure mode, especially when fracture or crush
produces the overall failure of the element.
- Plastic buckling produced in elements operates as a filter against large
  strains, reducing the danger of brittle fractures.
- Unreinforced masonry buildings usually perform very poorly.
Examining the failure modes of steel, reinforced concrete and masonry
structures, it can be observed that in case of steel structures, with the only
exception of the above-mentioned Pino Suarez building (see Chapter 2), no global
collapse has been registered but just some local damage has been suffered.
Contrary, the number of global failure cases is very high for reinforced concrete
and masonry structures. Therefore, steel structures remain a very good solution for
earthquake-prone countries. The image of a steel structure, standing undamaged
and surrounded by ruins during the Kobe earthquake (Fig. 3.14) is very illustrative
and full of meaning for this reality.

Figure 3.14 Undamaged steel structure during the Kobe earthquake
(courtesy of Fischinger et al, 1998)

The *conformation of non-structural elements* refers to the lessons to improve the behavior of partitions, equipments, etc.
-   Modern buildings are relatively flexible and must be designed to limit the damage in non-structural elements.
-   Recent moderate earthquakes emphasize the potential problem when property losses, resulting from damage of architectural and equipment systems may greatly exceed losses due to structural damage. The Loma Prieta and Kobe earthquakes are examples of these non-structural losses.
-   Main cause of partition damage is due to cracking, resulting from the distortions or drift in the primary structural framing system. The control of sway can avoid this kind of damage.
-   Damage to suspended ceiling systems in public event halls, sports arena, swimming pools or exhibition centers is reported especially after strong earthquakes (Kawaguchi, 2006).
-   Building contents could be severely damaged even during moderate earthquakes, if equipments are not adequately anchored to prevent falling, sliding or overturning (Fig. 3.15).

Figure 3.15 Overturning of building contents (FEMA, 2003)

3.4 LESSONS FOR MITIGATION OF SEISMIC RISK

The experiences of so many earthquakes produced in the last period have shown that the lessons can be related at urban and building levels.

The *urban lessons* refer to the improving of measures to limit the consequences of severe earthquakes on a large area scale.

-   It is required to avoid high density of population in areas near to known faults, especially in case of poor soil conditions. The worst examples are the San Francisco and Kobe earthquakes, where a densely urbanized city was developed in an area over a fault and liable to liquefaction site.
-   Special attention must be paid to the water system serving an earthquake-prone area. The Shinh-kang Dam was severely damaged during the Taiwan earthquake (Fig. 3.16), due to the fact that the fault surface line ran across through the dam; also the water gates were damaged. The resulting reservoir discharge did not inundate downstream, but the water was interrupted for some time.
-   Strategical centers for the post-earthquake emergency, like hospitals, fire stations and electrical facilities, police headquarters and so on, must be invulnerable to seismic effects and placed with direct access from the main road network. During the Mexico City and Californian earthquakes many hospitals were seriously damaged, so that they were not operational after the earthquake (Fig. 3.17).

- In several countries and many cities in the world there are a lot of buildings belonging to the cultural heritage, which are greatly exposed to seismic hazard. The majority of these buildings are not engineered, so they are lacking basic anti-seismic features and have never been fitted with adequate provisions against earthquake actions. Due to this fact, many historical monuments were damaged during the last strong earthquakes (Fig. 3.18). It is a duty of specialists in structural restoration to develop new methodologies to protect monumental constructions (Mazzolani, 2005, 2009).

- A great problem is the spreading of after-earthquake fires, mainly produced by the damage of gas pipelines. The 1906 San Francisco and, once more, the 1995 Kobe earthquakes, can be remembered like bad examples, the after-earthquake fires producing more important damage than from the earthquake itself (Fig. 3.19).

- Each area of potential disaster must have a regular road network with many connections. Especially during the Italian earthquakes, great problems arise due to very narrow streets (Fig. 3.20).

- It is required to assure the safety of engineering works such as bridges, dams or others, whose collapse can have negative effects on the performance of the urban systems during the seismic emergency. During Loma Prieta, Kobe, Izmit and Taiwan earthquakes a lot of first importance bridges collapsed, producing great difficulties in the accessibility to damaged areas (Fig. 3.21).

In order to evaluate the economic consequences of large earthquakes, a benefit for regional administrations and the owners is the availability of the data resulting from some scenarios of previous similar earthquakes, which can be useful for predicting appropriate measures to reduce the economical losses.

- A devastating earthquake is a relatively rare event and may occur once or maximum two times during the building life. It is possible that such an earthquake never affects the building. Therefore, in this case, according to the design philosophy, the design requirements are less severe than the largest possible ones and the structure is well prepared to support this special situation without collapse, but suffering important damage.

- The very recent losses produced during the last earthquakes in structures designed following the above design philosophy (especially in the cases of Northridge and Kobe earthquakes) have shown that the financial cost of the reestablishment of normal activity was so high that it was difficult to bear even by rich countries.

- This situation has forced the specialists to develop a new design philosophy, based on explicit and quantifiable performance criteria, considering multiple performance and earthquake hazard levels for earthquakes. The collapse of some poorly erected buildings and bridges, the failure of some portion of life lines and the simulation of post-fires can give a realistic idea about what can happen during a devastating earthquake.

Figure 3.16 Collapse of Shinh-kang Dam during Taiwan earthquake
(FEMA, 2003)

Figure 3.17 Collapse of San Fernando veteran hospital complex (USGS, nd)

Figure 3.18 Collapse of the Sant Agostino church, L'Aquila earthquake
(Mazzolani, 2009)

Figure 3.19 Fires produced after the San Francisco earthquake (USGS, nd)

Figure 3.20 Narrow streets of the Italian old cities

(a)   (b)

Figure 3.21 (a) Collapses of Loma Prieta Cypress Viaduct (USGS, nd); (b) Kobe
Hanshin express way (courtesy of Fischinger et al, 1998)

The *building lessons* refer to elaboration of a new design philosophy
considering that the damage is directly related to the adequacy of partition-
structure connections and to the brittleness or stiffness incompatibility of the
partitions.
-   Steel and wood panels with proper details perform very well.
-   Unreinforced and plaster-sheathed panels are very prone to damage by
    cracking, especially in case of tall flexible buildings.
-   Connections of precast concrete infilled panels must be careful to avoid
    damage.
-   Exterior cladding systems must be adequately detailed in order to allow
    interface relative movements. Special attention must be paid for exterior
    curtain walls, which are very sensitive systems to structure movements.
-   This new design methodology establishes the multi-level earthquake
    occurrence as occasional, rare and very rare, and the admissible
    consequence as no damage, reparable damage, and survival damage.
-   Code provisions consider only the minimum values for assurance against
    different levels of earthquakes, so some damage occurs and requires to be
    repaired. If an owner wishes to avoid any activity interruptions he can ask
    the designer to conceive a structure with increased values of strength and
    ductility, in order to reduce or, perhaps, even eliminate the potential
    damage produced by strong earthquakes.

- Special strategy must be developed for seismic zones with rare and moderate earthquakes, where some simplified design methodologies can be adequate to assure a reduced seismic risk.

3.5 LESSONS FOR SEISMIC DESIGN

3.5.1 Two Crucial Events in Seismic Design Philosophy

On April 1906 one of the greatest earthquakes shocked San Francisco, California, killing 1000 persons and causing a high damage for that time. A dramatic image of the consequences of this earthquake on the means of conveyance of this time, the horse carts, is presented in Figure 3.22. This event was very significant because it represented the starting of the first scientific attempt to understand the role of the existing geological Californian fault in generating earthquakes.

A group of creative engineers, responding to the observed damage of this earthquake, started to study, conceive and design a progression of structural solutions to solve the earthquake problem. This creative work has extended over a 100-year period and continues today.

A brief progression of key milestones in seismic design history follows (Elsessner, 2004):
- the initial seismic design for buildings was based on the similarity with static wind loads;
- in the early 1930s the concept of building dynamic response and the importance of building periods gained some interest;

Figure 3.22 Dramatic picture with horses killed during the 1906 San Francisco earthquake (USGS, nd)

- in the 1950s the dynamic concept was enhanced by the acceleration spectra method;
- from 1950 to 1980 analysis methods were developed and knowledge about non-linear behavior of structural components was enhanced;
- since 1980 to present, sophisticated computer programs have been, and continue, to be developed, to facilitate the design of complex structural systems and the study of their non-linear behavior.

On January 1995 a geological fault which had been quiet for hundreds of years under the sea near Kobe, in central Japan, suddenly began to break. Houses in Kobe and neighboring cities collapsed instantly, killing thousands of people and producing the most known damaging earthquake in the world history. Considering the evolution of the means of conveyance, from the horse cart in San Francisco to the high speed trains in Kobe, the image of the collapse of the Hanshin Expressway, shown in Figure 3.23, appears as dramatic as the one of the dead horses in San Francisco.

Where is the progress in seismic design, if the consequences of an earthquake remain the very same as hundred years ago?

The answer is related to the fact that during the period from San Francisco and Northridge-Kobe earthquakes, the majority of events occurred far from very urbanized areas. Therefore, the specialists cannot have the representation of the consequences of an earthquake shaking a densely populated city. The existing code provisions consider only earthquakes that occurred in normal conditions, because

Figure 3.23 Dramatic picture of the Hanshin Expressway collapsed during the 1995 Kobe earthquake (courtesy of Fischinger et al, 1998)

the recorded ground motions were performed far from sources, where an important attenuation exists. It was during the Northridge and Kobe earthquakes that for the first time earthquakes were recorded near to epicenter, and the differences between near-fault and far-fault ground motions were observed. These events were very important, because a new stage in seismic design philosophy for *exceptional earthquakes* started from them. Exceptional earthquakes are defined by Mazzolani (2002), Gioncu and Mazzolani (2003) in the follwing way:

- the action is higher than the code one;
- the action is not considered at all in the design process;
- the action is just formally considered in the design process, but in a wrong or not in complete way;
- the action, even if correctly considered during the design, after some time can act on a depreciated structure, due to the occurrence of existing damage, produced by ravage of time or by previous earthquakes;
- the design action has been correctly considered, but the execution of the structure has been bad, being the constructional imperfections largely greater than tolerance limits or the quality of materials not conform to the design pre-requisites for strength.

From the San Francisco earthquake and the Kobe earthquake about ninety years passed, during which so much has been studied and written in the field of seismic design that one may well reasonably wonder why, after such intellectual and financial efforts, it is possible to raise a level of damage so high as the one reached during the Kobe earthquake. In fact, the seismologists have recognized the faults generating earthquakes and they are able to determine, using today's modern methodologies for spatial supervision (such as GPS), the movements of tectonic plates. Also the structural engineers today have analysis methods, which, using the computer specialized programs, can evaluate the response of structures with a great accuracy for the most complicated actions. Therefore, the second question is fully justified:

Why, when the structure is located in a seismic zone, it is not possible to utilize some very reliable values for seismic actions to design structures in fully safe conditions?

The answer is related to the great difficulty to determine the earthquake hazard on the structure site and to transform it in some proper seismic actions for the evaluation of the structural response.

3.5.2 Three Lines of Defence

Reducing future earthquake losses depends on the strategy adopted at the national level. There are three lines of defence.

- Prediction of future earthquakes, referring to their location and importance for immediate, short, intermediate and long terms.
- Limitation of important building construction in the active seismic areas.
- Designing of earthquake-resistant structures in the active seismic areas.

The advanced methodologies in seismic design must be analyzed at the light of these lines of defence.

3.5.3   Improve Qualification of Seismic Risk

Recent earthquakes have demonstrated that the behavior of some structures was unacceptable (Gioncu and Mazzolani, 2002). This poor behavior can be explained for the buildings erected before the elaboration of the modern codes, but it is unacceptable for the damaged buildings which were designed and detailed in perfect accordance with these code provisions. These are the reasons why the structural engineers are sceptical and have many questions about today's design philosophy (Mazzolani, 1995, Kennedy, 2000):

- *Does the code give the correct loading conditions, reflecting in a proper way the actual ground motions on the structure site?*
- *Does the current design methodology consider the actual behavior of buildings from the point of view of material, structural and non-structural elements as well as structure properties?*
- *What makes the structure more safe but, at the same time, more economic?*

The last important earthquakes, such as Northridge and Kobe, gave rise to the starting of some very important programs of research activity in the USA, Japan and Europe, especially for steel structures. The results of this research provided much information about the behavior of structures during exceptional earthquakes. The main question resulting from this research is:

To what extent the lessons of these exceptional earthquakes produced in some special situations are valuable for other seismic areas?

This question was a challenge for the scientists and engineers, and progress in the last period led to the understanding of the complex phenomena, associated with an earthquake, which represents the advances in seismic design.

There are three main advances in building earthquake resisting structures, which must be examined.

(i) Advances in the knowledge of seismic actions. The knowledge of seismic actions is based on the last historic 300-years information and the last 50-years instrumentally monitored seismic records. Terms like fault, hypocenter, epicenter, tectonic plates, etc., were established only at the beginning of the 20th Century, but today, due to the common efforts of *seismology science* together with the new branch of *engineering seismology,* very important new terms have been added: interplate and intraplate earthquakes, active and moderate seismic areas, far-field and near-field earthquakes, etc. Although until now knowledge is poor in many cases, due to the difficulties to obtain proper information, the basic concepts are reliable and the following years will be devoted to complete the existing lacks. Learning from past earthquakes is a very important way to improve the knowledge, which must be enriched by systematic geological prospection and aerial supervision, using modern technologies such as Geographical Positioning Systems (GPS).

(ii) Advances in the design methodology. At the beginning of 20th Century structures were designed against earthquakes using the same methodologies as for wind loads. Only in the middle of this century the notions like dynamic resonance, spectra, seismic energy dissipation, ductility, etc, were introduced in the design

philosophy. Since the 1970s, a crucial change in seismic design has taken place, thanks to the availability of personal computers and the implementation of a great number of programs for structural engineering, which very easily perform static and dynamic analyses in elastic and elastic-plastic ranges. So, many kinds of analysis, such as modal, pushover, time history, etc., lead to a satisfactory prediction of the actual structure behavior under a given seismic loading. At the same time, lessons learned from the damaged structures after the last important earthquakes and recent experimental research are adding new knowledge, which improves the design methodologies.

(iii) Advances in earthquake protective systems, offering an alternative technique towards the conventional seismic design, based on control of seismic energy dissipation through the plastic hinges formation in the structure. In this direction, new techniques propose some innovative systems such as base isolation, changing the stiffness and/or damping in the structure (Housner, 1997, Smith, 2001, Torunbalci, 2003). The author's opinion is that this last methodology will be the future challenge of the structural protection against earthquakes. But today, these procedures are still very expensive and can be used only for very special structures. The great majority of structures remain to be protected by conventional design

The purpose of this book is to face the first two advances dealing with seismic actions and design methodologies.

3.6 REFERENCES

nd – no date: the web site is periodically modified

Ambraseys, N.N. (2002): Engineering seismology in Europe. In 12th European Conference on Earthquake Engineering, London, 9-13 September 2002, CD ROM 839.

Bertero, V.V. (1997): Performance-based seismic engineering: A critical review of proposals guidelines In Seismic Design Methodologies for the Next Generation of Codes (eds. P. Fajfar and H. Krawinkler), Bled. 24-27 June, 1997, 1-31

Burkhard, M. (2001): Seismotectonics and active faults.
http://www.ndk.ethz.ch/dowloads/publ/publ_B/15/Burkhard. pdf

Elsessner, E. (2004): Seismically resistant design: past, present, future. In 13th World Conference on Earthquake Engineering, Vancouver, 1-6 August 2004, Paper No. 2034

Fajfar, P., Krawinkler, H. (1997): Preface. In Seismic Design Methodologies for the Next Generation of Codes (eds. P. Fajfar and H. Krawinkler), Bled, 24-27 June 1997, IX-X

FEMA (2003): Earthquake hazard and emergency
http://training. Fema.gov/sesion/earthquake.pdf

Fischinger, M., Cerovsek, T., Turk, Z. (1998): EASY, Earthquake Engineering Slide Information System. A hypermedia learning toll. Electronic Journal of Information Technology in Construction, Vol. 3, str. 1-10

Gioncu, V., Mazzolani F.M. (2002): Ductility of Seismic Resistant Steel Structures. Spon Press, London

Gioncu, V. Mazzolani F.M. (2003): Challenges in design of steel structures subjected to exceptional earthquakes. In Behaviour of Steel Structures in Seismic Areas (ed. F.M. Mazzolani), Naples, 9-12 June 2003, Balkema, 89-95

Grossi, P. (2004): Source, nature and impact of uncertainties on catastrophe modeling. In 13th World Conference on Earthquake Engineering, Vancouver, 1-6 August 2004, Paper No. 1635

Hays, W.W. (1996): Post-earthquake investigations: A laboratory for learning. In 11th World Conference on Earthquake Engineering. Acapulco, 23-28 June 1996, CD ROM 448

Hays, W.W., Chaker, A.A., Hunt, C. S. (1999): Learning from disasters. Civil Engineering ASCE Magazine, Vol. 69, No. 12

Housner, G.W. (1997): Structural control: Past, present and future. Journal of Engineering Mechanics, Vol. 123, No. 9, 897-959

Kawaguchi K. (2006): Safety of large encloses and damage to neo-structural components. New Olympics. New Shell and Spatial Structures, IASS Beijing Symposium, 16-19 October 2006, CD-ROM

Kennedy, D.J.L. (2000): Be a skeptic, be happy. Invited speaker. Behaviour of Steel Structures in Seismic Areas, STESSA 2000 (eds. F.M. Mazzolani and R. Tremblay), Montreal, 21-24 August 2000

Mazzolani, F.M. (1995): Some simple consideration arising from Japanese presentation on the damage caused by the Hanshin earthquake. In Stability of Steel Structures (ed. M. Ivanyi), SSRC Colloquium, Budapest, 21-23 September 1995, Akademiai Kiado, Vol. 2, 1007-1010

Mazzolani, F.M. (2002): Structural integrity under exceptional actions: basic definitions and field of activity, In COST Seminar, Lisbon, 19-20 April 2002, 67-80

Mazzolani, F.M. (2005): Earthquake protection of historical buildings by reversible mixed technologies: The PROHITECH project. In International Conference on Earthquake Engineering, EE-21C, Skopje-Ohrid, 27 August-1 September 2005

Mazzolani, F.M. (2009): Lessons to be learned from the Abruzzo earthquake. STESSA 09 Conference, Philadelphia, 16-19 August, 2009, (oral presentation)

McClure, F.E. (1989): Lessons learned from recent moderate earthquakes. In Earthquake hazards and Design of Constructional Facilities in Eastern United States (eds. K.H. Jacobs and C.J. Turkstra), Annals of the New York Academy of Science, Vol. 558, 251-258

NGDC (nd): Geologic hazard photos Volume 2, earthquake events. http://www.ngdc.noaa.gov/seg/cdroms/geohazards_v2/document/647003.htm

NGDC (nd): Geologic hazard photos earthquake events http:// www.ngdc.noaa.gov/seg/cdroms/geohazard_v2/documents/

Nichols, C.R., Porter, D., Williams, R.G. (2003): Recent Advances and Issues in Oceanography. Greenwood Press, Westport, USA

Nunez, I.L. (2000): Compound growth or compound seismic risk of destruction? Some vulnerability lessons from the Izmit, Turkey, Earthquake of 17 August 1999. In 2nd Euro Conference on Global and Catastrophe Risk Management. Earthquake Risks in Europe, IIASA, Laxenburg, 6-9 July, 2000

Petroski H. (1985): To Engineer Is Human. The Role of Failure in Successful Design, Macmillan, New York

Petroski H. (2006): Success through Failure: The Paradox of Design. Princeton University Press

Roesset, J.M. (1992): Modelling problems in earthquake resistant design: Uncertainties and needs. In 10th World Conference on Earthquake Engineering, Madrid, 19-24 July 1992, Balkema, Roterdam, 6445-6483

Smith, K.G. (2001): Innovation in earthquake resistant concrete structure design philosophies: A century of progress since Hennebique's patent. Engineering Structures, Vol. 23, 72-81

Torunbalci, N. (2003): Earthquake protective systems in civil engineering structures: Evolution and application. In Earthquake Resistant Engineering Structures IV, ERES 2003, Ancona, 22-24 September 2003, 359-368

USGS (nd): Earthquake effects in Kobe, Japan
http://earthquake.usgs.gov/earthquakes/1995_01_16php

USGS (nd): Plate tectonics and people
http://pubs.usgs.gov/publications/text/tectonics.html

USGS (nd): Implication for earthquake risk reduction in the US from the Kocaeli, Turkey Earthquake of August 17, 1999
http://pubs.usgs.gov/circ/2000/c1193/

USGS (nd): Location of August 17, 1999, Turkish earthquake
http://quake.wr.usgs.gov/study/turkey/

USGS (nd): Monitoring earthquake shaking in Federal buildings
http:// pubs.usgs.gov/fs/2005/3052

USGS (nd): Marin County: surface fault rupture effect
http://geomaps.wr.usgs.gov/stgeo/quaternary/stories/marin_rupture.html

USGS (nd): Historic earthquakes: Niigata Japan, 1964 June 6
http://earthquake.usgs.gov/regional/world/events/1964_06_16.php

USGS (nd): The great 1906 San Francisco earthquake
http:// earthquake.usgs.gov/regional/1906/18april/index/php

USGS (nd): Multimedia Gallery, Loma Prieta
http:// gallery. usgs.gov/sets/1999_Loma prieta_California_Earthquake

Chapter 4

Advances in Conception about Earthquakes

4.1 FROM ENGELADOS TO GPS SYSTEM

4.1.1 Mythological Period

> *"...Greece is the Country where Philosophy, Democracy and **Seismology** were born..."* (Papazachos, 1994).

This sentence testifies that, where one of the oldest and most brilliant civilizations was developed, the seismic activity was the highest in the whole of Europe. Ancient Greek historians and geographers were among the first who made very careful macro-seismic observations and ancient Greek philosophers were also among the first who made interesting proposals for explaining the causes of earthquake generation. The ideas of people in the pre-philosophical period (before the 6th century BC) for the causes of earthquake generation have had a mythological character. Thus, according to the Greek and Roman traditions, *Engelados*, son of Tartarus and Earth and leader of Geants, was buried in Sicily under Etna by the goodness Athena, under the order of Zeus (Fig. 4.1). When Engelados moves and sighs in his grave, earthquake and eruption from Mont Etna occur. In the Japanese tradition it is believed that a monster *Namazu* (giant catfish) (Fig. 4.2) lives under the islands of Japan and causes earthquakes. Namazu liked to play pranks and could only be restrained by Kashima, a god who protected the Japanese people from earthquakes. So long as Kashima kept a mighty rock with magical power over the catfish, the Earth was still. But when he relaxed his guard, the catfish thrashed around, causing earthquakes. In India, the idea was that the Earth is held up by eight mighty elephants staying on the back of a turtle. When one of them grew weary, it lowered and shook its head causing an earthquake. Similar mythological explanations for earthquakes were present all over the world where these events occurred (ERI, 2001).

It is very important to mention that the first record of an earthquake was made in China in 1861 BC, carved into bamboo. A Chinese man named Zhang Heng, in 132 AD, more than 1700 years before Europeans, invented the first instrument properly designed to detect earthquakes (Fig. 4.3) (Agnew, 2002). The scope of this first seismograph was to detect as quickly as possible a strong earthquake, in order to allow the emperor to send immediate aid to the injured population (Wieland, 2001).

Figure 4.1 Battle between the Giants and Gods: Athens defeats Engelados

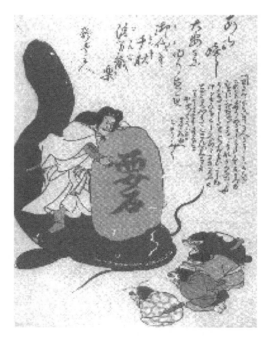

Figure 4.2 Namazu, the giant catfish, and Kashima, the God
(courtesy of Fischinger et al, 1998)

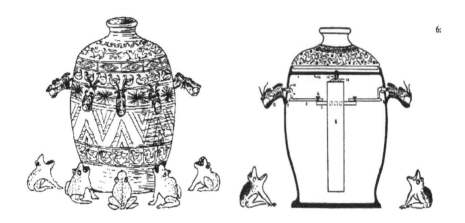

Figure 4.3 Seismoscope invented by Zhang Heng in 132 AD
(USGS, nd, RMS, 2006)

Using this instrument, the Chinese scientists evaluated the intensity of a documented earthquake for over 3000 years. Therefore, earthquakes in China have been recorded in the period from 1830 BC to the present. These records account for the damage effects of an earthquake by using an intensity scale which is similar to the Modified Mercalli Scale today (Bolton and Cole, 2006).

4.1.2 Period of Early Scientific Explanations

The classical period coincides with the Greek civilization, when the Greek philosophers attempted to give scientific explanations for all natural phenomena. *Thales* (624-546 BC) believed that an agitation of the great sea, which the Earth is floating on, is responsible for earthquake generation. *Pythagoras* (570-496 BC) considered that the heat of the Earth's interior as the cause of the earthquakes and *Epicur* (341-270 BC) believed that the erosion and transportation action of water generates earthquakes. The most influential explanations belong to *Aristotle* (384-323 BC) who attributed earthquakes to the wind blowing in underground caverns. The Aristotelian view of earthquakes remained the primary theory during the medieval periods of the countries of both Europe and Islam (Agnew, 2002).

The main period of the science development corresponds to the Renaissance, when a new spirit dominates and the scientists are basically interested in observational researches. The science moves from Greece to Western Europe. The new technology of chemical explosives suggested that earthquakes might be explosions in the Earth. *Agricola* (1494-1555) formulated the hypothesis that gasses are produced by chemical reactions between the central fire, the Earth and the moisture; these gasses, when submitted to high pressure, are forced to move the Earth's surface, producing faults in the crust and generating earthquakes. *Cardan* (1501-1576) supposed that earthquakes are due to the explosions produced by the chemical reaction of sulphur and nitrogen. Such theories for earthquake generation also explained volcanic events.

In 1620, *Sir Francis Bacon* studied the first crude map of the world in his famous *Novum Organum*, observing the strange similarity between the coasts of Africa and South America. He commented that this was no mere accidental occurrence. In 1596, the Dutch map maker *Abraham Ortelius* suggested that the Americas were torn away from Europe and Africa. In 1634 *Descartes* reached the conclusion that the creation of the Earth had been the result of ongoing natural processes, entering into conflict with the Church, which was, during those times, the supreme authority. Thus, many scientists during that period were forced to accept the Biblical history of Genesis, fearing expressing any alternative view.

The great opening to rational thought began in the 18th Century. In 1780 *Comte de Buffon* published his theory about the origin of the Earth, being the first publication beyond the limitation of the Bible. Concerning earthquake occurrences, the old theory of explosions of *Cardan* was accepted and dominated until it was expanded by *Lemery* (1645-1715). The new theory considered that these explosions were due to chemical reactions between iron and sulphur. In 1858, the Italian geographer *Antonio Pellegrini* attempted to demonstrate the fit of the American and African continents.

In China, a very strong seismic country, even the royalty was inspired to try to understand the seismic phenomenon, with the Emperor *Kangxi* (1654-1722) who attempted to describe the spatial distribution of earthquakes in China in an article written in 1720 (Bolton and Cole, 2006).

4.1.3   Early Modern Period

The greatest stimulus to seismological thinking came undoubtedly from the Lisbon earthquake of 1755, providing evidence of motion at a large distance. The English geologist and astronomer *Michell* (1724-1793) proposed for the first time that the wave propagation from a specific location caused such distant motions. For this reason, today *Michell* is regarded as the father of Seismology, which began to mature in 19th Century (Bertero and Bozorgnia, 2004).

During the 19th Century, the theory of the wave nature of earthquakes was for the first time set up by *Poisson* (1781-1840), *Mallet* (1810-1881), *Milne* (1800-1885) and *Stokes* (1819-1903), who theoretically predicted the possibility of the generation of two kinds of body waves: longitudinal and shear waves; but the cause producing these waves remained a mystery. In 1906, after the big San Francisco earthquake, *Reid,* Professor at the Johns Hopkins University in Baltimore, Maryland, studied the region in the vicinity of the San Andreas Fault in California and discovered that the Western side of the fault shifted toward the north-northeast before the earthquake took place. He also noted significant horizontal shearing in the fault. As a consequence, he considered that the

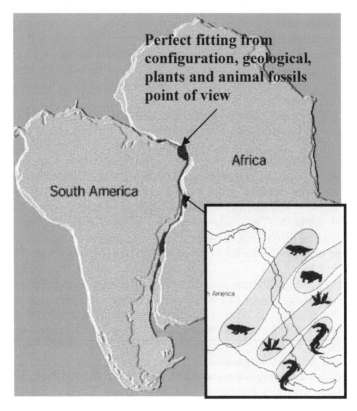

Figure 4.4 The well-known South America and Africa jigsaw (modified after Bokor, 2007)

earthquakes were caused by a sudden slippage in the fault, in the upper part of the Earth. According to Reid's view, knowing the elastic rebound theory, after an earthquake takes place, stress again starts to accumulate in the rocks. Thus, it means that, in a simplest form, the earthquake should take place at regular intervals (Ashida, 1996). Reid's theory was the first accounting for a proper explanation of earthquake generation. But this theory did not explain the formation of the faults.

A comprehensive theory about the actual position of continents was for the first time elaborated by the German meteorologist *Alfred Wegener* (1880-1930) in his book *The Origin of Continents and Oceans*, published in 1915 (Frater, 1998). He noted the remarkable fit of the South American and African continents (Fig. 4.4) and clearly identified geological occurrences on matching the coastlines of these two continents from the point of view of geological structures (mountain belts and mineral deposits) as well as of common plants and animal fossils. The break-up of lithosphere produced the two different continents, Africa and South America (Fig. 4.5). Many scientists of 18th and 19th Centuries, such as *Benjamin Franklin, James Hutton, Alexander von Humboldt*, etc., studied the problem of the Earth's

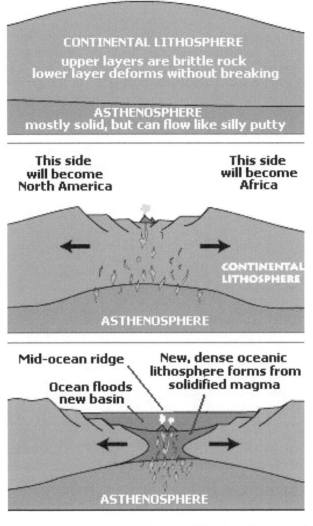

Figure 4.5 Break-up of continental lithosphere (USGS, nd)

geologic crust together with plant and animal life, showing the similarities between the Africa and Asia coasts, which were not considered as accidental coincidences.

Wegener was not the first to notice that the continental patterns fit together like a jigsaw puzzle, but he was able to make this idea as a part of modern Earth science. *Wegener* considered that, just like the pieces of a jigsaw puzzle, the world's continents could imaginably be fitted together to form, 225 million years ago, one super-continent named *Pangaea* (from Greek, *all land),* surrounded by a universal ocean, *Panthalassa* (from Greek, *all sea*) (Fig. 4.6). But Pangaea was, geologically speaking, a rather short-lived phenomenon. At the end of this period, 200 million years ago, some cracks began to occur in this super-continent. The first separation into two continents occurs 135 million years ago, the Northern continent *Laurasia* has been separated from the Southern continent *Gondwanaland*. Laurasia was formed by the collision of two small continents: Laurentia (now North America) and Baltica (which now comprises a major part of Europe and European Russia). Gondwanaland was already a super-continent in its own right extending from the equator to the South Pole. The separation began when the Indian isle began moving towards the North.

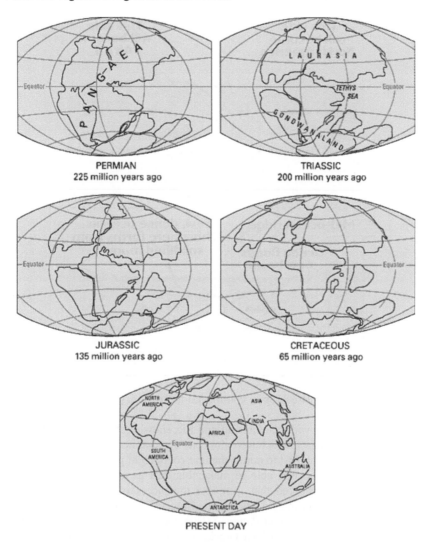

. **Figure 4.6** Wegener's Earth, from Pangaea to the present
(USGS, nd, Bokor, 2007)

About 65 million years ago, the Atlantic and Indian oceans were formed, the Eurasia continent began to have a clear contour, India continued its travel towards Eurasia, and Australia began to separate from Antarctica. About 50 million years ago, India, then an island continent, collided and subducted Asia, giving birth to the Himalayan Mountains. The last image of Figure 4.6 shows the present continental configuration.

Wegener's theory is based on the old observations that the Earth's rocky crust is broken into massive slabs named *Tectonic plates,* floating over the interior molten basalt and the actual conformation of continents is the result of the movement of these plates. This theory failed to explain the nature of the forces propelling the plates, because the simple interpretation that continents float through the ocean floor is not enough. Because Wegener was basically a meteorologist, many of his arguments related to the solid Earth are erroneous from the perspective of present-day knowledge (Uyeda, 2002).

4.1.4 Modern Period

The classical geologists contested the *Theory of Continental Drift* for a long time. But the new hypothesis of the early 1960s explained several puzzling sets of observations.   The theory was confirmed after the examination of deep-sea expeditions to the mid-ocean ridges performed by *Maurice Ewing*. In 1961 *Hess* and *Dietz* developed the *Theory of Ocean Ground Expansion.* The ocean ridges, where magma is constantly extruded into the ocean floor, push the plates apart. Therefore, the mantle convection drives the plate motions. The zone where this process takes place is referred to as a mid-oceanic ridge; the most known is the Mid-Atlantic Ridge, where Eurasian and North American continental plates are drifting apart at a rate of around 2.5 centimeters each year. In 1967 *Jason Morgan* proposed that the Earth's surface consists of 12 rigid plates that move relative to each other. In the same year, *Xavier Le Pichon* published a synthesis showing the location and type of plate boundaries as well as their direction of movement. These discoveries in the period 1966-1970 led to developing the scientific *Theory of Tectonic Plates*, which was the basic hypothesis during more than 50 years. Numerous methods tested this theory and now almost all seismologists accept it.

But the history of hypotheses about the formation and migration of continents continues nowadays, when new theories are developed. The *Earth Expansion Theory*, which considers that the dimensions of the Earth increase, the oceanic rifts are simply cracks in the skin of the planet produced by its growing and, therefore, the distances between the continents are increasing. Another theory is the *Shock Dynamic Theory*, which discovered that all the continents move by keeping one unique central point, located in the North part of Madagascar. This theory proposes the hypothesis that the Earth was catastrophically struck by a giant meteorite 180 million years ago, creating a giant crater and producing a dynamical shock. The meteorite impact was like a hammer blow for the brittle crust, which cracked as a glass globe.

4.1.5   A New Vision about Plate Tectonics: Super-continents Cycle Theory

The most convincing evidence for the theory of plate tectonics comes from the seafloor spreading. Millions of years ago, the super-continent Pangaea developed major rifts along which the continents separated into their present configuration.

But what happened before Pangaea?

Studying the geology and geophysics of old mountain ranges, such as the Appalachians and Urals, the sea-level history, the rock types and their magnetic orientation, scientists have proposed that a *Super-continents Cycle* exists, according to which super-continents form and break up at the intervals of tens of millions of years. This theory is largely based on the Wilson's Cycle, which occurs as part of the breakup and subsequent reassembly of the super-continents (Nichols et al, 2003).

Wegener's theory was generalized in 1967 by Tuzo Wilson, who observed that the evolution of the Earth is cyclical. This theory is born as an answer to the question:

... did the Atlantic Ocean open, close, and reopen?

This theory is known as *Wilson's Cycle Theory* (Everything, 2002). In the frame of this theory, Wegener's great Pangaea super-continent (formed about 300 million years ago) was only the last actual step in these cycles. The theory is based on the observation that cyclical opening and closing of the ocean basins occurs during the history of the Earth's crust formation (Fig 4.7) (Geology 202, 2004).

At some late time, continental rifting begins again for the next cycle. Until today, there may have been more than six occurrences of these cycles; each one taking 300 to 500 million years (Bokor, 2007):

- First super-continent *Ur*   -3.0 billion years;
- Assembly of *Kenorland*   -2.5 billion years;
- Assembly of *Columbia*   -1.8 billion years;
- Assembly of *Rodinia*   -1.1 billion years;
- Breakup of *Rodinia*   -760 million years;
- Assembly of *Pannotia*   -600 million years;
- Breakup of *Pannotia*   -550 million years:
- Assembly of *Pangaea*   -300 million years;
- Breakup of *Pangaea*   -200 million years.

What will happen after Pangaea?

Figure 4.8 shows the presumed future world (about 250 million years from now), when a new super-continent will be formed, by the disappearance of the Atlantic Ocean, the collisions of North and South Americas with Africa and Australia with Antarctica, and the closing of the Indian Ocean, the Pacific Ocean, will take over the majority of the world's surface.

The main lesson of the Wilson's Cycle Theory, if it is accepted, refers to the fact that today the continental configuration approach starts only after the breakup of the Pangaea super-continent, ignoring previous cycles. For a complete scenario of the configuration today of the existing faults, the plate movements for all the cycles must also be considered (provided that it could be possible, knowing that the present knowledge about the remote past is very limited).

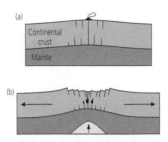

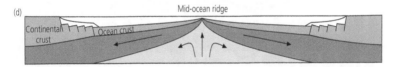

Continental rifts, such that the crust stretches and faults occur

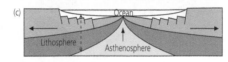

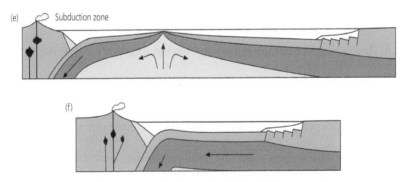

Seafloor spreading begins, forming a new ocean basin and the ocean widens and it is flanked with sediment passive margins

Subduction of oceanic plate begins on one of the passive margins, closing the ocean basin and standing of the continental mountain building

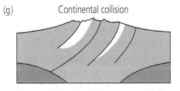

The ocean basin is destroyed by a continental collision, which completes the mountain building process

Figure 4.7 Wilson's Cycle (modified after Geology 202, 2004)

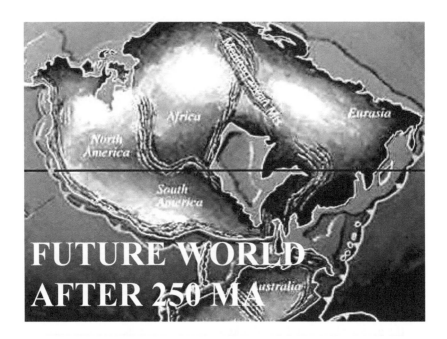

Figure 4.8 After Pangaea – future world (modified after Bokor, 2007)

4.1.6. Revolution in Seismology: Global Positioning System (GPS)

For Seismology, since 1990, there are several factors contributing to advances in development of knowledge. First, the creation around the world networks of seismic stations to detect ground motions provides a large amount of seismic data throughout the world. Second, the development of computer technologies allows seismologists to analyze the massive quantities of new data and to model the tectonic plate movements. But the real revolution after 1990 is due to the development of space-based geodetic investigation, the *Global Positioning System* (GPS). This system is a constellation of 24 satellites (Fig. 4.9a), which is used for navigation and precise geodetic position measurements. These satellites are placed in orbits at about 3.75 times the Earth radius (NASA, 2005). Daily position estimates are determined from satellite signals, which are recorded by almost 1000 GPS receivers on the ground. Horizontal velocities due to the motion of the Earth's tectonic plates and deformations in plate boundary zones are represented in maps (Fig. 4.9b). The GPS system has greatly contributed to the knowledge of regional tectonics and fault movements through its ability to measure, at sub-centimeter precision, the relative positions of points on the Earth's surface (Jackson, 2001). Re-measurement at a later date then gives the changes in relative positions, values which can be converted into the relative direction of tectonic plate movement, the velocity of this movement and the produced stresses. So, today, due to GPS, one

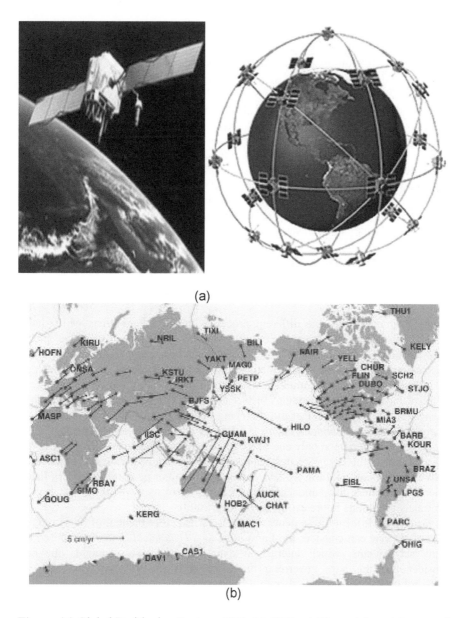

(a)

(b)

Figure 4.9 Global Positioning System, GPS: (a) GPS satellite and the orbits around
Earth (USGS, nd); (b) Measured directions and velocities of movements
(NASA, 2002)

can understand how the faults move in horizontal and vertical directions. The
quality and abundance of seismological data, together with the ability of the
computers to use them, give the perspective to investigate in detail the fault

slipping and the rupture processes which would have been impossible to see ten years ago.

Based on GPS observations, one can notice that the drift of continents will continue. In the frame of Wilson's Cycle Theory, a new conformation of continents must be considered. In the future, after 50 million years, the Atlantic and Indian oceans will extend their surfaces, in detriment of the Pacific Ocean, where the tendency of closeness of North American plate to Eurasian plate could be observed. In the next future, the Californian area will separate from the North American plate along the San Andreas fault; a part of Eastern African plate will move towards East; the Mediterranean area will be reduced, but the Arabian plate will move away from the African plate. Finally, Australia will move in the North direction, finishing its travel by a collision with the Eurasian plate. A new great continent will rise by the collision of the existing ones. After this, a new separation will occur, in the frame of the cyclic evolution of the Earth.

4.2 TOWARDS A NEW VIEW IN SEISMOLOGY

4.2.1 Earth Science as a Fundamental Interdiscipline

Earth Science explores the different processes and transformations which have produced the Earth and looks at how this world is likely to evolve in the future. This science is fundamentally interdisciplinary, because it requires the basic disciplines of physics, chemistry, mathematics and biology for its scope, which can be pursued only by the simultaneous application of various mixtures of these different modes of understanding. In fact the Earth Science, more than any other science, is the embodiment of the emerging new sciences, where the focus is on Complex Systems. *Complex Systems* is a science which focuses on systems in their entirety, rather than upon only one of the system's component parts. In the past, the primary goal of the Earth Science has been to understand the form and function of individual parts of the Earth System. In contrast, a major goal of the modern Earth Science is to more fully understand the Earth System as a whole, so that it may accurately predict its future evolution (Peltier, 2000).

What does it mean these general aspects for the development of Seismology? First of all, one can consider the Earth crust as a *Dynamical System* composed by many sub-systems, which interact in dynamical processes. Second, one can consider that the laws governing the other parts of Earth System (such as Meteorology) can be used also in Geophysics. Therefore, the following new theories about the systems must be considered:

-   *System Theory.*
-   *Self-Organized Dissipative System Theory.*
-   *Bifurcation Theory.*
-   *Chaos Theory.*

In the following, the main aspects of these theories, in relation to Seismology Science, will be presented.

4.2.2 System Theory and Earth Science

During the past three decades, scientists, philosophers and mathematicians have been working to construct a theoretical framework for unifying the many branches of the sciences, natural and social. The outcome of this effort is the *System Theory*, which provides a framework for understanding both natural and human-constructed environments and it is an attempt to formulate common laws which apply to every scientific field.

The System Theory is a trans-disciplinary/multi-perceptual theory which studies structure and properties of systems in terms of relationships, from which new properties of the whole emerge (Wikipedia, nd). This theory focuses on arrangement and relations among the parts, connecting them into the whole. It means that the whole is more than the sum of the parts and it is directly related to the new *Science of Complexity* (Casti, 1994). The Theory of Complexity can be interpreted in the elementary meaning that, if a system has more components, they are joined in such a way that it is difficult to separate them. This theory tries to elaborate computational methodologies for the quantification of the System Theory. Due to the dynamical nonlinear behavior of the complex systems, the Complexity Theory is related to *Evolutionary System, Self-Organized Dissipative Systems, Bifurcations, Catastrophes and Chaos Theories* (Casti, 1994).

In order to have a holistic science to study the Earth, the System Theory approach has been widely accepted as a framework by science communities (Turcotte and Malamud, 2002). So, the *Earth Science* (also known as Geoscience or Geophysics) built a branch of the general theory, in which the Earth is treated as a system, which evolves as a result of cooperation of some constituent sub-systems, the main topics being the Atmosphere (with Climatology and Meteorology), the Biosphere (with Biogeography), the Hydrosphere (with Oceanography) and Geosphere (with the fields of Geophysics and Seismology). Fundamental for the Earth Science approach is the need to emphasize the interaction among these sub-systems as a dynamical process which extends over spatial scales from microns to the size of planetary orbits, and over time scales of milliseconds to billions of years. Therefore, within the concept of the Earth as a complex and dynamic entity involving the disciplinary spheres for land, air, water and life, there is no process or phenomenon which occurs in complete isolation from the other elements of the Earth System.

4.2.3 Self-Organized Dissipative Systems and Mantle Convection

The history of conceptions about the *evolutionary systems* is represented by a succession of changing of paradigms. The first main paradigm was the Newtonian one, which starts with Brahe's, Kepler's and Galilei's observations about the mathematical models for the solar system. Newton, in his masterpiece *Philosophiae Naturalis Principia Mathematica* (published in 1687), called also the *Principia* for short, was searching for universal principles which could be used to explain the physical world around him. In this process, he arrived at three laws, the well-known Newtonian Laws of Motion. The first law states that a body in motion remains in motion until a force starts acting on the body. The second law, $F = Ma$, equates the force with the rate of change momentum. The third law states that for each action there is an equal and opposite reaction. These three laws are as insightful as they are simple. Even today, in the results of more recent mechanical theories, such as quantum mechanics and relativistic mechanics, the Newtonian Laws are very widely used. It is very hard to overstate the importance of Newton's contributions to the advancement of the science (Sell and You, 2002).

Newton introduced the concept of *determinism*, because his laws imply that anything which happens at any future time is completely determined by what happens now, and moreover that everything now was completely determined by what happened at any time in the past. The determinism is the philosophical proposition stating that every event is causally determined by an unbroken chain of prior occurrences (Wikipedia, nd). According to the deterministic model of science, the universe unfolds in time like the working parts of a perfect machine, without randomness or deviation from the predetermined laws.

"... We may regard the present state of the universe as the effect of its past and the cause of its future. An intellect which at any given moment knew all of the forces that animate nature and the mutual positions of the beings that compose it, if this intellect were vast enough to submit the data to analysis, could condense into a single formula the movement of the greatest bodies of the universe and that of the lightest atom; for such an intellect nothing could be uncertain and the future just like the past would be present before its eyes." (Laplace) (Gioncu, 2005)

This *intellect* has been later dubbed as Laplace's Demon (Garrette, 2003).

In the last time the Newtonian paradigm is contested by the new developments in the System Theory.

The first doubt about the universality of the Newtonian paradigm comes from the Thermodynamics Laws, which introduce the time as an important parameter in the study of systems. Contrary to the Newtonian concept, where all the phenomena are reversible, in Thermodynamics the most important characteristic is the irreversibility.

Secondly, Leibnitz's aphorism:

"Natura non facit saltum" (Nature does not make jump)

based on the Newtonian paradigm, is denied by the surrounding reality, characterized by sudden changing, discontinuous variations or mutations. Even the earthquake events cannot be framed in the phenomena which respect the Newtonian paradigm. Therefore, now it is widely understood that:

"Natura facit saltum" (Nature makes jump).

Studying these phenomena, Prigogine (during the 1970s) introduced a new paradigm about the nonlinear behavior of Thermo-dynamical Systems (Prigogine and Stengers, 1979). The behavior of a complex system is presented in Figure 4.10, where two different phase types can be perceived. The first is the *Near to equilibrium phase,* where all the phenomena are framed in the linear behavior, the influence of fluctuations is much reduced and the rules of Thermodynamics referring to the closed systems can be applied. In this field, systems respond to fluctuations by returning to equilibrium. Contrary, the second one is the *Far from equilibrium phase,* where it exists only if an open system is considered in the field of nonlinear behavior. Systems far from equilibrium may respond to fluctuations by evolving to new states. The main characteristic of this phase is the dependence of its environment. Due to this fact an internal self-organization of the system can be noticed.

Therefore, the new Prigogine paradigm replaced the early Newtonian paradigm. The new paradigm is based on the observation that the dynamics of a system in nonlinear range can tend to increase the inherent order of the system.

Self-organization is a process of attraction and repulsion, in which the internal organization of a system, normally an open system away from equilibrium, increases in complexity, without being guided or managed by an outside source (Wikipedia, nd). This concept in early days was identified as a system which changes its basic structure as a function of its experience and environment. The present definition considers the organism and environment taken all together. The concept of self-organization is central for describing the biological system (evolution and morphogenesis), but it can be found in many other disciplines, both in the natural sciences (physics, chemistry and biosphere), mathematics (computer science and cybernetics) and in the social sciences (human society and economics), which tend to produce organization.

These systems, also named as *Dissipative Systems,* are the thermodynamically open systems operating far from the equilibrium in an environment, where energy and matter are exchanged (CSCS, 2006, Wikipedia, nd). These systems tend to dissipate energy as they interact with the environment and they are able to maintain identity being open to flows of energy, matter or information from the environment. The energy arrived from the outside is quickly dissipated by the system situated far from the equilibrium. At the same time, these systems can produce internal energy and exchange it with the external environment (Green, 2000).

In the frame of these theories, there are some new aspects as bifurcation and chaos, which will be discussed in the following sections.

What does it mean this general theory for understanding the occurring of earthquakes?

Earthquakes are complex, destructive phenomena and it is extremely difficult to model them. The occurring of earthquakes must be related to the plate tectonics, which is one of the most successful theories in the history of natural sciences, having revolutionized all the Earth science. The idea of continental fixity has been

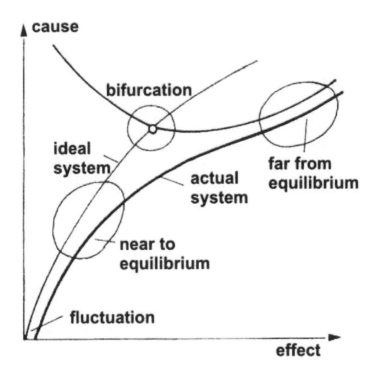

Figure 4.10 Behavior near and far from equilibrium (Gioncu, 2005)

replaced by the idea of continental mobility, due to the convection in the mantle (Anderson, 2003) (see Chapter 2, Figure 2.1). Scientists today generally agree that the plate-driving forces result from the hot softened mantle which lies below the cold rigid plates. So, both the Earth's surface and its interior are in motion. Circular motion of the mantle carries the plates in the direction of rotation.

The Earth's surface is covered by about twelve large tectonic plates which relatively move each other and respect the underlying mantle. The plate tectonic can be considered as a successful example of a far-from-equilibrium self-organized system, powered by heat (as an external source of energy) and gravity from the mantle and organized by dissipating in and among the plates (Anderson, 2001a). Figure 4.11 shows a cross-section of the Earth, where the equilibrium between the sea-floor spreading in the Mid-Atlantic and East Pacific Rise is equilibrated by the subduction in the East and West Zones of Pacific Ocean, forming a self–organized system among plates and mantle convection. The surface plates are the active element, the convicting mantle is the passive element.

When a tectonic plate is placed on the convecting system, it organizes the convective flow and the plates themselves become a dissipative self-organized system.

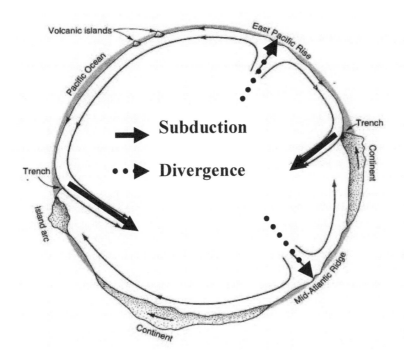

Figure 4.11 Cross-section of the Earth (modified after Uyeda, 2002)

Mantle convection is extremely sensitive to changes in the surface conditions. The collision of two plates can reverse the mantle flow. Both plate tectonics and balanced convection represent a balance between driving and resisting forces and the change in any force is immediately transmitted throughout the system, causing a global reorganization. Therefore, the plate tectonics can be interpreted as a non-equilibrium process in the sense that a small perturbation in the system can cause a complete reorganization in the configuration and sizes of plates and in the platform of mantle convection. Such perturbations include stress changes in a plate and plate to plate collision or separation. Changes in boundary conditions are transmitted essentially in instantaneous way throughout the system and new plates with new boundaries form. These changes in topside system formed by plates and mantle cause mantle convection to reorganize and the topside is fractured into new plates due to the extension or compression. Therefore, the plate tectonics, as a very illustrative example of far-from-equilibrium of a self-organized system, may respond by evolving to a new state. In the frame of this concept it is clear that the conformation of plates is not permanent. The actual conformation is a temporary alliance of major and minor plates. The reorganization process episodically changes direction and speed of plates and redefines the plate configuration.

Episodical plate reorganizations are inevitable and can explain the permanent movement of the Earth's plates (Anderson, 2001a,b, Sankaran, 2002).

4.3 BLOCK MODELS AND SEISMICITY

4.3.1 Bifurcation Theory

Bifurcation (named also Self-organized Criticality, Wikipedia, nd) is the phenomenon in which the system reaches the crisis at the bifurcation point (Fig. 4.10) and results in a system splitting into two possible behaviors for a small change of one parameter. The important new possibility is that the self-organized state can suddenly appear beyond this crisis point, far from the equilibrium in the nonlinear range.

Fluctuations. In thermodynamics, a fluctuation is a perturbation in the equilibrium state, which changes the bifurcation point in an accelerated behavior. Near to equilibrium, these fluctuations have no important effects (Fig. 4.10). In exchange, far from the equilibrium, these fluctuations produce a great changing in system behavior, not proportional to the sizing of these fluctuations. The smallest fluctuation can lead to a radically new behavior of the system. To predict the future behavior, it should be necessary to measure the fluctuations with infinitely accurate precision, which is an impossible task both in principle and in practice (Coveney, 2003). Many phenomena in nature exhibit anomalously large fluctuations, exceeding what cannot be explained as a consequence of statistically independent random events. Therefore, a suitable determination of the nonlinear systems is out of the question.

Evolution in cascade (Fig. 4.12). The Complex Systems are characterized by a cascade succession of bifurcation points. The system evolution is composed by slow evolutions, followed by bifurcations or accelerated changing. In some cases, jumps from a stable position to another position occur. This is a very characteristic situation for the earthquake occurrence, when some imperceptible deformations are accumulated for a long time in the plates due to the mantle convection and, suddenly, a perceptible important modification of the relief or the loss of plate equilibrium produces the earthquake.

4.3.2 Block Structure Models

As was shown in the previous sections, the Earth crust is divided into blocks (tectonic plates), which move in different directions with different velocities. So, these blocks are stressed, mainly along their boundary, which are represented by the faults of the Earth crust, where they interact with the neighboring ones. The stresses are mainly concentrated in the contact zone. It is possible to define a limit level of pressure supported by rocks in contact (mainly given by friction under pressure), for which a rock rupture or a slippage between the rocks occurs along a fault plane. When the level of stress remains below this limit pressure, the blocks are in equilibrium. When the limit level is exceeded in some part of the fault plane, a stress-drop occurs (in accordance with the dry friction model) with important

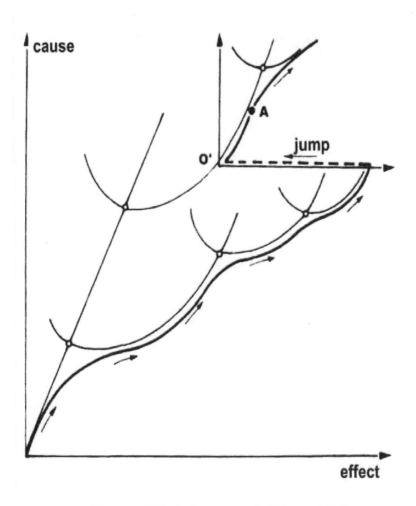

Figure 4.12 Evolution in cascade (Gioncu, 2005)

displacements, producing an earthquake. Immediately after the earthquake and some time after, the affected parts of the fault plane are in a state of creep, characterized by faster growth of inelastic displacements, lasting until the stress falls below some other level and a new stress state begins, due to the external existing forces.

A seismic region is considered as a system of absolutely rigid blocks of the lithosphere (solid part of the Earth's crust), which are separated by thin boundary zones (for instance South American Plate, Fig. 4.13a).

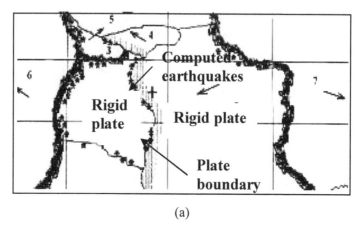

(a)

Figure 4.13 (continues)

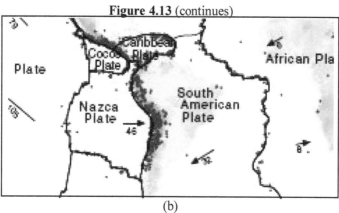

(b)

Figure 4.13 Block structure of South American Plate: (a) Model of block and resulting earthquakes; (b) Produced earthquakes (modified after Rozenberg et al, 2002)

So, after this model, an earthquake is the result of the loss of the block system equilibrium (Soloviev, 2001, Soloviev and Maksimov, 2001, Digas et al, 2001, Rozenberg et al, 2002). The motion of a block is defined so that the system is in a static state of equilibrium, being stressed by the forces resulting from convective currents in the Earth mantle. When the ratio of the stress to the pressure exceeds a critical level in some part of a fault zone, a slip between two adjacent blocks occurs. This slip represents a loss of equilibrium and a jump occurs until a new position of equilibrium, far from the original one. This new position is reached when stress and pressure along the fault find a new equilibrium condition. The jump is a dynamic process and it generates the earthquakes, the oscillations round the new position producing waves, which propagate in the crust until the surface.

After this stress relaxation and due to the fact that the external forces, produced by convective currents, continue to exist, a new accumulation of deformation takes place. So, new equilibrium losses can be produced at the same or in a new position along the fault.

Using these concepts, the solution of dynamic equilibrium equations is obtained by discrimination, by introducing a time step. Computational experiments show (Digas et al, 2001) that the block models of the lithosphere dynamics are quite time-and memory-consuming on sequential computers, so that it does not allow the simulation of the dynamics of complicated systems with a large number of blocks. However, the approach applied for modeling admits the parallelization of calculations on a multiprocessor machine, using a parallel algorithm for numerical simulation. This fact makes it realistic to pass to a system of tectonic plates in the global scale (Rozenberg et al, 2002).

The South American region, composed by South America, Caribbean, Cocos and Nazca plates, was studied by Digas et al, (2001) and Rozenberg et al, (2002). The surrounding plates, North America, Africa, Antarctica and Pacific, are treated as boundary blocks. Between the obtained synthetic results and the recorded earthquakes a very good correspondence is obtained (Fig. 4.13)

Using the concept of block equilibrium and parallel technologies, some very interesting applications were performed. The blocks system of the Vrancea (Romania) subduction zone has been studied by Soloviev (2001) and Rozenberg et al (2002). Figure 4.14 shows the main structural elements of the Vrancea blocks

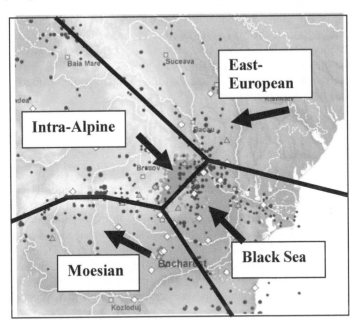

Figure 4.14 Simulation of Vrancea (Romania) earthquakes
(modified after Rozenberg et al, 2002)

system and the block structure approximation (composed by four blocks, East-European, Moesian, Black Sea and Intra-Alpine micro-plates). The temporal distribution of the large synthetic earthquakes in a period of 7000 years (interval corresponding to the main tectonic process of this region) contains over 9000 events with magnitude between 5.0 and 7.6. The maximum value of the synthetic magnitude is 7.6. Note that in the 20th Century four earthquakes with magnitude 7.0 or more occurred in the Vrancea region.

The Theory of Block Dynamical Equilibrium has been very successful in explaining numerous problems in Seismology, but at the same time, it is very clear that it is only a first order of approximation for the blocks movements. The theory can be criticized in the following aspects:

- The plate dynamic equilibrium is considered only for the conformation of tectonic plates today, ignoring the historical steps in formation of this conformation. For instance, the very important collisions among some plates, as the Italian, Chinese or Indian ones, with the Eurasian plate, which occurred in the remote past, are ignored.
- Plates are not truly perfectly rigid and the deformations cannot be concentrated only along the plate boundaries. In many cases the deformations affect some crust parts far from the plate edges. So, the world's plate boundaries in many cases are diffuse and not concentrate, as it is considered in the theoretical approach.
- The stress along the boundaries is considered as only compression or shear. But, in the real situation, the stress state is more complex. For instance, for the subduction, the shear between the two plates is accompanied by back-arc extension.

All these questionable aspects show that the results obtained using this theory must be used with wise precaution.

4.4 CHAOS THEORY AND SEISMOLOGY

4.4.1 Chaos Theory

The behavior of the nonlinear dynamic systems far from equilibrium can be described only by the *Chaos Theory*. The word "chaos" has been used since antiquity to describe various forms of randomness. In the ordinary nomenclature, the term chaos defines a condition of disorganization and suggests a lack of order or direction. For complex phenomena there are two essential factors: *time* and *nonlinearity*. The theory uncovers a new uncertainty principle, which governs how the real world behaves and explains why time goes in only one direction. Nonlinearity causes small changes on the level of organization, producing large effects at the same or different levels.

The first experimenter in chaos was the meteorologist Edward Lorenz. In 1960 he was working on the problem of weather prediction using a set of twelve nonlinear equations to model the weather, discovering the great sensitivity of these equations on the initial conditions. Just a small change in the initial conditions can drastically change the long–term behavior of a nonlinear system. For instance, with a starting of initial conditions with 2.000000 values, the final result can be entirely different from the same system with a start of 2.000001 values. Starting from this revelation, it is impossible to predict the weather accurately. In 1963, Lorenz published a paper describing what he had discovered in a meteorological journal, because he was a meteorologist, not a mathematician or physicist. As a result, Lorenz's discoveries were not acknowledged until years later, when others rediscovered them. Many scientists believe that the 20th Century will be known for only three revolutionary theories: relativity, quantum mechanics and *Chaos*.

A comparison of a linear and nonlinear system is shown in (Figure 4.15). In a linear system the left or right fluctuations have no great influence. For this system, Newtonian laws show how it is possible, given an initial point in phase space, to plot out the trajectory for all future times. In other words, given the full specification of the system, it is possible to determine its future behavior. Any external force or perturbation will produce a predictable change, while tiny external fluctuations have a negligible effect.

In exchange, for a complex nonlinear system, the fluctuations produce increasing effects as the parameter increases in value, but the fluctuations have periodical characteristics. At the same value of parameter, the curve parameter-effect breaks in two in a point named *bifurcation*. The fluctuations increase, but remain in the periodical field. Raising the growth rate a little more, a new bifurcation occurs in four different values. As the parameter rose further, the curve bifurcated again. The bifurcation comes faster and faster, until suddenly, *chaos* appears.

In this field it is impossible to predict the behavior of the *Complex Nonlinear System*, because it strongly depends on the initial value from which the chaos occurs.

The great influence of a small perturbation in the initial conditions can grow to have unpredictably large consequences. The meteorologist Lorenz called this phenomenon the *butterfly effect*. The sensitivity to these initial conditions is described in the following metaphor:

"... A flap of a butterfly's wings in Brazil starts a tornado in Texas..."

showing, in a philosophical signification, the impossibility to determine the behavior of nonlinear complex system in the high field of parameters. Even the deterministic systems, which are not stochastic or random systems, could present, under some conditions, the chaotic motions.

So, a fundamental question arises, however in this pessimistic sentence: Is it possible to predict the main factors influencing the chaotic motions, even if no exact description is possible? Even if there are still much more unsolved than solved problems, the answer is positive. There are some properties of the chaotic phenomena, discovered by the new *Theory of Chaotic Dynamics*, which allow the prediction of the chaotic motions. One of them is the deterministic feature of the parameters of the initial conditions, studied in the frame of the *Theory of Deterministic Chaos* (Moon, 1987, Schuster, 1988).

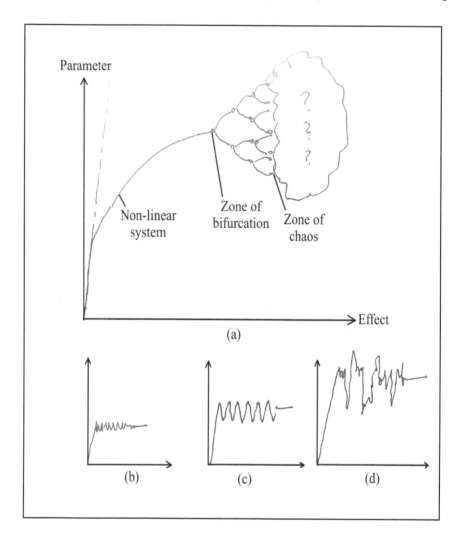

Figure 4.15 Nonlinear behavior and chaos: (a) Changing of periodical fluctuations in chaotic movements; (b) Periodical fluctuation; (c) Increasing of fluctuation after bifurcations; (d) Chaotic fluctuations

In the last few decades, physicists became aware that even the system studied by classical mechanisms can behave in an intrinsically unpredictable manner. Although such a system may be perfect in principle, its behavior is completely unpredictable in practice. This phenomenon is called *deterministic chaos*. To explain its origin, it is necessary to come to the concept of linear and nonlinear behavior of systems. Linear equations of equilibrium are easily solved, but the nonlinear ones are in general very hard or impossible to solve. Therefore, until the beginning of the 20th Century most nonlinear problems in classical mechanics were linearly approximated. Linearity basically means that the effects are proportional to the cause. Contrary, the essence of nonlinearity is that the effects are not proportional to the cause. Small causes may have large effects. The processes which are very sensitive to small fluctuations are called chaotic. This is because their behavior is in general very irregular, so that they give the impression of being random, even though they are driven by deterministic forces (Heylighen, 2002).

A very illustrative model for this deterministic chaos of dynamical systems is the nonlinear behavior of a double pendulum (Fig. 4.16a). In horology, a double pendulum is a system of two pendulums on a common mounting which move in anti-phase (Wikipedia, nd). The motion of a double pendulum is governed by a set of coupled ordinary differential equations. Above certain energy, its motion is changed. The pendulum behavior for small oscillations is very well known. But in the field of nonlinear behavior, for large oscillations, the motions are chaotic (Fig. 4.16b) (Weisstein, 2007, Wikipedia, nd). The most important result obtained from

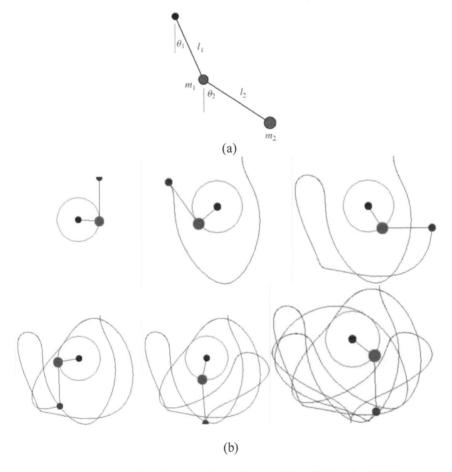

(a)

(b)

Figure 4.16 Double pendulum behavior (after Weisstein, 2007)

analyzing the double pendulum behavior is the great sensitivity to the initial conditions. Two identical double pendulums, but with very small differences (in order of 0.00001) in the initial conditions, have completely different oscillations (Wheatland, 2007).

4.4.2 Applications of Chaos Theory in Seismology

The relation between Chaos Theory and Seismology is given by the Burridge-Knopoff model (Saito and Matsukawa, 2007). In case of the interplate earthquake type, where an ocean lower plate goes downward beneath an upper land plate (Fig. 4.17 shows the case of tectonic plates of Japan), the fault is composed by some segments. Burridge and Knopoff introduced a model which exhibited some characteristics similar to the dynamics of an earthquake fault. They were interested in the role which friction plays with regard to the earthquake mechanism. The basic configuration of this model would consist of some blocks (modeling the fault segments) connected to the upper plate by linear springs. The blocks are also connected to each other by linear springs. The frictional forces act between the blocks and the lower plate. The upper plate is driven with constant velocity. When the friction cannot hold the blocks, they will move forward to a certain distance to release the energy in the springs. Due to the coupling between the blocks, the number of blocks involved in a single slip event has a very broad distribution. If only one of the blocks slips, the event is small. If a great number of the blocks are involved in slipping, the event is large.

During the slipping, a vibration of blocks around the initial position occurs. For a symmetric model, as in the original Burridge-Knopoff one, the system presents a

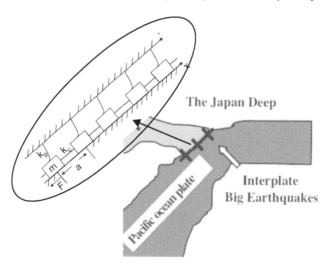

Figure 4.17 Burridge-Knopoff model: The interplate fault model (modified after Saito and Matsukawa, 2007)

regular cyclic behavior. Contrary, in presence of asymmetry of the system (different friction forces, or different spring rigidities), a chaotic behavior (as for the double pendulum) occurs (Huang and Turcotte, 1990, Santos, 2006, Xia et al, 2005). Therefore, all the phenomenological aspects determined for deterministic chaotic dynamics are valuable for earthquakes.

How far can this theory be applied for understanding earthquake phenomena?

As shown above, this question is related to the possibility to use the recorded ground motions for the prediction of the next earthquakes on the same site. This problem is born from the observation that considerable differences have been obtained among ground motions recorded in the same place during different earthquakes, originating more or less from the same source zone. At the same time, it can happen that very similar buildings situated at some dozens of meters from each other behave in a completely different way, one collapsed due to important damages and the other did not suffer any damage.

If the above block theory gives an image about the places where earthquakes occur, the magnitude of them remains a very difficult problem. Indeed, the determination of the next event characteristics is a crucial problem, and the question is whether there are coherent and sure methodologies to solve this design aspect.

The recorded ground motion of a particular earthquake gives the possibility of its very good description, but it is questionable whether it could be considered as a prediction for the next earthquakes occurring on the same site. This remark is supported by the observation that considerable differences have been obtained among ground motion records at the same place during different earthquakes, originating more or less from the same source zone (Gioncu and Mazzolani, 2002). This is due to the fact that each event is offering new surprises in the ground motion characteristics and in the buildings' vulnerability, showing the great complexity of the phenomenon.

The Sumatra earthquake can be considered as an example of these observations. The 26 December 2004 earthquake, with a magnitude of 9.0, produced a terrible tsunami, killing thousands of people. After three months, on 28 March 2005, practically in the same place, an new earthquake with a nearer magnitude of 8.7 occurred, but without any tsunami effect. Both in a deterministic or probabilistic way, this new earthquake should have produced a new tsunami, but with a reduced intensity. Why did it not occur?

The greatest earthquakes occur in case of subduction plates, where the two plates are locked over some width, resulting in both uplift and horizontal shortening of the overlying plate margin. The uplift is produced by a large flexion of this plate (Fig. 4.18a). Once the accumulating stress exceeds the strength of the fault, the locked zone fails and a great earthquake occurs. During the rupture, the stored strain is released, resulting in horizontal extension and vertical uplift. Underwater these displacements can cause important tsunami, especially due to the produced blow. The great earthquake occurred according to the mechanism shown in Figure 4.18b, explaining the great magnitude and the important tsunami.

In case of a second earthquake produced in the same place, only a slipping between the two plates (without extension) should occur (Fig. 4.18c), because the.

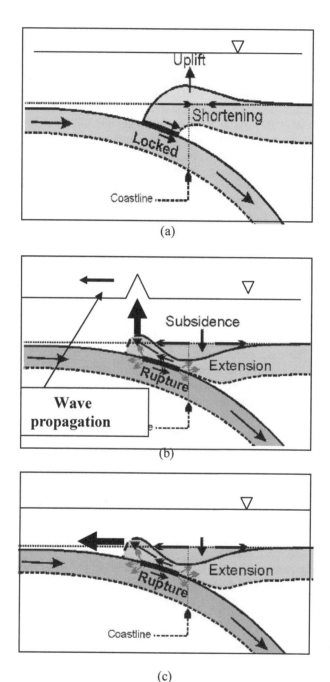

(a)

(b)

(c)

Figure 4.18 Process of subduction producing tsunami (a) Flexion of plate;
(b) Vertical blow on sea surface and wave propagation, Sumatra, 26.12.2004;
(c) Horizontal slipping, Sumatra, 28.03.2005 (modified after GSC, 2008)

time was very short for the accumulation of the plate flexion. The absence of vertical blow explains the lack of tsunami at practically the same earthquake magnitude

This is an example showing the great influence of initial conditions on the development of an earthquake.

The earthquakes are produced by the local sudden movements of some tectonic plates which belong to a very complex system of world tectonic plates. The behavior of the Earth crust belongs to a nonlinear dynamic system, with many similarities with the atmospheric dynamics. If for this climate problem one must admit the limitation of the variability statistic, the same restriction must be considered for seismic problem modeling.

The earthquake starts in an existing fault due to the slipping of the two borders or for a new rupture. The source is characterized by the fault type, depth and area of slipping or rupture. The main characteristic of the movement of the tectonic plates is the shock acting as an impulse. Unfortunately, it is not possible to record the event directly at the source. But it is clear that the restoring of the new equilibrium produces some oscillations, with properties depending on the characteristics of the source. They are longer with two or more important pulses in case of slipping or with only one pulse in case of rupture. The source movements generate the waves, which cross the crust and the site soil. The characteristics of source, traveled path and site characteristic form the initial conditions of soil vibrations. In accordance with the Theory of Chaos, the chaotic ground motion, being the result of a nonlinear system very sensitive to the differences in these initial conditions, will present a great variability of the main characteristics.

A very significant case for this variability is the Vrancea (Romania) seismic zone, where frequent intermediate earthquakes (depth between 90 to 160 km) occur practically in the same place. The last very important ones happened on 4 March 1977, M = 7.2, 16 August 1986, M = 5.0 and 30 August 1986, M = 7.0, 30 May 1990, M = 6.7 and 31 May 1990, M = 6.1 (Pustovitenko et al, 1994). The different positions of epicenters of these earthquakes were situated at some kilometers distance from each other, but gave rise to very different rupture directions. The 1977 and 1986 earthquakes were oriented mainly in the Southwest direction, while in 1990 the rupture orientations were to Northeast. The most important records of these earthquakes were registered at the INCERC in Bucharest, situated at 160 km from the source, on the site with very bad soil conditions. Figure 4.19 presents the spectra respectively obtained for these events on the same site (Gioncu, 1995). One can see that the soil conditions did not have very important effects during low magnitude earthquakes, the maximum spectral accelerations being obtained for T < 1.0 sec. Contrary, the characteristics of spectral accelerations for the 1977 and 1986 earthquakes, having a greater magnitude than the other ones, were completely changed, a very important amplification being obtained for T = 1.6 sec. The 1986 and 1990 earthquakes were very similar in magnitude (M = 7.0 for 1986 and 6.7 for 1990), but some differences in initial conditions dramatically changed the ground motion effects. Thus, in the light of Theory of Chaos, one can undersand

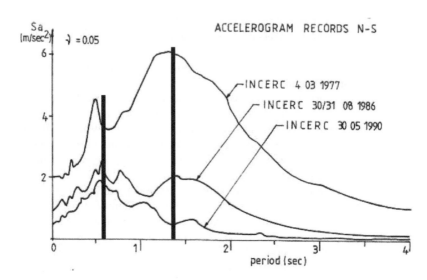

Figure 4.19 Comparison among the Vrancea (Romania) earthquakes recorded in the same place by INCERC Bucuresti (Gioncu, 1995, Lungu and Moldoveanu, 1997)

why this simple changing of initial conditions completely modified the ground motion response.

Therefore, according to the Chaos Theory, some conclusions can be drawn:

-    All earthquakes start off as tiny earthquakes, which happen frequently, but only a few of them result in large earthquakes, due to small perturbations with unpredictable large consequences.
-    There is nothing special characterizing these tiny earthquakes which happen to grow into large ones.
-    Even if the occurrence of an earthquake can be predicted (see next section), one could never be able to foresee the accurate value of the magnitude of the next earthquake.
-    The interval between tiny and large earthquakes is highly variable.

4.5 EARTHQUAKE: PREDICTING THE UNPREDICTABLE?

An earthquake prediction must specify the geographical area where it will occur, the time interval within which it will happen and the expected magnitude. In the light of the previous sections, the prediction of future earthquakes would require measuring the initial conditions with literally infinite precision, but this task is impossible in practice. However, considerable advances have been made in past decades in understanding the plate motions and rupture dynamics, which make it possible to predict earthquakes to a certain extent.

Earthquakes have killed more than 3 million people during the period from 1900 to today. Earthquakes are so deadly because they strike without warning. When more reliable methods for predicting earthquakes can be found, earthquake warnings can be given in advance, saving human lives (Gray, 1996). There is a great interest to predict location and time of large earthquakes, in order to mitigate their devastating potential. A great deal is known about *where* earthquakes are likely, due to the fact they occur in predictable areas such as at tectonic plate boundaries. Contrary, there is not a current reliable way to predict the days or months *when* an event will occur in any specific location and also the most important aspect, the *predicted magnitude* remains an open question, about which there are many debates among the specialists.

The *Earthquake Prediction* intends to forecast an earthquake of a given magnitude range, occurring in a localized region and at a specific time window (Wikipedia, nd). Earthquake prediction at the present time is not an exact science and the forecast of the earthquake occurrence cannot be very accurate. There are many controversies in trying to predict an earthquake, due to some successful cases, followed at the same time by a greater number of unsuccessful cases of failure in determining the occurrence of strong earthquakes.

Among the successes, one must mention the well-known earthquake prediction of the 1975 earthquake of Haicheng, a very densely populated town in China, where an evacuation warning was issued the day before the earthquake of M 7.3 magnitude. As expected, because of the great magnitude and the presence of a lot of non-engineered constructions, the damage in Haicheng and in the neighboring area was extensive, but very few lives were lost. The prediction was based on a number of geophysical observations of precursor events as well as the abnormal animal behavior. The increasing in foreshock activity triggered the evacuation warning. This success gave hope that earthquake prediction was eventually possible.

Unfortunately, most earthquakes do not have such obvious precursors. In fact, one year after, the 1976 Tangshan earthquake with M 8.0 magnitude, with the highest number of deaths from an earthquake last century, suddenly reduced the enthusiasm of Chinese scientists. Although precursor events were observed as well as geophysical and geochemical anomalies were detected, these precursor events occurred over a very widely spread area, making it extremely difficult for scientists to focus on a particular region and, therefore, to issue a short-term prediction or a warning. There were some remarkable differences between the particular precursor events of the Tangshan earthquake and those of other predicted earthquakes. Furthermore, no significant foreshocks were observed. With this earthquake, the hope in accurate predictability of earthquakes evaporated, showing that there is a long way to setup a sure methodology to predict earthquakes.

The above two examples give a measure of the validity of the earthquake prediction at the present time. However, some improvement is now in progress for obtaining better predictions. After these two earthquakes, the main obvious lesson is that there are some earthquakes which can be predicted and others not. The target of scientists is to prepare adequate methodologies for both cases.

To understand the earthquake prediction, five different types have been assigned by scientists (Keilis-Borok, 1996, Wieland, 2001, Carayannis, 2007):

- *Long-term prediction*, involving a time of decades or more and having just a general interest, without usefulness for public safety;
- *intermediate term prediction*, which falls into a time span of a few weeks to some months, with a limited interest for practical usefulness;
- *short-term prediction,* giving specific information on the earthquake location and time prediction within days, weeks or months (not years), which would be useful for any kind of public safety and evacuation planning;
- *immediate term prediction*, which considers prediction at the level of one day;
- *early warning*, involving a time of seconds or no more that one minute, function on distance from source.

The possibility of prediction is explored separately for these different types.

For *long-term predictions*, statistical analysis is one method for predicting earthquakes, when, looking at the history of earthquakes in a given region, one can see that there exists a recurrence in earthquake occurrence. For large events one may apply the concept of the earthquake cycle. The advantage of this analysis is that it would give people several years for the preparation to withstand the earthquake. For existing buildings, one can determine the ones which are not able to resist the predicted earthquake and must be reinforced. For new buildings, this methodology gives the possibility to build earthquake–resistant structures. The disadvantage is that the statistical analysis cannot give a short-term prediction of an earthquake with a specific range for the expected magnitude.

For *intermediate term*, some specific methodologies are developed, based on the concept that the Earth's crust is not perpetually in a critical state, but under a given stress condition, exceeding some limits, it works as a nonlinear system (see the previous section concerning Theory of Chaos). From the viewpoint of mechanics, the earthquake is the failure or the instability (bifurcation) phenomenon of a focal medium, accompanied by a rapid release of strain energy. The mechanical problem is complicated by the existence in the rocks of cracks (faults) or weakened zones, with different size, shape and orientation, under high pressure and temperature. In the frame of these aspects, there are many approaches for the prediction of earthquakes, but two of them seem to be the most important.

An international team of scientists from Russia, USA, Western Europe, Japan and Canada, under the leadership of professor Keilis-Borok (Keilis-Borok, 1990, 1996, Keilis-Borok and Shebalin, 1999, Keilis-Borok and Soloviev, 2003, Keilis-Borok et al, 2007), an outstanding Russian seismologist and mathematical geophysicist, now resident in the USA, proposed a methodology in which the establishment of precursors of strong earthquakes plays a crucial role in the prediction. It was observed that, before a strong earthquake, the earthquakes of medium magnitude range become more intense and irregular; they become more clustered in space and time and the range of their correlation probably increases (Hough, 2005). The prediction is aimed at the determination of *Time of Increased Probability* (TIP), which is the time interval within which a strong earthquake has to be expected. Algorithm is designed to determine the TIP, which can be considered as the alarm time. Using these algorithms, some earthquakes in the USA and Japan were announced with reliable prediction (Rotwain and Novikova, 1999). The Keilis-Borok team in June 2003 predicted the 6.5 magnitude earthquake which struck Central California on 22 December and also in July 2003 predicted the 8.1 magnitude earthquake which occurred in September on Hokkaido. Previously, this team predicted the Northridge earthquake (Science Daily, 1994). But the scientific community is fairly skeptical about these successes in prediction, considering that a larger number of successful predictions is necessary to guarantee that the developed methodology is correct.

Another methodology for earthquake predictions has been developed in China under the leadership of professor Yin (Yin et al, 1994, 1995, 2000, 2002), using the so-call *Load-Unload Response Ratio* (LURR). This parameter can be defined from the constitutive curve of the focal zone for the controlling-state parameters. In case of the earthquake prediction, this constitutive curve can be drawn for one factor defining precursor parameters (see the next details concerning the precursors of an earthquake). The LURR parameter is the ratio between the slopes of curve for loading and unloading behavior. If the nonlinear system is stable, this parameter is 1. When the system is close to instability, the LURR value significantly differs from 1. For this reason, it can quantitatively indicate the degree of the imminence of instability. The temporal variation of LURR in many seismic areas in China, Japan and the USA has been calculated and analyzed, showing in all cases that this ratio is close to 1 in the periods of stability and significantly higher than 1 before the occurrence of earthquakes with magnitude higher than 6.0.

For *short-time prediction,* it is very important to determine the geophysical precursors of the earthquake (Gray, 1996). The monitoring of these geophysical precursors enables scientists to make short-term predictions of an earthquake. There are a variety of geophysical precursors. However, only five of them are the most important; they are: the P-wave changes, ground uplift and tilt, random emission, electrical resistivity of rocks and the water level fluctuations. In order to monitor the changes in the Earth's magnetic field, the *Demeter* (Detection of Electro-Magnetic Emissions Transmitted from Earthquake Regions) micro-satellite was launched in 2004 (Wikipedia, nd).

Another prediction for very short-time earthquakes is given by *animal behavior* (Kirschvink, 2000, Wikipedia, nd, Carayannis, 2007). Animals can detect the P-wave or ultrasonic wave generated by an earthquake, even if the waves are too small for humans' senses, or when there are some changes in low-frequency electromagnetic signals. When this happens, animals start behaving agitatedly and nervously. The Chinese began a systematic study of this unusual animal behavior and these studies had an important role in the prediction of the 1975 Haicheng earthquake. Scientists said snakes could sense a quake from 120 km away, up to five days before it happens, the reptiles being perhaps the most sensitive to earthquakes. Thus, the crate of some reptile farms, where a camera monitors the snake behavior, is one solution for short-term prediction.

In case of *early warning*, the effectiveness depends on the distance from source to site, because it is based on the difference between the velocity of P and S waves, the P one being the first arriving to the site. Thus an early warning system for earthquakes is possible only in case of earthquakes produced from deep sources and excludes the ones of crustal fracture in the near source zones. The warning time is rather small, ranging from seconds to a maximum of about one minute. There are some examples of using this system. For Mexico City, situated about 320 km from the source (subduction of the Cocos plate beneath the North American plate), the warning time is 60 to 75 seconds. For Bucharest, situated at 130 km from the Vrancea epicenter, the warning time is about 25 seconds, while for Taipei (Taiwan), about 10 to 17 seconds.

What is it possible to do during this very short warning time? The following activities are possible: Evacuation of buildings (for time greater than 30 seconds) or putting the inhabitants in a safer position, shutting-down the critical systems (nuclear and chemical reactors), stopping the high-velocity train, putting the strategic facilities into safe location, etc.

One must mention that, from the design point of view, only the long time prediction, based on statistical analysis, can be considered in order to eliminate or to limit the economical losses. The other predictions give the possibility to alarm the authorities in order to prepare actions for limiting the loss of life.

4.6 STATISTICAL SEISMOLOGY AND THE THEORY OF MULTI-SOURCE DATA FUSION

Statistical seismology is a relatively new field, which applies statistical methodologies to earthquake data in an attempt to raise new questions about earthquake mechanisms and to make some progress towards earthquake characteristic prediction (Vere-Jones, 2006, Vere-Jones et al, 2005). But the main question to be agreed is: can the physics of earthquakes be a statistical problem (Turcotte, 1999).

The physics of earthquakes covers a broad range of topics. Some of them are quite well understood, but others not. Among the understood ones, the general origin of earthquakes is well accepted as the displacement of tectonic plates along the pre-existing faults. But many aspects still remain unsolved in the problem of the dynamic rupture occurrence and of the prediction of earthquake characteristics, because there are important differences between theoretical results and in-field recorded values.

Considering the seismic data, there is a historical period until 1900 during which only qualitative information was available and the last period of 50 years in which seismic records have been available, being instrumentally monitored. For this last period, there are many data for low seismic events, few values for moderate earthquakes and, a very few for large earthquakes. Thus, a very important question arises concerning the validity of the probabilistic methodology in absence of sufficient valuable data. Due to this situation, the deterministic methodologies have an important position in determining the seismic actions. Supporters of the probabilistic approach defend its use, whereas the deterministic followers deny the probabilistic way. The weakest aspect of the deterministic approach is the tendency to give a definite numerical answer mainly based on the understanding of the phenomenological aspects by proposing design values related to them, despite the complexity of the phenomenon governed by the Theory of Chaos and the enormous number of unknowns.

The criticism to the deterministic approach refers to the fact that the proposed design values of seismic actions are the maximum ones, which probably could never be reached during the structure life. Contrary, the probabilistic approach is criticized for the reduced number of records at the same site related to the same source.

In this dispute, some examples can be very suggestive, like for instance the subduction fault in Northern California through Oregon and Washington into British Columbia and the New Madrid area in Eastern North America. In the first case, during the past decades since the seismic activity has been followed, no earthquakes with magnitude greater than 6.0 have been recorded. But, geological studies have shown that some giant earthquakes have struck the region in the past 7000 years. The last one was produced in January 1700 with a magnitude $M = 9.0$, very close to maximum possible value. In the second case, during the period 1811-1812 three very strong earthquakes of 7.6 to 7.9 shook the Missouri US area. After these former events, the zone remains almost quiet. According to the probabilistic approach, these situations could be ignored, but interpreting the geological aspects through the deterministic approach, the possibility of occurrence of a very important earthquake has to be considered. In exchange, as a lot of recorded ground motions exist for both the earthquakes produced along the San Andreas (US) and in the Vrancea (Romania) zones, the Statistical Seismology can be very usefully applied in these cases. A different situation corresponds to the case of the Balkan earthquakes, where the locations of the epicenters are very diffuse, two events never being recorded in the same place.

These very different aspects show that there are situations where the positions of sources are very well identified; the *Statistical Seismology* can be very useful to define the design earthquakes in this case. But in case of diffuse earthquakes, a different approach must be used to predict these earthquakes. This is the task of the new developed *Theory of Multi-Source Data Fusion* (Semerdjiev, 1999, Leebmann and Kiema, 2000, Yager, 2004). This new field of statistical analysis is devoted to obtain information of better quality by data fusion originated from different sources. This is a general theory, which can be used in very different fields of engineering. In the seismic field, this theory is based on the evaluation of tectonic, geological, seismological and geotechnical data, for different sources. If these characteristics of some sources are similar, the number of valuable data for statistical analysis increases. For instance, the San Andreas, North Anatolian and New Zealand faults have the same characteristics: crustal shallow interplate and strike-slip type. Therefore the data used for the more studied Californian earthquakes can be used also for the North Anatolian earthquakes, but not for other seismic zones. A very frequent mistake is to use the well-known El Centro record, with characteristics corresponding to strike-slip earthquakes, for designing buildings located in Italy, Romania or Greece, with very different earthquake characteristics. Some seismic areas of North America, Europe, Asia and Australia present shallow intraplate, normal dip-slip type. The differences in the earthquake records are only the consequence of the site conditions, which can be separately analyzed. So, the data obtained in a given place can be used also for other places with the same earthquake type, provided that the site effects can be eliminated.

4.7 REFERENCES

nd – no date: the web site is periodically modified

Agnew, D.C. (2002): History of seismology, Chapter 1. International Handbook of Earthquake & Engineering Seismology. Part A (eds. H.K. Lee, H. Kanamori, P.C. Jennings, C. Kisslinger), Academic Press, Amsterdam, 3-11

Anderson, D.L. (2001a): Topside tectonics.
http:// www.gps.calthech.edu/~dla/tectonics-drift4.1.pdf

Anderson, D.L. (2001b): Plate tectonics as a far-from equilibrium self-organized dissipative system. American Geophysical Union, Fall Meeting 2001

Anderson, D.L. (2003): Plate tectonics: The general theory
http://www.gps.caltech.edu/~dla/DLAPlate Tectonics.pdf

Anderson, D.L. (2008): Mantle convection
http://www. mantleplumes.org./Convection.html

Ashida, M. (1996): Fault premise. The Sciences, September/October, 15-18

Bertero, V.V., Bozorgnia, Y. (2004): The early years of earthquake engineering and its modern goals. In Earthquake Engineering. From Engineering Seismology to the Performance-Based Engineering (eds. V.V. Bertero and Y. Bozorgnia), CRC Press, Boca Raton, 1.1–1.17

Bokor, L (2007): The supercontinent cycle hypothesis. Wolverhampton University Report
http://www.frutex-cowsite.com/Geography?Supercontinent_Cycle_theory.doc

Bolton, P., Cole, S. (2006): Earthquakes and a brave new China. Bendfield Hazard Research Centre Report.
hppt://benfieldhrc.org/activities/issue6/pages/eq_china.htm

Carayannis, G.P. (2007): Earthquake prediction in China.
http://www.drgeorgepc.com/EarthquakePredictionChina.html

Casti, J.L. (1994): Complexification: Explaining a Paradoxal World Through the Science of Surprise. Harper Perennial, New York

Coveney, P. (2003): Chaos, entropy and the arrow of time.
http://www.fortunecity.com/emachines/e11/entropy.html

CSCS, Center for the Study of Complex Systems (2006): Dissipative structures.
http://www.cscs.umich.edu/~crshalizi/notebene/dissipative-structures.html

Digas, B.V, Melnikova, L.A., Rozenberg, V.L. (2001): Numerical simulation of dynamics of tectonic plates: Spherical block model. 6th Workshop on Non-linear Dynamics and Earthquake Prediction, ICTP, 15-27 October 2001

ERI, Earthquake Research and Information (2001): Earthquake Myths and Folklore.
http:// www.ceri.memphis.edu/public/myths.shtml

Everything (2002): Wilson cycle.
http://everything2.com/index.pl/node_id=1392372

Fischinger, M., Cerovsek, T., Turk, Z. (1998): EASY, Earthquake Engineering Slide Information System.A hypermedia learning toll. Elecdtronic Journal of Information Technology in Construction, Vol. 3, str. 1-10

Frater, H. (1998): Natural Disasters. Cause, Course, Effects, Simulation. Multimedia Program on CD-ROM

Garrette L. (2003): Laplace's Demon. Hypography, Science for Everyone.
http://www.hypography.com/topics/Laplaces_Demom_112215.cfm

Geology 202 (2004): The Wilson cycle.
http://www.earth.northwestern.edu/people/seth/B02/new_2004/wilson_cycle.html

Gioncu, V. (1995): Development and design of seismic-resistant steel structures in Romania. Behavior of Steel Structures in Seismic Areas, STESSA 94 (eds. F.M. Mazzolani and V. Gioncu), 26 June-1 July 1994 , E & FN Spon, London, 3-27

Gioncu, V. (2005): Instabilities and Catastrophes in Structural Engineering (in Romanian) Editura Orizonturi Universitare, Timisoara

Gioncu, V., Mazzolani, F.M. (2002): Ductility of Seismic Resistant Steel Structures. Spon Press, London

Gray, Ch. (1996): A review of two methods of predicting earthquakes. The Undergraduate Engineering Review,
http://tc.engr.wisc.edu/UER/uer96/author3/content.html

Green, D.G (2000): Self-organization in complex systems. In Complex Systems (eds. T.R.J. Bossomaier and D.G. Green), Cambridge University Press, 15-50
http:// books.google.com/books?isbn = 0521462452

GSC, Geological Survey of Canada (2008): Geodynamics. Simplified subduction thrust earthquake cycle.
http://gsc.nrcan.gc.ca/geodyn/eqcycle_e.php

Heylighen, F. (2002): Deterministic chaos
http://pespmc 1.vub.ac.be/CHAOS.html

Hough, S.E. (2005): Earthquakes: Predicting the unpredictable? Geotimes, March.
http://www.agiweb.org/geotimes/mar05/feature_eqprediction.html

Huang, J., Turcotte, D.L. (1990): Are earthquakes an example of deterministic chaos? Geophysical Research Letters, Vol. 17, No. 3, 223-226

Jackson J. (2001): Living with earthquakes: Know your faults. Journal of Earthquake Engineering, Vol. 5, No. 1, 5-123

Keilis-Borok, V.I. (ed) (1990): Intermediate-term earthquake prediction: Models, algorithms, worldwide tests. Physics of Earth and Planetary Interiors, Special issue, 61,

Keilis-Borok, V.I. (1996): Intermediate-term earthquake prediction. Proceedings of National Academy of Science, Vol. 93, April, 3748-3755

Keilis-Borok, V.I., Shebalin, P.N. (eds) (1999): Dynamics of Lithosphere and Earthquake Prediction. Physics of Earth and Planetary Interiors, Special Issue, 111, 179-330

Keilis-Borok, V.I., Soloviev, A.A. (eds) (2003): Nonlinear Dynamics of the Lithosphere and Earthquake Prediction, Springer-Verlag, Heildelberg

Keilis-Borok, V.I., Shebalin, P., Gabrielov, A., Turcotte, D. (2007): Reverse detection of short-term earthquake precursors.
http://arxiv.org/pdf/physics/0312088

Kirschvink, J.L. (2000): Earthquake prediction by animals: Evolution and sensory perception. Bulletin of the Seismological Society of America, Vol. 90, No. 2, 312-323

Leebmann, J., Kiema, J.B.K. (2000): Knowledge representation in technical information systems for earthquake loss mitigation. 2nd Euro Conference on Global Change and Catastrophe Risk Management, Laxenburg, July 2000

Lungu, D., Moldoveanu, T. (1997): Introduction. Design of Structures in Seismic Zones (eds. D.Lungu, F.M.Mazzolani, S.Savidis). Tempus Project 01198,1-60

Moon, F.C. (1987): Chaotic Vibrations. An Introduction for Applied Scientists and Engineers. John Wiley & Sons, New York

NASA (2002): Datbase
http://sideshow.jpl.nasa.gov/mbh/series.html

NASA (2005): Global positioning satellites. Global positioning orbits
http://hyperphysics.phy-astr.gsu.edu/hbase/gps.html

Papazachos, B.C.(1994): Seismology in Greece, European Seismological Commission, XXIVth General Assembly, Athens, 19-24 September 1994, 39-54

Peltier, W.R. (2000): Earth evolution and global change. University of Toronto
http://www.crsng.gc.ca/programs/real2000/Global-change.pdf

Prigogine, I., Stengers, I. (1979): La Nouvelle Alliance. Metamorphose de la Science. Gallimard, Paris

Pustovitenko, B.G., Kapitanova, S.A., Gorbunova, I.V. (1994): The complex rupturing in the deep focal sources of Vrancea region. In European Seismological Commission, XXIVth General Assembly, Athens, 19-24 September 1994, 627-633

RMS (2006): The 1976 Great Tangshan earthquake. 30-year retrospective
http://www.rms.com/Publications/1976Tangshan.pdf

Rotwain, I., Novikova, O. (1999): Performance of the earthquake prediction algorithm CN in 22 regions of the world. Physics of the Earth and Planetary Interiors, 111, 207-213

Rozenberg, V.L., Soloviev, A.A., Ermolieva T,Y..(2002): Earthquake generators: application of parallel technologies. In 4th Workshop on Symbolic and Numeric Algorithm, Synasco 02 (eds. D. Petcu, V. Negru, D. Zaharia, T. Jebeleanu), Timisoara, 9-12 October 2002, Mirton, Timisoara, 314-326

Saito, T., Matsukawa, H. (2007): Size dependence of the Burridge-Knopoff model. Journal of Physics, Vol. 89, 012016

Sankaran, A.V. (2002): Mantle convection results from plate tectonics: Fresh hypothesis reverses current views. Current Science, Vol. 82. No. 7, 785-787

Santos, I.M. (2006): Modeling and numerical study of nonsmooth dynamical systems. PH Thesis, Universitat Politecnica de Catalunya

Schuster, H.G. (1988): Deterministic Chaos. An Introduction. VCH Verlag, Weinheim

Science Daily (1994): Earthquake can be predicted months in advance.
http://www.sciencedaily.com/releases/2004.htm

Sell, G.R, You, Y. (2002): The evolution of evolutionary equations. In Dynamics of Evolutionary Equations, Vol. 143, Springer Verlag

Semerdjiev, T. (ed.), (1999): Multi-Source Data Fusion, ProCon Ltd, Sofia

Soloviev, A. (2001): Model of block structure dynamics and its application to study lithosphere block dynamics and seismicity. 6th Workshop on Non-Linear Dynamics and Earthquake Prediction, International Centre for Theoretical Physics, 15-27 October 2001

Soloviev, A., Maksimov, V. (2001): Block models of lithosphere dynamics and seismicity. Interim Report, IR-01-067, International Institute for Applied Systems Analysis, Laxenburg, Austria

Turcotte, D.L. (1999): The physics of earthquakes: Is it a statistical problem? ACES Inaugural Workshop, Australia, http://www.quakes.uq.edu.au/ACES_WS1_proc/PDF/2.1_1.pdf

Turcotte, D.L., Malamud, B.D. (2002): Earthquake as a complex system. In International Handbook of Earthquake & Engineering Seismology (eds. W.H.K. Lee et al), Chapter 14, Academic Press, Amsterdam, 209-229

USGS (nd): The break-up of Pangeea. http://vulcan.wr.usgs.gov/ ivingWith/vulvancPast/Notes/breahupof_pangeea.html

USGS (nd): Early monitoring http://earthquake.usgs.gov/learn/eqmonitoring/eq-mo.php

USGS (nd): This dynamic Earth http://pubs.usgs.gov/gip/dynamic/dynamic.html

USGS (nd): The earthquake surface in 3D-Sereo system andtopographic maping

Uyeda, S. (2002): Continental drift, sea-floor spreading and plate/plume tectonics. In International Handbook of Earthquake & Engineering Seismology (eds.W.H.K. Lee et al), Chapter 6, Academic Press, Amsterdam, 51-67

Vere-Jones, D. (2006): The development of statistical seismology: A personal experience. Tectonophysics, 413, 5-12

Vere-Jones, D., Ben-Zion, Y., Zuniga, R. (2005): Statistical seismology. Pure and Applied Geophysics, Vol. 162, 1023-1026 http://en.wikipedia.org/wiki/Plate_tectonics

Weisstein, E.W. (2007): Double pendulum. http://scienceworld.wolfram.com/physics/DoublePendulum.html

Wheatland, M. (2007): The double pendulum http://www.physics.usyd.edu.au/~wheat/dpend_html

Wieland, M. (2001): Earthquake alarm, rapid response and early warning systems: Low cost system for seismic risk reduction. Disaster Reduction Workshop, Reston, 19-22 August 2001

Wikipedia (nd): Earthquake prediction http://en.wikipedia.org.wiki/Earthquake_prediction

Wikipedia (nd): Demeter satellite http://wikipedia.org/wiki/Demeter_satellite

Wikipedia (nd): Plate tectonics http://en.wikipedia.org/wiki/Plate_tectonics

Wikipedia (nd7): Determinism. http://en. Wikipedia.org./wiki/Determinism

Wikipedia (nd): Satellite constellation.
http://en.wikipedia.org/wiki/Satellite_constelation
Wikipedia (nd):Self-organization
htpp://en.wikipedia.org/wiki/Self-organization
Wikipedia (nd): Dissipative systems
http://en.wikipedia.org/wiki/Dissipative_system
Wikipedia (nd): Self-organized criticality
http:// en.wikipedia.org/wiki/Self-organizedcriticality
Wikipedia (nd): Double pendulum
http://en.wikipedia.org/wiki/Double_pendulum

Xia, J., Gould, H., Klein, W., Rundle, J.B. (2005): Simulation of the Burridge-Knopoff model of earthquakes with variable range stress transfer. Physical Review Letters, 248501.1-4

Yager, R.R. (2004): A framework for multi-source data fusion. Information Science, Vol. 163, Nos. 1-3,175-200

Yin, X.C., Yin, C., Chen, X.Z. (1994): The precursor of instability for nonlinear systems and its application in earthquake prediction: The load-unload response ratio theory. Nonlinear Dynamics and Predictability of Critical Geophysical Phenomena (eds. W.I. Newman and A.M. Gabrelov), AGU, 55-60

Yin, X.C., Chen, X.Z., Song, Z.P., Yin, C. (1995): A new approach to earthquake prediction: the load/unload response ratio (LURR) theory. Pageoph, Vol. 145, No.3-4, 705-715

Yin, X.C., Wang, Y.C., Peng, ,K.Y., Bai, Y.L., Wang, H.T., Yin, X.F. (2000): Development of a new approach to earthquake prediction: load/unload response ratio (LURR) theory. Pure and Applied Geophysics, Vol. 157, 2365- 2383

Yin, X.C., Mora, P., Peng, K., Wang, Y., Weatherley, D. (2002): Load-unload response ratio and accelerating moment/energy release critical region scaling and earthquake prediction. Pure and Applied Geophysics, Vol. 159, 2511-2523

Chapter 5

Tectonic Plates and Faults

5.1 PLATE TECTONICS

5.1.1 Major Plates

Today it is accepted that the Earth is covered by some rigid plates which move across its surface, over and on a partially molten internal layer. Using geological terms, the plates form the *lithosphere*, which is the Earth's solid rock. The rigid lithosphere can be considered to float on the ductile *asthenosphere*, which flows. So, the lithosphere (surface of the Earth) is broken up into what are called *tectonic plates*; *Plate tectonics* (from the Greek *tecton*, meaning one who constructs and destroys) being the theory of geology developed to explain the phenomenon of continental drift. This theory defines the tectonic plates and their boundaries. In function of their surface, the *tectonic plates* are divided into major and minor plates. Tectonic plates can include continental crust, oceanic crust or both. The distinction between continental crust and oceanic crust is based on the density of constituent materials. The continental crust is composed primarily of granite, so it is relatively light. An average thickness may by around 30-40 km, while the thickest part is about 70 km in the zone of the Himalayas and Tibet. The oceanic crust is denser than the continental one, being composed of basalt. It is relatively thin, about 7 km thick. As a result, the oceanic crust generally lies below the sea level, while the continental crust is situated above the sea level. There are 52 important tectonic plates (USGS, nd, Wikipedia, nd). The 14 *major tectonic plates* are shown in Figure 5.1. Among them, the seven most important plates are:
-	African plate covering Africa (continental plate).
-	Antarctic plate covering Antarctica (continental plate).
-	Australian plate (known also as Indo-Australian plate) covering Australia (continental plate).
-	Eurasian plate covering Europe and Asia (continental plate).
-	North American plate, covering North America and North-East Siberia (continental plate).
-	South American plate, covering South America (continental plate).
-	Pacific plate (the biggest), covering the Pacific Ocean (oceanic plate).

In addition to the above great tectonic plates, the following can be also considered as major plates: Arabian plate (continental plate), Caribbean plate, Cocos plate, Juan de Fuca plate, Nazca plate and Philippine plate (all being oceanic plates).

It is very important to underline that the tectonic plates do not coincide with the continent forms.

5.1.2 Minor, Micro and Ancient Plates

There are 38 *minor tectonic plates* (Bird, 2003): Aegean Sea, Altiplano, Amurian, Anatolian, Burma, Banda Sea, Balmoral Reef, Birds Head, Caroline, Conway Reef, Easter, Futuna, Galapagos, Juan Fernandez, Kermadec, Manus, Mariana, Maoke, Molucca Sea, New Hebrides, Niuafo'ou, North Andes, North Bismarck, Okhotsk, Okinawa, Panama, Rivera, Sandwich, Scotia, Shetland, Somali, Solomon Sea, South Bismark, Sunda, Timor, Tonga, Woodlark and Yangtze plates.

To these major and minor plates, one must add some hundreds of *micro tectonic plates*, formed during the continental drift, as the Iberian, Adria, Turkish, Black Sea, etc., microplates.

The presented configuration corresponds to today's situation, ignoring the history of formation of each plate. There are also some *ancient plates*, which disappeared during the genesis of the actual configuration. For instance, the major Eurasian plate is divided in two parts, European and Asian plates, in contact along the Ural Mountains, due to the fact that, about 400 million years ago, they were different continents. The Farallon and Kula plates almost totally subducted the North American plate, the Juan de Fuca and Gorda plates being the remains of this plate. The China plateau is the result of the collision of an ancient minor plate with the major Asian plate.

5. 2 PLATE BOUNDARY TYPES

5.2.1 Relative Plate Movements

Tectonic plates are able to move because of the relative density of oceanic lithosphere and the relative weakness of the asthenosphere, which allows the tectonic plates to easily move towards a subduction zone. Therefore, the plates are in constant, but very slow, movement. The recent development of GPS has proven that plates are moving almost exactly as the plate tectonics postulate. Figure 4.8 shows the relative plate motions, with different directions and velocities. The fastest-moving plates are the oceanic ones, with the Cocos plate having the highest velocity at 8.6 cm/year. The giant Pacific plate is the next, moving at 8 cm/year. The slowest-moving plates are the continental ones, because of their deep roots acting to slow their movement. The slowest is the Eurasian plate with 0.7 cm/year, followed by the North American plate with 1.1cm/year.

Another very important aspect is the direction of the plate movement. In Figure 5.2 one can see that these directions are very different, being divergent, convergent or quasi parallel along some lines.

5.2.2 Plate Boundaries

When plates meet, the consequence depends on the density of the respective plates. The essential hypothesis of the plate tectonics is that the interior of the plate is rigid and that the deformations are restricted to the *Plate boundaries*, forming

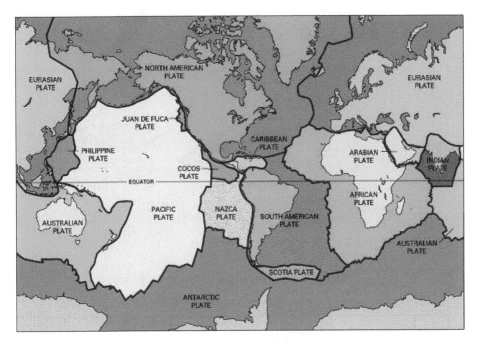

Figure 5.1 The main tectonic plates (USGS, nd, Wikipedia, nd, Nelson, 2003)

Fault networks. The relative movements are produced along the boundaries with velocity varying across the Earth, but with an average of the order of tens of centimeters per year. In the region where the hot molten lava comes out, the movement of the two boundaries is divergent and so some fault in the form of a rift is created. Contrary, in the region where the cold rock mass goes into the Earth, the movement of the two boundaries is convergent and, due to the collision along the fault, some volcanoes and mountain chains are formed. Considering these relative movements, different boundary types exist (Fig. 5.2):

-   *Divergent boundaries,* when two plates move away from each other, due to the ascension of molten lava, which fills the opened fault between the two plates.
-   *Convergent boundaries,* when two plates move slower in the front or in contrary directions.
-   *Transform boundaries,* when two plates move side-by-side along the same fault with different velocities or in opposite directions.

Figure 5.3 shows the plate boundaries which are active at the present time. In the following, each boundary type will be detailed, with the most important examples.

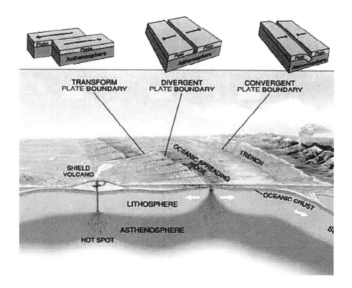

Figure 5.2 Plate boundary types (USGS, nd)

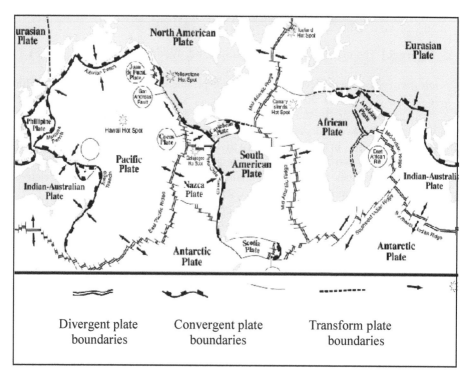

Figure 5.3 World plate boundaries (modified after NYSED, 2001)

5.2.3 Divergent Boundaries

In plate tectonics, the *Divergent boundaries* (also known as constructive or extensional boundaries) are a linear fracture between two tectonic plates which are moving away from each other, rising convection currents (Nelson, 2003, Wikipedia, nd). The rising current pushes up on the bottom of the lithosphere, lifting it and flowing laterally beneath it. The magma (liquid rock) seeps upward to fill the cracks. In this way, a new crust is formed along the boundary. There are two divergent boundary types.

Oceanic divergent boundary (Fig. 5.4a) occurs beneath oceanic lithosphere, the rising convection producing a mid-oceanic ridge. Extensional forces stretch the lithosphere and produce a deep fissure. When the fissure opens, pressure is reduced on the super-heated mantle material below. It responds by melting and the new magma flows into the fissure. The magma then solidifies and the process repeats itself. The *Mid-Atlantic Ridge* (Fig. 5.4b) is a classic example of this type of plate boundary (Wikipedia, nd). It is the result of the separation of some tectonic plates: the North American plate from the Eurasian plate in the North Atlantic and the South American plate from the African plate in the South Atlantic. These plates are still moving apart, so the Atlantic is growing at the ridge, at a rate of about 5-10 cm per year in the East-West direction. The volcanic country of Iceland, which straddles the Mid-Atlantic Ridge, offers scientists a natural laboratory about the process of spreading between North American and Eurasian plates. There are also other important oceanic divergent boundaries

Continental divergent boundary occurs beneath a thick continental plate (Fig. 5.5). Generally, the pull-apart is not vigorous enough to create a clean, single break through the thick plate. As the two plates pull apart, normal faults develop on both sides of the rift and the central blocks slide downwards. Early in the rift forming, streams and rivers will flow into the sinking rift valley to form a long linear lake or a shallow sea. The East Africa Rift Valley is a classic example of this plate boundary. This rift is in a very early stage of development, but the Northern part of the rift, the Red Sea, is in a more complete stage of evolution, showing the separation of Saudi Arabia from the African continent. Geologists believe that, if this spreading continues, the new Somalia plate will break away and Eastern Africa will become an island, as happened for Madagascar a long time ago.

5.2.4   Convergent Boundaries

In plate tectonics, the *Convergent boundaries* (also known as destructive boundaries) are the actively deforming region where two tectonic plates move towards one other (Nelson, 2003, Wikipedia, nd). This motion of two plates produces the subduction of one plate beneath the other, or the collision. In the first case the less dense plate rides over the denser plate, which is subducted. In the second case, if the densities are similar, the plate's collision occurs. Generally, the oceanic plates are more dense that the crust plates. In function of these differences in density, there are the following convergent boundaries.

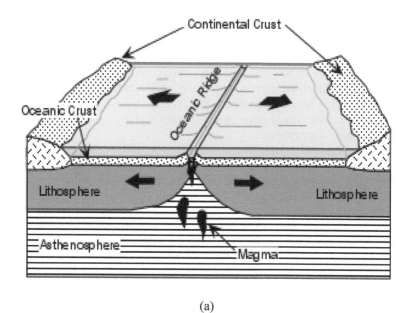

(a)

(b)

Figure 5.4 Divergent boundaries: (a) Oceanic ridge; (b) Oceanic divergent boundaries (USGS, nd)

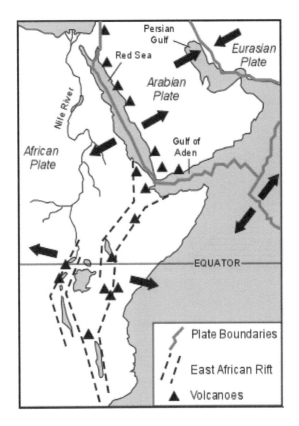

Figure 5.5 Continental divergent boundary: African Rift Valley (USGS, nd)

Ocean-ocean convergent boundaries (Fig. 5.6) occur when two convergent plates are oceanic ones and one of them subducts under the other, resulting in trenches and some chains of islands. The most known case is the Marianne trench (paralleling the Marianne Islands), where the Pacific plate converges against the Philippine plate. Other cases are the island arcs produced at the ocean surface, such as the Japanese, Aleutian or Caribbean islands. The most dangerous result of this convergence is the possibility of forming the tsunami phenomenon. The 2004 Sumatra tsunami is the best example of this risk (see Fig. 4.18).

Ocean-continent convergent boundaries (Fig. 5.7a) occur when an oceanic plate with higher density subducts a continental plate, forming trenches thousands of kilometers long and 8 to 10 km deep. The main ocean-continental boundaries are presented in Figure 5.7b. Off the coast of South America, along the Peru-Chile trench, the oceanic Nazca plate is pushing into and being subducted under the continental plate of South America. In turn, this plate lifts up, creating the Andes Mountains, the backbone of the continent. This subduction zone is one of the deepest and largest in the world, in some cases being 650 km from the surface. Another very important subduction zone is the Western zone of Canada, where the Juan de Fuca plate subducts the North American plate.

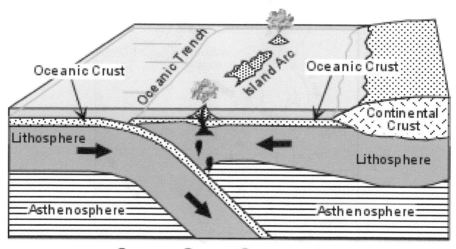

Ocean - Ocean Convergence

Figure 5.6 Convergent ocean-ocean boundaries: (USGS, nd)

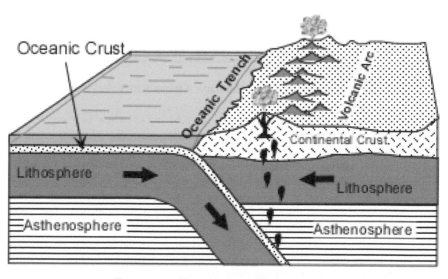

Ocean - Continent Convergence

(a)

Figure 5.7 (continues)

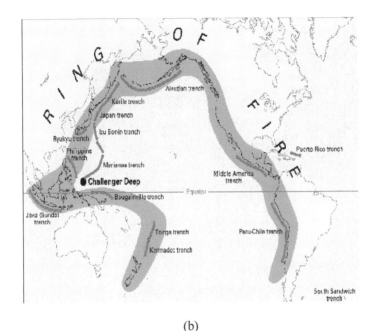

(b)

Figure 5.7 Convergent oceanic-continental boundaries: (a) Subduction of oceanic plate; (b) Main convergent oceanic-continental boundaries (USGS, nd)

Continental-continental convergent plates (collision) (Fig. 5.8a), when two continental plates, having similar density, collide into each other. The crusts of these plates tend to buckle, forming mountain belts. Currently, the highest mountains in the world represent the result of this collision. The Himalayan Mountains resulted from the collision, 10 million years ago, of the plate containing India with the Eurasia plate (China-Tibetan plateau) (Fig. 5.8 b).

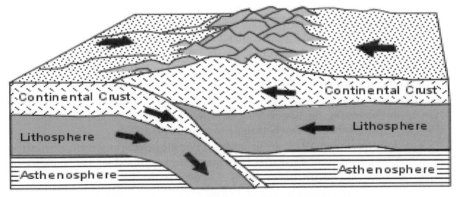

Continent- Continent Convergence
(a)

Figure 5.8 (continues)

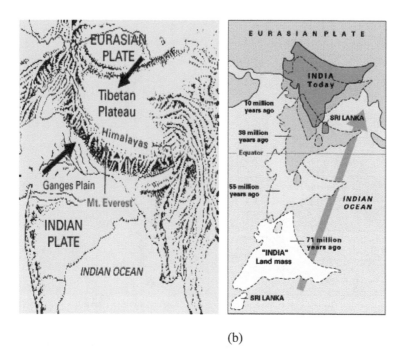

(b)

Figure 5.8 Convergent continental-continental boundaries: (a) Collision between two continental plates; (b) India-Eurasia Plates collision (USGS, nd)

Similar collisions were produced between European and Asian plates forming the Ural Mountains, and Adrian and European plates forming the Alps.

5.2.5 Transform Boundaries

Transform boundaries are locations where two plates slide along each other. There are two transform boundary types, one oceanic and one continental.

Most transform boundaries are found on the *ocean floor*, where they often offset spreading ridges to form a zigzag plate boundary (Fig. 5.9a), being transversal to the divergent oceanic boundaries, such as the Mid-Atlantic, Nazca or Antarctic ridges (Fig. 5.9b). The name of transform boundaries is proposed to underline that they transverse the mid-oceanic ridges. Generally, this boundary type is not interesting for Engineering Seismology, due to the fact that usually it does not affect inhabited areas, belonging to oceanic plates.

Few transform boundaries occur on land. For the *continental zone*, one of the most famous examples of these transform boundaries occurs along the boundary of North America and Pacific plates and it is known as the *San Andreas Fault* (Fig. 5.10). It is interesting to notice that the sliding of the San Andreas Fault occurs in some part at depth, without surface traces, but in some others part its movements are marked at the surface by valleys and lakes. Another very important case is the *North Anatolian Fault* (see Fig. 3.4). A comparison between these two transform

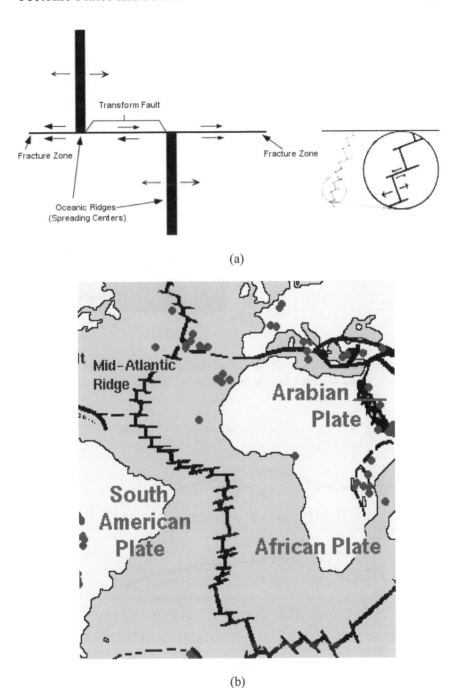

(a)

(b)

Figure 5.9 Oceanic transform boundaries : (a) Zigzag plate boundaries;
(b) Mid-Atlantic transform boundaries of tectonic plates (USGS, nd)

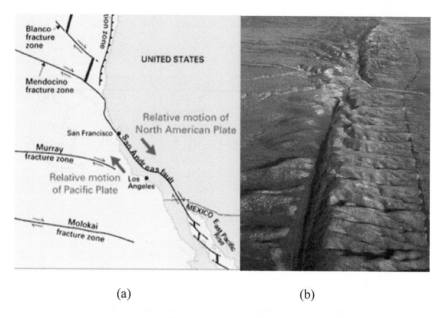

(a)   (b)

Figure 5.10 San Andreas Fault: (a) Active faults; (b) Aerial view of the San
Andreas Fault in Central California   (USGS, nd)

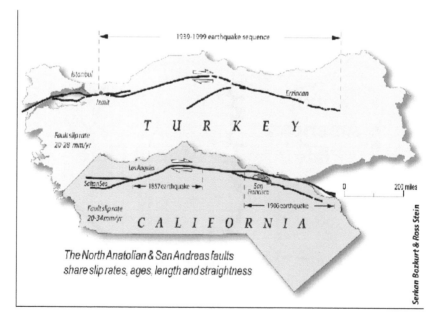

Figure 5.11 Comparison of two transform faults (USGS, nd)

boundaries is presented in Figure 5.11. One can see a very significant similitude between these continental transform boundaries. Similar continental transform boundaries occur in New Zealand, being known as the *Alpine Fault*.

5.3   DIFFUSE ZONES AND FAULT TYPES

5.3.1 Zones with Diffuse Faults

An important category of seismic zones, extending through the very densely inhabited areas, is the one occurring in the large areas along the plate boundaries or in areas between boundaries of tectonic plates. The occurrence of earthquakes violates the plate tectonic model, which considers that the earthquakes are concentrate on the narrow bands of plate boundaries.

The characteristics of faults in diffuse zones are very complex due to the multitude of fault types. Looking to the world map of seismic zones (see Fig. 2.2), one can see that there are zones where the model of tectonic plates is not accurate. Nevertheless, this spreading is mainly due to (Fig. 5.12):
- Extension of seismic zones under subducted plates;
- Collision between two continental plates;
- Fracture of weak crustal plates between the boundaries.

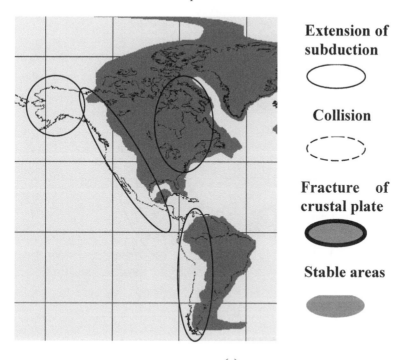

(a)

Figure 5. 12 (continues)

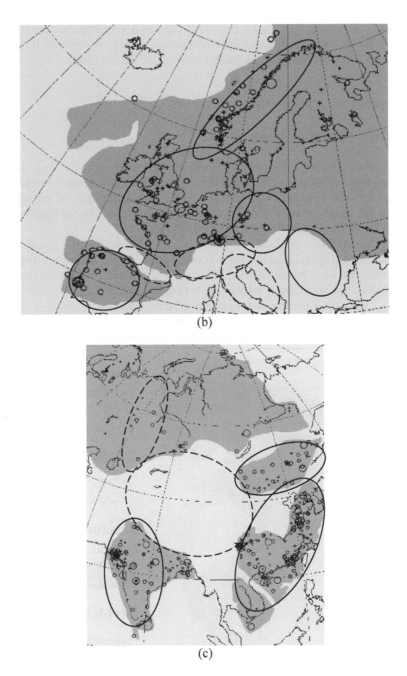

(b)

(c)

Figure 5.12 World zones with diffuse boundary zone: (a) North and South
America; (b) Europe; (c) Asia (modified after Mooney et al, 2004)

5.3.2 Plate Boundary Zones: Extension of Seismic Zones

The areas affected by the ocean-continent convergence boundaries (see Figure 5.7a) are larger than the zone of plate boundaries, containing very complex geological formations of the subduction zone (Fig. 5.14). The extension of seismic zone depends on subduction slope. The reduced slope produced a long extension of seismic zones. So, the subduction process induces plate boundary zones, characterized by the presence of multiple faults, distributed on the large area, with the potential for some crustal earthquakes.

A very illustrative example is the subduction of Kula Plate under the North American Plate (Fig. 5.14a). After the almost complete subduction, the remainder plate is the Juan de Fuca Plate (Fig. 5.14b). The spreading of faults in the case of this subduction refers to the diffuse zone of Cascadia of Western Canada. At some time in future, this seismic zone will generate a huge subduction earthquake, with magnitude greater than M 9.0. Geological evidence indicates that one huge earthquake strikes this Canadian coast every 300-800 years.

Another illustrative example of this diffuse zone is the seismic area of Los Angeles City, crossed by the San Andreas fault (see Fig. 2.13a), around an important zone with a network of diffuse faults. It must be noticed that the well-known 1994 Northridge earthquake is produced by a secondary fault of this network. A similar example is the 1995 Kobe earthquake, occurred in a fault which not belonging to the tectonic plate boundaries (see Figure 2.14a). The faults from these diffuse zones have characteristics very different from the ones corresponding to the tectonic plate boundaries

The formation of the seismic Californian zone is dominated by the complete subduction of the ancient Farallon and Pacific plates under the Western coast of the North American Plate (Fig. 5.16a), creating the San Andreas fault (Wikipedia, nd). The remains of the Farallon plate are the Juan de Fuca plate, subducting under the Northern part of North American plate, the Cocos plate subducting under the Central America, and the Nazca plate, subducting under the South American plate. It is very important to notice that the very shallow angle of subduction creates a large extension of seismic zone (Fig. 5.16b), characterized by the presence of many parallel faults (see Fig. 3.5) and the mountains in the subduction areas (as the impressive American Rocky Mountains in the Western North America).

In the same time it very interesting to understand the formation of the volcanic Andes Mountains in South America, as the result of subduction of Nazca plate under the South American plate, forming the Peru-Chile trench.

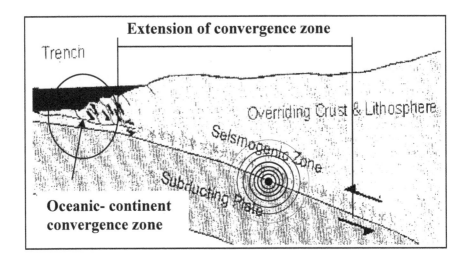

Figure 5. 13 Extension of seismic zone (modified from Lecture 14, 2003)

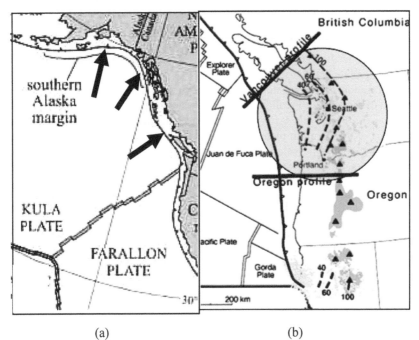

(a)   (b)

Figure 5. 14 Subduction under North American Plate: (a) Kula Plate;
(b) Extension of seismic Cascadia zone (USGS, nd)

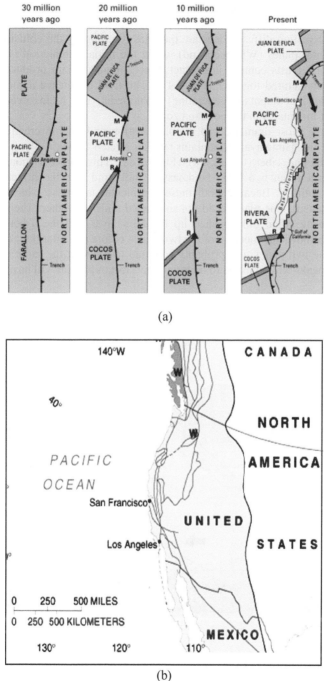

(a)

(b)

Figure 5.15 Californian seismic zone: (a) Subduction of Farallon Plate under
North American Plate; (b) Extension of seismic zone (USGS, nd)

5.3.3 Collision between Two Continental Plates

The most interesting diffuse zones are the ones resulting by the collision between two continental plates (Wikipedia, nd), when the initial subduction is destroyed, the continental crusts coming in contact direct, mountains are produced, and the two continents sutured together (Fig. 5.17). The field of science named *Orogeny* (from Greek, for mountain generating) (Wikipedia, nd) is involved with the formation of mountains in the collision zone.

Among the most dramatic and visible creation and visible creation of plate tectonic are the Himalayan Mountains, which stretch 2900km along the border between India and Tibet (USGS, nd), (Fig. 5.18a). This immense mountain range began to form when two large landmasses, India and Eurasia, driven by plates movement, collided. Because both these continental landmasses have about the same rock density, one plate could not subducted under the other. The presence of impinging plate could only pushed up the Himalayas and the Tibetian Plateau Figure 5.18b shows the meeting of the two plates before and after their collision. In the future the two plates will probably continue to move at the same rate. So the Himalayan peaks will continue to raise and the Tibetan plateau will have grown.

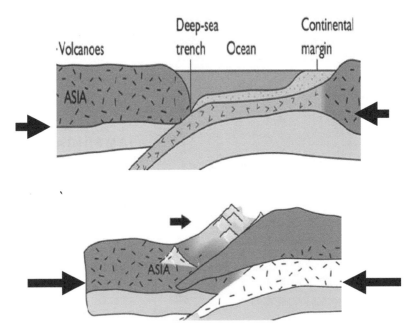

Figure 5. 16 Collision between two continental plates (after Oceanography, 2008)

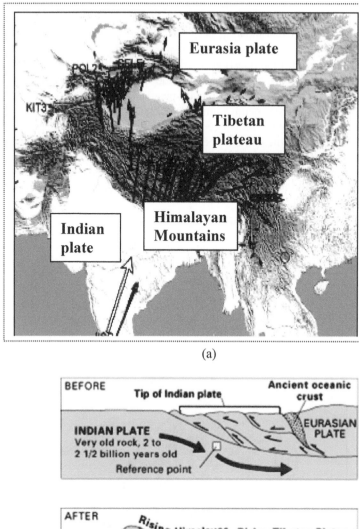

(a)

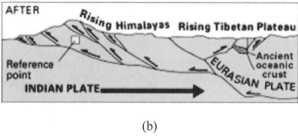

(b)

Figure 5. 17 Collision of Indian and Eurasian plates: (a) Forming the Himalayan
Mountains and Tibetan plateau; (b) Meeting of the two plates (USGS, nd)

Another very important zone where collision earthquakes are present is the territory of Italy (Fig. 6.9a) (Rosenbaum et al, 2004, Valensise et al, 2003, Carminati et al, 2004, Chiarabba et al, 2005, Solarino and Cassinis, 2007, Battaglia,et al, 2004). It is mainly due to the presence of the Adria microplate. During the last few decades the question whether the Adria Plate is a promontory of the Africa Plate, or whether it moves as an independent microplate, have been addressed by numerous authors. The Italian peninsula is formed by different sectors deriving from different geodynamical processes. There are three different orogenic belts characterizing the Italian geology (Solarino and Cassinis, 2007):

(i)   In Northern Italy, the collision between the Adria Plate and the European Tectonic Plate gave rise to the Alps Mountains.

(ii)   In the central part, the collision between the Adria Plate and the European Plate, represented by the Tyrrhenian Sea, produced the Apennines Mountains.

(iii)   The Southern zone is characterized by the crustal subduction of the Ionian Plate belonging to the African Plate under the Tyrrhenian plate belonging to the European Plate, along the Calabrian Arc.

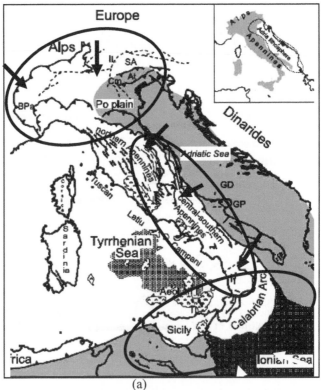

(a)

Figure 5.18 (continues)

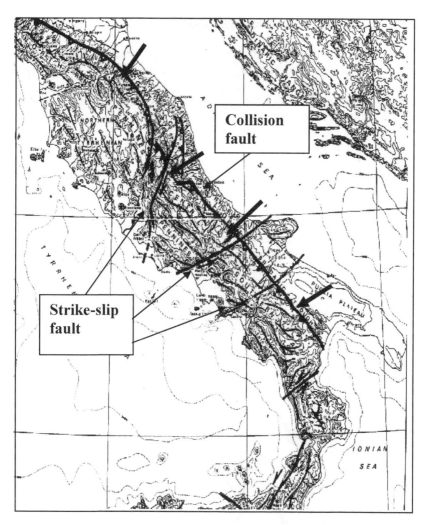

Figure 5.18 Formation of the Alps and Apennines Mountains: (a) Collision of African and Eurasian Plates through Adriatic Micro-plate (modified after Chiarabba et al, 2005); (b) Faults along the Apennines (modified after Monaco and Monaco, 2005)

The process of orogenesis is continuing until now the Alps are rising somewhere between some millimeter to a centimeter each year. This permanent stress produces permanent ground motions along the created faults, especially along the Apennines Mountains.

5.3.4   Fracture of Weak Compressed Crust

Produced exclusive in the Earth's crust, the faults in diffuse zones due to the fracture of crust have different features in comparison with the tectonic plate boundaries: no divergence, no subduction, as in the case of tectonic plate boundaries. There are the following fracture types:

- Fracture in the compressed crust.
- Fracture in the tension crust.
- Fracture in the sheared crust.

In the followings, the differences between these fracture types will be presented.

Fracture of weak compressed crust (Fig. 5.19). Active plates and microplates are stressed by forces applied in the adjacent plates, forces applied at the plate boundaries due to the motions of plates. If the compressed crust presents a weakness a fracture of crust occurs, forming a reverse fault (Fig. 5.20a). In case of low-angle (average < 45 degree), the fault is called as *thrust fault* and it consist often of fault planes parallel to sedimentary layers, forming ramps. The hanging wall moves up relatively to the footwall (Fig. 5.20b), producing a shortening of the crust. It is very important to notice that, after some time, an erosion of hanging wall peak occurs (Fig. 5.20c). Figure 5.21 shows the map of Eastern USA and adjacent Canada, with compression stresses produced by the Mid-Atlantic fault

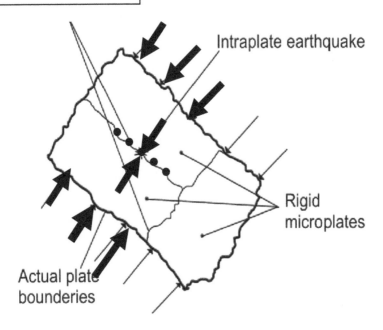

Fracture of weak compressed crust

Intraplate earthquake

Rigid microplates

Actual plate bounderies

Figure 5.19 Fracture of weak compressed crust

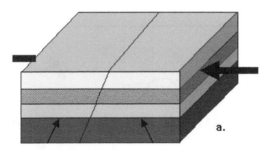

Hanging wall   Foot wall

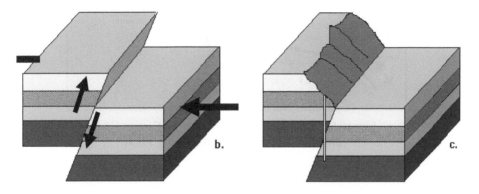

Figure 5.20 Reverse or thrust faults: (a) Crustal block before faulting; (b) After faulting; (c) Erosion of reverse fault (modified after Dutch, 1997)

Fracture in tension crust (Fig. 5.22). If a crust weakness exists, the plume rising from the mantle, producing tension stresses and fractures in the continental crust. So, *normal fault* forms when the crust is stretched, tensioned or pulled apart (Fig. 5.23). After a certain amount of elastic deformation, the tension in crust becomes too much and brittle fracture must cause a gap between the two crustal blocks. But, as this gap could never exist in nature, the two blocks must stay in contact across the fault plane. To do this, in many cases, instead of having just one normal fault, a whole series of fault plane can occurs, the crust pulls apart, giving rise to a *rift fault* (Fig. 5.24), with two parallel normal faults. In this fault, the uplifted segment is called *horst block* and the down-dropped segment, *graben block*. The best examples of these fault types are the Dead Valley (North America), the Upper Rhine (Western Europe, Fig. 5.25) and the Lake Baikal (Asia) (the deepest continental rift on the Earth).

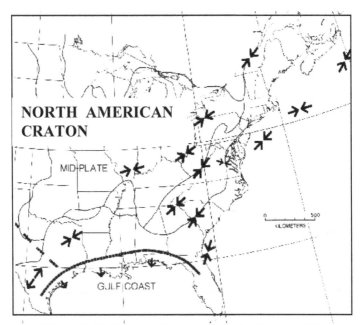

Figure 5. 21 Compression stress map in Eastern USA
(modified after Zoback and Zoback, 1989)

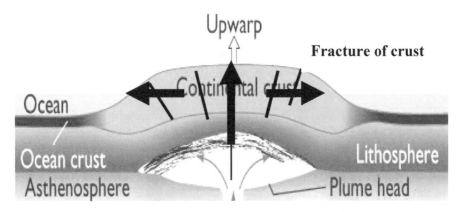

Figure 5.22 Tensioned crust caused by rising mantle plume (modified after
EPS 50, 2009)

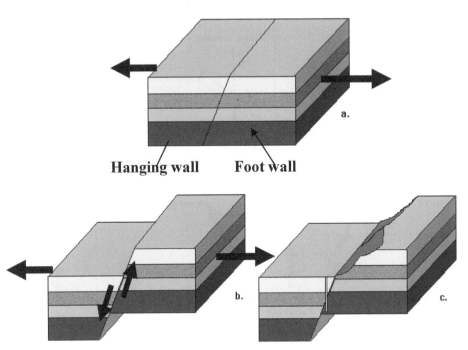

Hanging wall   **Foot wall**

Figure 5. 23 Normal fault: (a) Crustal block before faulting; (b) After faulting; (c) Erosion of normal fault (modified after Dutch, 1997)

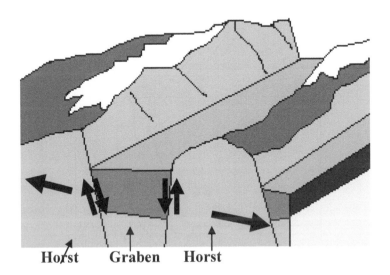

Horst   Graben   Horst

Figure 5.24 Rift produced by normal faults (modified after Dutch, 1997)

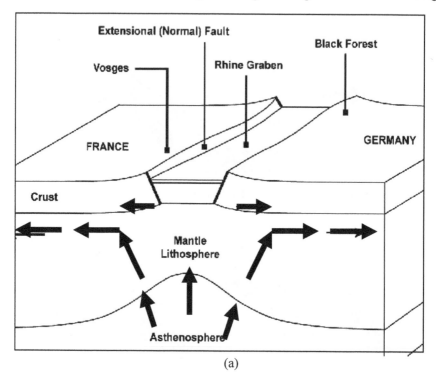

(a)

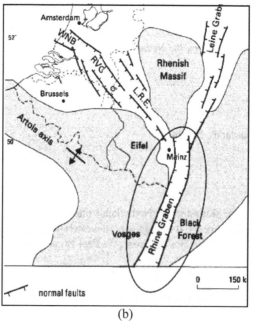

(b)

Figure 5.26 Rhine's rift: (a) Formation of the rift (modified after Wikipedia, 2009); (b) Rhine's graben (modified after Geluk et al, 1994)

Fracture of sheared crust. In some cases, during the plume rising, instead to produce normal faults, the shear faults occur. There are two types for shear faults (Schettino, 2007):

- *Symmetrical pure shear type* (known as McKenzie model) (Fig. 5.26a), which considers the ductile behavior of the lower crust and the brittle behavior of the upper crust. Due to the shear stresses, a symmetrical system of normal faults results.

- *Asymmetrical shear type* (known as Wernicke model) (Fig. 5.26b), which is a low-angle fault that crosses through the lower crust and determines an asymmetric tectonic structure of faults in the upper crust.

From the fault point of view, in both cases, a whole series of fault plane can occur by the rotation of a series of micro-blocks (Fig. 5.28), separated by fault planes.

The main example for this crustal fault is the New Madrid-Missouri (central USA) fault system, producing the largest historically documented earthquake in the United States (Fig. 5.29)

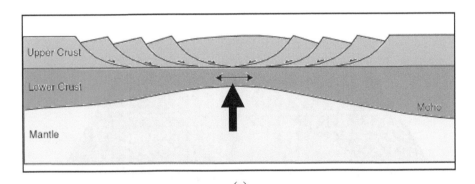

(a)

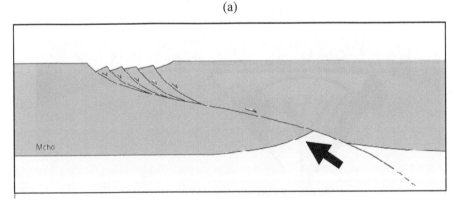

(b)

Figure 5.26 Sheared crust: (a) Symmetrical pure shear type;
(b) Asymmetrical shear type (modified after Schettino, 2007)

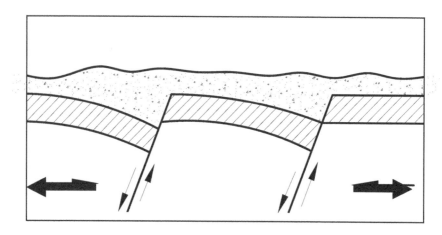

Figure 5. 27 Rotation of micro-blocks (Gioncu and Mazzolani, 2006)

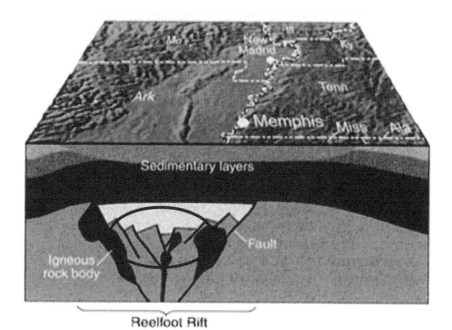

Figure 5.29 Central Mississippi Valley (USGS, nd)

5.4 WORLD'S MAIN PLATE BOUNDARIES

5.4.1 Plate Boundaries Affecting the Continental Earthquakes

In spite of the great efforts made by seismologists of different countries to solve the problem of the existing faults, reliability of their locations and characteristics are far inferior to what it should be expected in practice. This situation is mainly due to the fact that the processes of the fault movements are until now poorly known. Only in the last few years, due to the GPS technique, the progress can be qualified as notable.

Figures 5.3 presents an image of all boundaries types over the world. One can see that, for continental areas, the divergent boundaries, with the exception of Eastern Africa, have no influence, being situated on the ocean floor. The great majority of boundaries are convergent and generally situated at the continent coasts, with the exception of the Arabian and Indian areas, where they cross the Asian Continent. The transform boundaries are rare, only the Western North American and the North Anatolian area are affected by this boundary type.

Examining the tectonic plate boundaries it can be observed that three major belts dominate the movements of tectonic plates: the Circum-Pacific Ring, the Alpide-Himalayan Belt and the Mid-Atlantic Ridge. In the following, these plate boundaries, their characteristics and the earthquake types occurring along them will be presented.

5.4.2 Circum-Pacific Ring

The *Circum-Pacific Ring* (Fig. 5.27) (known also as the *Ring of Fire,* due to the presence of some very active volcanoes), is composed of segments with different characteristics. In the last century, over 90% of the world's earthquake energy release occured along this ring.

In the Eastern part, the Pacific Plate is in contact with the North American Plate along the continental coast. The side of Central America separates two secondary plates, the Cocos and the Caribbean Plates. The Southern part of the ring is formed by the contact between the Nazca and the South American Plates. The Western part of the ring is dominated by the interaction between the Pacific Plate and a strip of the North American, Philippine, Eurasian and Indo-Australian Plates. In the Northern part the Pacific Plate has common boundaries with the North American Plate, while in the Southern part with the Antarctic Plate.

The following segments characterize the *Eastern Pacific plate boundary*:

Canadian segment is composed by the North part, where the contact between Pacific and the North American Plates produces the movements, and the South part, where the interaction of three plates, the Pacific, the Juan de Fuca and the North American plates exists. All these movements are produced by the Pacific plate subduction beneath the North American plate.

Figure 5.30 Circum-Pacific ring (USGS, nd)

Californian segment is dominated by the San Andreas Fault and its ramifications. This fault forms a 1200km long boundary between the enormous tectonic plates of the Pacific and the North American. The Pacific plate slips in a Northwest direction with a rate higher than that of the North American plate, resulting in a transverse boundary. The rupture is located in the crust at 15-20 km. The South part is situated in the region of Los Angeles, where a diffuse network of faults exists.

Mexican segment is divided in two parts. In the North part, a subduction between the Pacific and the North plates occurs, while in the South, the most active part is due to the subduction of the Cocos plate under the North American plate.

Central American segment is dominated by the differences in movements between the North American, the Cocos, the South American and the Caribbean plates. The subduction fault between the Caribbean and the North American plates dominates the Northern boundary. The continental segments refer to Venezuela and Columbia, where some important strike-slip faults exist. The Western part is due to the subduction of the Cocos plate under the Caribbean one.

South American segment is dominated by the fault resulting from the subduction of the Nazca plate beneath the South American plate. The main characteristic of this subduction is that the depth is shallow in the North and very deep in the South (until 600 km). Due to this aspect, the earthquake influence occurs over a wide boundary zone.

The following segments characterize the *Western Pacific plate boundary*:

Japanese segment. In the North and central parts, the Eastern part of Japan is influenced by the Pacific plate subduction under the North American plate, while the Western part by contact of the North American plate with the Eurasian plate. In Southern Japan, the Philippine plate subducts the Eurasian plate.

Taiwanese segment. The Taiwan Island is located in the complex zone of junction between the Eurasian and the Philippine plates. Contrary to the Japanese coast, the Eurasian plate subducts beneath the Philippine plate.

Philippine and Guinea segments is due in the Northern part to the interaction between the Indo-Australian and the Philippine plates, whereas in the Southern part, to the interaction between the Pacific and the Indo-Australian plates.

New Zealand segment. New Zealand lies astride the junction of the Pacific (at the East) and the Indo-Austalian (at the West) plates. The confluence of these plates comprises several tectonic types. At the Northeast of the North Island, the Pacific plate subducts the Indo-Australian plate. At the Southwest of the South Island, the roles are reversed as the Indo-Australian plate subducts the Pacific plate. These two different subduction zones are jointed by a transverse fault.

The *Northern Pacific Plate boundary* is characterized by the subduction of the Pacific Plate under the North American Plate.

The *Southern Pacific Plate boundary* results from the contact with the Antarctic Plate.

5.4.3 Alpide (Alpine)-Himalayan Belt

The *Alpide (Alpine)-Himalayan belt* (Fig. 5.28) is a mountain range, being, after the Pacific Ring, the second for ground motion activities, but the largest and the most varied one. Due to the fact that this belt is mainly continental, large areas with diffuse faults characterize it. The extension is from the Mid-Atlantic Ridge to the Pacific zone. About 5 to 6% of the world's earthquake energy and 17% of the world's largest earthquakes have occurred along this belt. The first segment is composed mainly by the Mediterranean area, resulting from the subduction of the African plate beneath the Eurasian one. The Asian segment is due to the contact of the Arabian and Indo-Australian plates against the same part of the Eurasian plate.

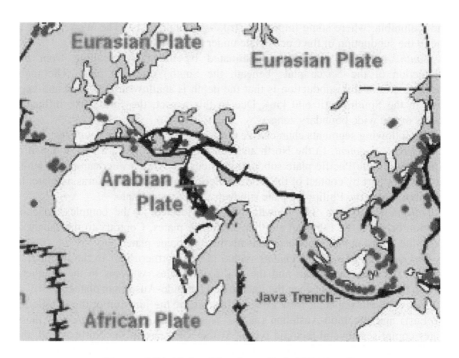

Figure 5.31 Alpine-Himalayan Belt (USGS, nd)

Azores-Maghrebian segment starts from the Azores triple junction among the North American, Eurasian and African plates. The first part, The East Azores fault is a transverse one, followed by the subduction fault. The Maghrebian fault is again a subduction fault between African and Eurasian plates, crossing the continental zone of Morocco, Algeria and Tunisia.

Calabrian and Hellenic Arcs. The Calabrian Arc is at the South of Italy (Sicily and Calabria) and the Albanian coast is continuous with the Hellenic Arc situated at the South of Crete and Cyprus Isles. These faults are due to the subduction of the African plate under the Eurasian one.

Anatolian segment. Anatolia is a well-known earthquake-prone region, being crossed by the East Anatolian fault (continuation of the Cyprus fault), the West Anatolian fault (continuation of the Crete fault), the North and the Northeast Anatolian faults. These faults are characterized by transverse type due to the movements in different directions of both Anatolian and Eurasian plates.

Arabian segment movement dominates the Eastern Mediterranean region. The 1000 km long Red Sea is a divergent boundary between African and Arabian plates. The contact of the Arabian plate with the Eurasian plate at the North and Northeast is composed by subduction faults. In the Southeastern part, the Arabian plate is in contact with the Indo-Australian plate forming a transverse boundary.

Zagros-Makran segment (South Iran and Pakistan) is produced by the subduction of the Indian plate beneath the Southern Eurasian plate.

Himalayan segment is dominated by the collision and subduction of the Indo-Australian plate beneath the Eurasian plate, forming the Himalayan Mountains.

Indonesian segment is the last part of the Alpine-Himalayan belt, ending in the Circum-Pacific Ring. The first part is located along Sumatra Island, where the Indo-Australian plate subducts beneath the Eurasian plate.

5.4.4 Mid-Atlantic Ridge

The last important belt is the Mid-Atlantic Ridge (Fig. 5.31). It is an underwater mountain range of the Atlantic Ocean, which runs from Iceland to Antarctica: this is the longest mountain range on Earth, separating the both North American from the Eurasian plates and South American from the African plates (Fig. 5.31). According to plate tectonics, this ridge runs along a divergent boundary. Fortunately, this belt traverses only one inhabited territory, the Iceland island (Fig. 5.32).

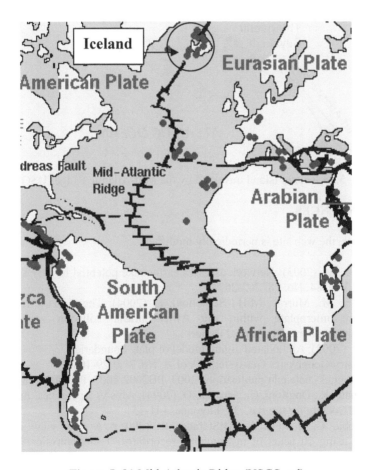

Figure 5. 31 Mid-Atlantic Ridge (USGS, nd)

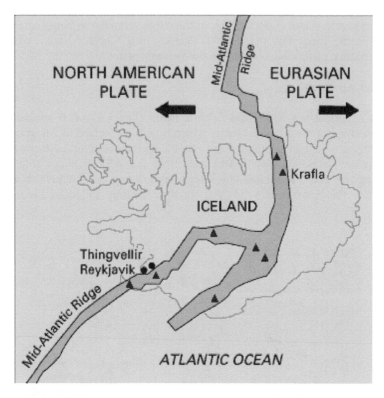

Figure 5.33 Iceland Island traversed by the Mid-Atlantic Ridge (USGS, nd)

5.5 REFERENCES

nd – no date, the web site is periodically modified

Battaglia, M. (2003): Network assesses earthquake potential in Italy's Southern Alps, Eos, Vol..84, No. 28, 262-263

Battaglia, M., Murray, M.H., Serpelloni, E. (2004): The Adriatic region: An independent microplate within the Africa-Eurasia collision zone.Berkeley Seismological Laboratory, Annual Report

Bird, P. (2003): An updated digital model of plate boundaries. Geochemistry Geophysics Geosystems, Vol. 4, No. 3, 1029-1049. http://element.ess.ucla.edu/publications/2003_PB2002/2003_PB2002.htm

Carminate, E., Doglioni, C., Scrocca, D. (2004): Alps vs Apennines. Italian Geological Society for the IGC 32, Florence, 141-151

Chiarabba, C., Jovane, L., DiStefano, R. (2005): A new view of Italian seismicity using 20 years of instrumental recordings. Tectonophysics, Vol. 395, 251-268

Dutch, S. (1997): Faults and earthquakes. http:// www.uwgb.edu/DutchS/EarthSC202Notes?quakes.htm

EPS 50 (2009): Lecture 28. Tectonics & Crustal evolution
http://eps.berkley.edu/course/eps50/documents/lecture28/tectonics.pdf

Geluk, M.C., Duin, E.T., Dusar, R.H.B.,Van den Berg, M.W. (1994): Stratigraphy and tectonics of the Roer Valley Graben. Geologie en Mijnbouw, 73,129-141

Gioncu, V., Mazzolani, F.M. (2002): Ductility of Seismic Resistant Steel Structures. Spon Press, London

Gioncu, V., Mazzolani, F.M. (2003): Challenges in design of steel structures subjected to exceptional earthquakes, in Behaviour of Steel Structures in Seismic Areas, STESSA 2003 (ed. F.M.Mazzolani), Naples, 9-12 June 2003, Balkema, Lisse, 89-95

Gioncu, V., Mazzolani, F.M. (2006): Influence of earthquake types on the design of seismic-resistant steel structures. Part 1: Challenge for new design approaches. Part 2: Structural responses for different earthquake types, in Behaviour of Steel Structures in Seismic Areas, STESSA 2006 (eds. F.M. Mazzolani and A. Wada), Yokohama, 14-17 August 2006, Taylor & Francis, London, 113-120, 121-127

Lecture 14 (2003): Earthquakes in Latin America
http://www.utdallas.edu/~pujana/latin/Lecture/%20LA%20Earthquakes.pdf

Monaco A., Monaco, R. (2005): Urbanistica e Rischio Sismico. Sistemi Editoriali, Napoli

Mooney, W.D., Schulte, S., Detweiler, S.T., (2004): Intraplate seismicity and the discrimination of nuclear test events using an updated global earthquake catalogue. 26th Seismic Research Review – Trend in Nuclear Explosion Monitoring, 439-448

Monaco, A., Monaco, R. (2005): Urbanistica e Rischio Sismico. Sistemi Editoriali, Napoli

Nelson, S.A. (2003): Global tectonics. EENS 111.
http://www.tulane.edu/-sanelson/geol111/pltect.htm

NYSED (2001): Earth Science Reference Tables
http:// www. emsc.nysed.gov/osa/reftable/reftablearch/esrtchart.htm

Oceonography (2008): Origin of oceanic basins.
http://www.tulane.edu/~bianchi/Courses/Oceanography/chap%203.ppt

Rosenbaum, G., Lister, G.S., Duboz, C. (2004): The Mesozoic and Cenozoic motion of Adria (central Mediterranean): A review of constraints and limitations. Geodinamica Acta, Vol. 17, No. 2, 25-139

Schettino, A. (2007): Plate tectonic modelling: Tools and methods. Chapter 2. Plate boundaries.
http://www.serg.unicam.it/Boundaries.html

Solarino, S., Cassinis, R. (2007): Seismicity of the upper lithosphere and its relationships with the crust in the Italian region. Bullettino di Geofisica Teorica e Applicata, Vol. 48, No.2, 99-114

USGS (nd): Active volcanoes, plate tectonics and "Ring of fire"
http://www.crystalinks.com/rof.gif

USGS (nd): Uncovering hidden hazards in the Mississippi Valley
http://quake.wr.usgs.gov/prepare/factsheets/HiddenHazs/

USGS (nd): Farallon plate.
http://pubs.usgs.gov/publications/text/Farallon.html
    USGS (nd): Understanding plate motions
http://pubs.usgs.gov/publications/text/understanding.html
    USGS (nd): Major tectonic plates of the world.
http://geology.er.usgs.gov/eastern/plates.html
    USGS (nd): Earthquake hazard program, Northern California
http://quake.wr.usgs.gov/research/deformation/modeling/people/shinji.html
    USGS (nd): Location of August 17 1999 Turkish earthquake
http://earthquake.usgs.gov?research/geology/turkey/images/turkey_loc.html
    USGS (nd): Cascadia subduction zone.
http:// erthquae.usgs.gov/research/structure/crust/cascadia.php
    Valensise, G., Amato, A., Montone, P., Pantosti, D. (2003): Earthquakes In
Italy: past, present and future. Episodes, Vol. 26, No.3, 245-249
    Wieland, M. (2001): Earthquake alarm, rapid response and early warning
systems: Low cost system for seismic risk reduction. Disaster Reduction
Workshop, Reston, 19-22 August 2001
    Wikipedia (nd): Divergent boundary
http://en.wikipedia.org/wiki/Divergent_boundary
    Wikipedia (nd): Continental collision
http:/en.wikipedia.org/wiki/Continental_collision
    Wikipedia (nd): Orogeny
http://en. Wikipedia.org/wiki/Orogeny
    Wikipedia (nd): List of tectonic plates.
http://www.answers.com/topic/list-0f-tectonic-plates
    Wikipedia (nd): Plate tectonocs
http:// en. Wikipedia.org/wiki/Plate_tectonics
    Wikipedia (nd): Mid-Atlantic ridge.
http://en.Wikipedia.org/wiki/Mid-Atlantic_ridge
    Wikipedia (nd): Convergent boundary.
http://en.wikipedia.org/wiki/Convergent_boundary
    Wikipedia (nd): Earthquake
http://en.wikipedia.org/wiki/Earthquake
    Wikipedia (nd): Farallon plate.
http://en.wikipedia.org/wiki/Farallon_Plate
    Wikipedia (nd): Mid-Atlantic ridge.
http://www.wikipedia.org/wiki/Mid-Atlantic_Ridge
    Wikipedia (nd): Upper Rhine Plain
http://en.wikipedia.org/wiki/Upper_Rhine_Plain
    Zobak, M.D., Zoback, M.L. (1989): In situ stress, crustal strain and seismic
hazard in Eastern North America. Earthquake Hazards and the Design of
Constructed Facilities in the Eastern United States (eds. K.H. Jacob, C.J. Turkstra),
Annales of the New York Academy of Sciences, Vol. 558, 54-65

Chapter 6

Faults and Earthquakes

6.1 ANATOMY OF AN EARTHQUAKE

6.1.1 Uniqueness of Earthquakes

An *earthquake* is a phenomenon resulting from the sudden release of stored energy in the Earth's crust which creates seismic waves. At the Earth's surface, earthquakes may manifest themselves by a shaking or displacement of the ground and sometimes tsunamis, which may lead to loss of life and destruction of property (Wikipedia, nd).

The great number of observations and ground motion records for past earthquakes worldwide, together with ground response analyses, have indicated that this tremendously complicated natural phenomenon of earthquakes is very difficult to understand. Despite the considerable amount of data available so far and a relevant literature on these topics, nobody can say with enough confidence that exact knowledge and rigorous statement exist about these events which occur below the Earth's crust due to unforeseeable forces at an unpredicted time.

Each earthquake is unique, being the result of the effect of many factors (defined by type and value), influencing the ground motions and generating very complex phenomena. Examining the recorded ground motions, a chaotic movement can be observed, without showing any rule at first glance. But, like a cardiologist who understands the heart movements looking at the cardiogram, the seismologist can detect the anatomy of an earthquake. So, contrary to the first impression, the ground motions are the result of the overlapping of the effects of a limited number of basic factors, having different importance and giving their variety in the recorded movements. Due to the fact that an interaction between these factors exists, this complex aspect of ground motions can be identified.

The human and economic losses are generally considered as the main characteristic of an earthquake, which is classified according to its magnitude; but we can observe that this aspect cannot only be considered as a representative factor. In fact, there are some cases in which very strong earthquakes in magnitude have not produced important losses, like for example the 1960 Chile earthquake, with magnitude 9.5, and the 1964 Alaska earthquake, with magnitude 9.2, because they both occurred in zones with low population density. Contrary there are other cases, in which even moderate magnitude earthquakes caused very large damage which was foreseeable, because they occurred in urbanized zones: the 1994 Northridge earthquake, with magnitude 6.7, the most damaging earthquake in the US history, and 1995 Kobe earthquake with magnitude 6.9, the most devastating earthquake in world history. Considering these different events, one can look to an earthquake as a local phenomenon having some distinct features, which differ from each other due to its peculiar typology. Starting from the basic features, given by the source, the overlap of different influence factors can change these features by creating amplification or attenuation of the ground motions, by increasing or reducing the movement duration, etc. Therefore, the main characteristics in a site can differ very much from the initial ones and this is the main reason why, in some cases, a moderate magnitude earthquake can be metamorphosed into a devastating one.

6.1.2 Mechanical and Thermo-dynamical Processes Producing Earthquakes

There are two types of phenomena producing earthquakes:
  - The first phenomenon is due to a *mechanical process* occurring in the Earth's crust, where the equilibrium is based on the friction between two plates and the earthquake is the result of the loss of this equilibrium, or, when new faults are forming, is due to the brittle fracture of rocks.
  - The second phenomenon, dominated by a *thermo-dynamic process,* is not due to friction between two plates, but is produced at a deep depth where temperatures and pressures are very high and there is only one plate.

Due to the first phenomenon, the earthquakes occur at shallow depths of 0 to 50 kilometers in the Earth's crust, where the friction forces between tectonic plates normally assure the stability of these plates, being in contact along the fault. Some dislocations exist due to previous movements or other dislocations can occur at the end of a fault already produced, being characterized by a new brittle rupture in the continuation of the existing one.

The second phenomenon gives rise to a particular category of earthquakes which is largely ignored by the scientific community. The earthquakes occur at depths of 50 to 300 kilometers beneath the Earth's crust and are caused by different processes. Extreme heat and pressure in the subduction zone cause the metamorphoses of rocks or change them into different forms. During the change, the rocks release water, which essentially lubricates the fault, facilitating it to move. The mechanism whereby earthquakes are produced, because of the high pressures occurring below 100 kilometers of depth, remains until today an enigmatic and controversial problem (Persh and Houston, 2004). But the main unsolved problem is to understand the earthquake features produced below 300 kilometers, where the slab disappears in the mantle. It is very important to mention that the deepest recorded earthquake exceeds 600 kilometers.

6.1.3 Source Types

A large number of earthquakes daily occur on the Earth. So, earthquakes can strike any location at any time. But the history shows that they principally occur in some specific zones of the Earth. Many earthquakes occur in narrow regions around the plate boundaries. This is evident by comparing the Figure 2.2, showing the distribution of earthquakes in the world, with Figure 5.1 indicating the main boundaries of the main tectonics plates. But this is not a general rule, because a lot of earthquakes occur far from these boundaries and others are located in the crustal zones or under the Earth crust.

In function of the earthquake position, there are the following main earthquake types (Fig. 6.1):

- *Crustal interplate earthquakes,* are produced along the tectonic convergent plate boundaries, where a fault exists due to previous plate movements. Along the fault, shear friction forces between the two plates equilibrate the gigantic tectonic plate forces and the elastic strain energy is stored. When the shear force capacity is reached, sudden slip releases the large elastic strain energy stored in the interface rocks. So, in case of interplate earthquakes, the slipping occurs along these tectonic plate boundaries and their positions are well defined. Only in case of existing parallel faults, the activated one is an unknown.

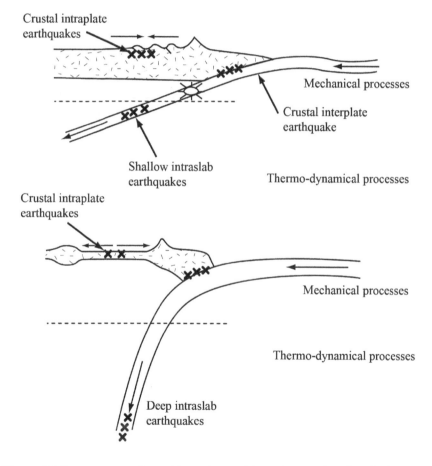

Figure 6.1 Source types: Crustal interplate and intraplate earthquake, shallow and deep intraslab (inslab) earthquakes (modified after Stern, 2002)

-Crustal intraplate earthquakes, are produced in the interior crust of a tectonic plate, far from the boundaries, in the diffuse zones (see Section 5.3). In this case, the ground movements are caused by the previous rupture, or by a new one, due to the reaching of the rock bearing capacity. Thus, generally, an intraplate earthquake occurs in the same seismic area but not in the same place, as in the case of interplate earthquakes. So, the position of an intraplate earthquake is, generally, very difficult to determine beforehand.

-Intraslab (inslab) earthquake, affected by thermal phenomena, the earthquake occurs in slab (subducted oceanic plate), at shallow or deep depth. It is situated in a slab, in a zone under the Earth crust, where the solid rock begins to be transformed into molten-lava, due to the high temperatures and pressures. Therefore, very special features, different from the cases presented above, characterize this source type.

Generally, a seismic zone is composed by one or more source types. The best case to exemplify this fact, is given by the European territory, which is composed by four distinct zones (Fig. 6.2). The first zone is the Mediterranean basin, where the Northern zone (along the Alps) is dominated by crustal interplate earthquakes and the Southern zone (along North Africa, Sicily and Greece) by crustal interplate and intraslab earthquakes. The second zone is the Central Europe, characterized by crustal intraplate earthquakes (France, Belgium and Germany). The third zone is the Northern area (UK and Norway), which is dominated also by the intraplate earthquakes. The last important zone is the Carpathian one (Romania), characterized by intraslab earthquakes.

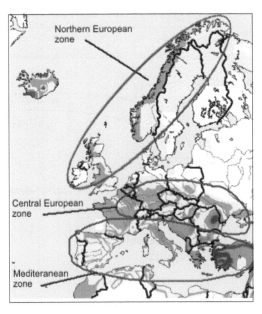

Figure 6.2 European seismically active zones (Gioncu, 2006)

6.2 CRUSTAL INTERPLATE EARTHQUAKE

6.2.1 Crustal Interplate Earthquake Types

The crustal interplate earthquake has its source at the convergent and transformed boundaries of the Earth's crust, under form of collision, subduction or strike-slip earthquakes (Fig. 6.3) (Gioncu, 2006, Gioncu and Mazzolani, 2003, 2006).

6.2.2 Crustal Subduction Earthquakes

Crustal subduction earthquakes occur in the zones where the oceanic plate is colliding with and descending beneath the continental plate. Therefore, this earthquake type always occurs along continental coasts and is the most important source of the world's earthquakes, because crustal subduction is the cause of the majority of the produced earthquakes. The Earth's largest earthquakes occur in these subduction zones. At the margin of the Pacific Ocean, along the Ring of Fire, where these earthquakes occur frequently, hundreds of millions of people and trillions of dollars of economical infrastructure are at risk from these earthquakes.

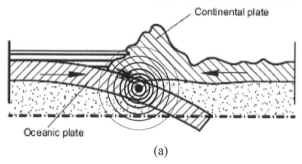

(a)

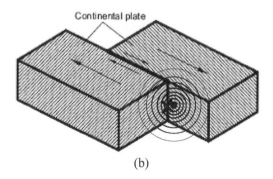

(b)

Figure 6.3 Crustal interplate earthquake types: (a) Subduction; (b) Strike-slip
(Gioncu, 2006)

Seismic zones corresponding to the subduction are the most dangerous, because it happens near the surface. These subduction sources produced also the largest world's tsunamis during the last period, like the 1960 Chile earthquake (M 9.5), 1964 Alaska (M 9.2), 1952 Kamchatka (M 9.0) and 2004 Sumatra (M 9.0) events. The history mentions that on January 1700 one of the largest earthquakes (known as the Cascadia earthquake) occurred along the Western Canadian coast of North America, due to the sliding of the Kula and Juan de Fuca plates under the North American plate (see Figures 5.14). The same subduction process produced in 1964 the second known mega earthquake in Alaska (M 9.2). The geological evidence indicates that 13 great earthquakes occurred in this region during the last 6000 years. While large earthquakes have not been observed in the last short period (150 years), there is compelling evidence that they will happen sometime in the future; it represents a considerable hazard for people living in the very urbanized zones of this region. For these reasons, these faults are very carefully examined by seismologists.

Figure 6.4 shows the main world's subduction zones marked by lines with sags. They are: Aleutines, Cascades, Central America, Andes, South America, Kurile, Mariana, Lau Basin, New Zealand (at the borders of the Pacific Tectonic Plate along the Ring of Fire) and North Africa, Sicily, Crete and Indonesia (at the borders of the Euroasian Tectonic Plate with the African and Indian Tectonic Plates, respectively, along the Alpide-Himalayan belt). In all these zones crustal subduction earthquakes are very frequent.

The Western coasts of both the American continents are constantly affected by earthquakes due to the subduction of different tectonic plates. One of the most important zones, for exemplification of this earthquake type, is the coast of South America. Figure 6.5a shows the region where the Cocos Plate in the North, the Nazca Plate in the Centre (with the majority of important earthquakes) and the Antarctic Plate in the South subduct the Southern American Plate. The Nazca Plate subduction forms the very long Peru-Chile Trench and the Andes Mountains (with its highest elevations over 6000 m representing the second largest orogen after the Himalaya). The 1960 Chile earthquake (M 9.5), the strongest recorded in the world, had a rupture length along the fault line of about 1300 km. It is very important to observe in Figure 6.6b that there are two subductions, the first along the Peru-Chile Trench and the second inland of the continent, corresponding to the collision type which will be discussed in the next section. This particular characteristic of subduction zones produces very large diffuse areas, especially in the Peru zone and in the Northern areas of Chile.

The second very important subduction activity, due to the subducting of the Pacific Plate beneath the North American Plate, is along the Aleutian and Alaska basins (Fig. 6.6), where the second strongest earthquake was recorded: the 1964 Alaska M 9.2. The Aleutian–Alaskan arc is extended for over 4000 km (Black, 2000, Burns 2005)

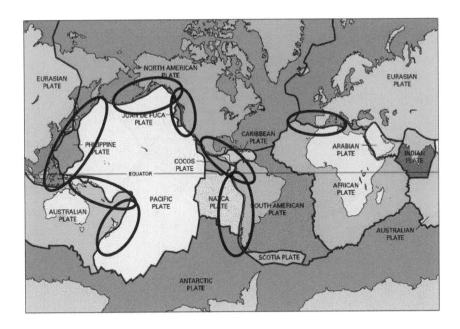

Figure 6.4 Subduction zones with crustal earthquakes (modified after USGS, nd)

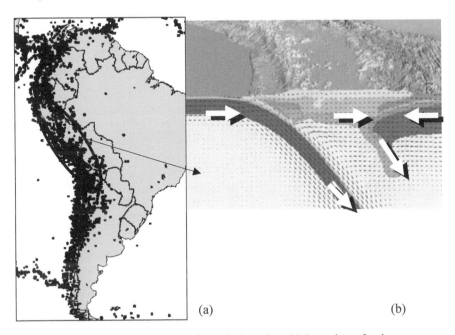

(a)   (b)

Figure 6.5 Earthquakes of South America: (a) Location of epicenters;
(b) Subduction in the South American plate (after Nature, 2005)

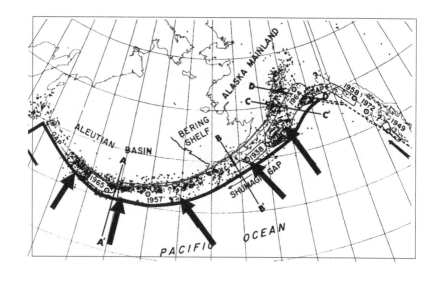

(a)

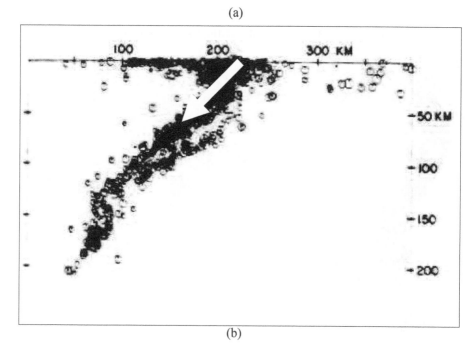

(b)

Figure 6.6 Aleutian-Alaskan arc: (a) Distribution of earthquakes;
(b) Subduction of Pacific Plate (modified after Black, 2000)

6.2.3 Crustal Strike-slip Earthquakes

Figure 6.7 presents the world's zones where crust strike -slip earthquakes occur. One can see that, in comparison with subduction earthquakes, the areas are reduced.

This earthquake type occurs when the two tectonic plates slide and grind against each other along a transform fault. Despite the fact that few lateral movements occur on land, they have produced the largest damaging earthquakes (see San Francisco, Loma Prieta, Kobe and Kocaeli earthquakes).

The most famous earthquakes occurred along the Californian San Andreas Fault (see also Figure 5.10), the main of which are shown in Figure 6.8a, with two main areas, San Francisco and Los Angeles zones. In some case, the strike-slips produced slipping at the surface of some meters, in many cases, the strike-slips remain blind (Fig. 6. 8b).

Another important strike-slip earthquake occurred in Turkey's North Anatolian Fault (Fig. 6.9).It is very significant to understand how the earthquakes occur in these fault systems, being supervised by GPS. Geodetic analysis shows a relative motion of Anatolia to Europe. While Europe relatively moves from left to right, Anatolia relatively moves in contrary direction (Doglioni et al, 2002).

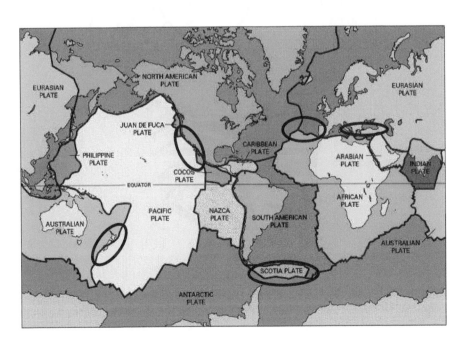

Figure 6.7 Crustal strike-slip earthquakes (modified after USGS, nd)

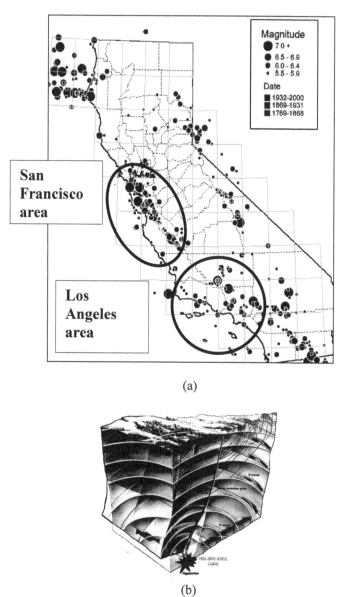

(a)

(b)

Figure 6. 8 Californian earthquakes: (a) Locations of earthquakes (after CGS, 2007); (b) Crustal blind strike-slip earthquake (Gioncu, 2006)

Therefore, it is evident that a strike-slip movement occurred along the North Anatolian fault, which runs for about 1400 km in the Aegean Sea, being a surface fault.

Another zone with strike-slip earthquakes is the Alpine fault in South Island of New Zealand (Fig. 6.10). It is situated on the Fire Ring, where the Pacific and Australian plates push each other sideways. The Alpide fault is the largest active fault in New Zealand and extends over 650 km, the most active part of the fault is the central section with earthquakes shallower than 15 km.

There are also two areas with strike-slip earthquakes, around the Scotia plate and Middle of Atlantic Ocean, but these ones are situated in oceanic zones, uninteresting for earthquake Engineering

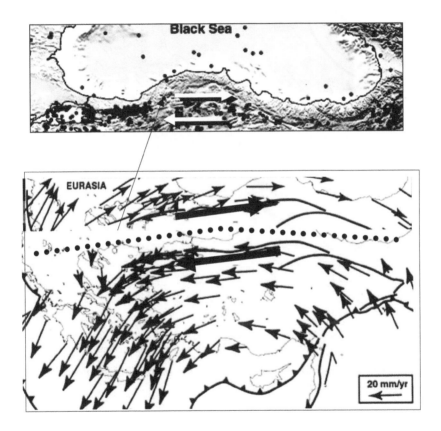

Figure 6.9 Anatolian strike-slip earthquakes: (a) Location of epicenters (after Taymaz et al, 2007); (b) Measured direction of movements (Doglioni et al, 2002)

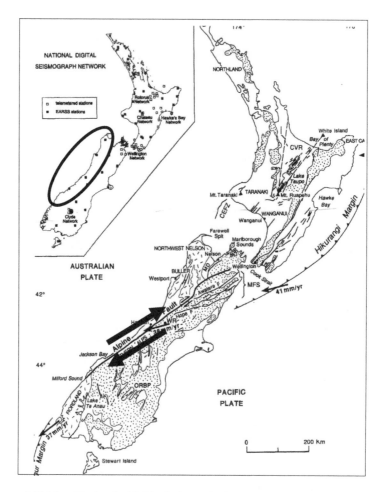

Figure 6.10 New Zealand: Alpide fault producing strike-slip earthquakes (modified after Anderson and Webb, 1994)

6.3 CRUSTAL INTRAPLATE EARTHQUAKES

6.3.1. Crustal Intraplate Earthquake Types

The crustal intraplate earthquake has its source in the interior of a plate, far from the plate boundaries (Iio and Kobayashi, 2002, Gangopadhyay and Talwani, 2003). It occurs in the upper crust, which in the Tectonic theory is regarded as a purely elastic medium. As shown in Chapter 5, tectonic plate boundaries are characterized by the presence of very well-known active faults, their characteristics being very well established, due to the permanent seismic activity of these regions. No comparable simple description exists yet for quantifying active intraplate earthquakes in diffuse zones, known as *Stable continental interiors*. In contrast with the interplate boundaries, the interior of the plate appears to tectonically behave as a rigid block, but with some faults in uncertain positions. Due to their low activity, the geological and geophysical indicators and the seismic data are limited for this type of earthquake. Broad mountain belts and rifts with diffuse seismicity and strong variations in the crust structure show that the studies of intraplate earthquakes are more complex than the interplate ones.

The difficulties come from the fact that the well-known Tectonic Plate Theory starts from the division of the Pangaea super-continent into the actual continents, ignoring the fact that, before this, some cycles of gathering and spreading of continents occurred (see Wilson's Cycle Theory, Section 4.1.5), with different configurations which are known today. Therefore, the mountain belts and the diffuse seismic zones in the inland of actual continents are the result of this pre-Pangaea history. In addition, the tectonic plates are not perfectly rigid and some inland faults exist in the weakest zones of the crust. The best model for the diffuse zones must consider that they are composed by some rigid micro-plates, as a result of (see Section 5.3.1):

(i) Extension of seismic zones under subducted plates;

(ii) Collision along two continental plates or of some ancient tectonic activity;

(iii) Faults due to some weaknesses of the crust.

In fact, the tectonic plates look like a broken plate. Therefore, the so-called Stable Continental Interiors do not exist in the reality. So, an earthquake can occur in almost any world zone.

6.3.2 Earthquakes along the Extension of Subduction Plates

Most of the world's great earthquakes occur in the ocean-continent convergence boundaries, in the extension of subduction zones (see Figs. 5.7a and 5.13). A subduction zone is a large convergent boundary where two tectonic plates collide. They are constantly shifting and moving, so when they subduct, one pushes beneath the other. Subduction zones create geologic formations such as ocean trenches and mountain ranges, as well as phenomena like earthquakes and volcanoes. Because subduction zones are gently inclined at shallow depths, they have the largest seismogenic area, exceeding the plate boundary zones.

While scientists have a general understanding of plate boundaries, many problems of seismic zones remain obscures, mainly regarding the dimensions of the subduction areas.

The Figure 6.11 shows the two large subduction areas of South American plate, and the earthquakes produced in these areas. One can see that the seismogenic zones are large extended in the continental territory.

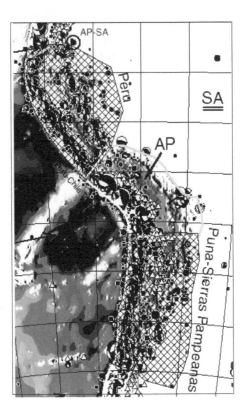

Figure 6.11 Earthquakes extension in South America (after Bird, 2003)

6.3.3 Crustal Collision Earthquakes

The continental collision is a variation of the fundamental process of subduction, where the subduction zone is destroyed. This form of continental genesis happens when the consummation of an ocean basin is complete and the continental margins of the ocean enter into collision. When two continental plates, with the same thickness and weight, converge, instead of having a true subduction, as presented in the above section, the edges crumple, the two continents suture together, a great uplift occurs and the formation of a large mountain chain results (Fig. 6.11). This process of mountain building is studied by *Orogeny* (meaning mountain generating from the ancient Greek) (Wikipedia, nd), which is a science studying the processes of crustal uplift, folding and faulting which form the systems of mountains. The Orogeny usually produces long linear mountains, known as an *orogenic belt*. Due to the fact that the convergence between the continents continues, the mountains grow and the earthquakes in collision faults continue to be frequent events.

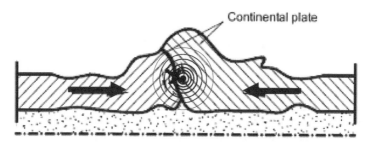

Figure 6.12 Crustal collision earthquakes (Gioncu, 2006)

Considering these aspects, it is clear that the collision earthquakes, produced along the suture fault, occur always in the inland of continents.

There are some main collision zones: the Himalayas, Alps and Adria zones, Ural Mountains and Eastern China.

The result of one the most known collision is the greatest mountain range of the modern world, the Himalayan Mountains, as a result of the collision of the Indian plate with the Eurasian plate (Fig. 6.13 see also Figure 5.8). One can see that the seismicity of the Tibetan Plateau is very high and is large extended in the Asian continent.

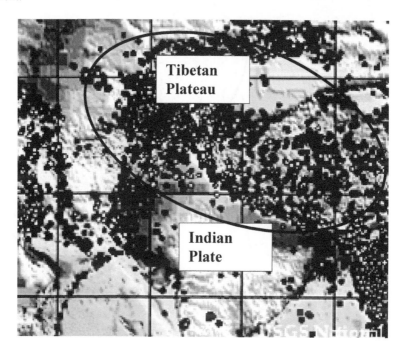

Figure 6.12 Seismicity of Tibetan Plateau (USGS, nd)

The second very important collision area is the result of multiple collisions and subduction in the zone of the Adria plate (see Fig., 5.18). The very dense location of epicenters in Italy is presented in Figure 6.14a. It is very interesting to observe that the majority of sources are situated in the upper rigid crustal zone (no more than about 30 km); only in the zone of the African Plate subduction, are the sources very deep (Fig 6.14b), what corresponds to the intermediate interslab earthquakes (see the Section 6.4).

In the Eastern zone of the Adria Plate a network of faults is formed due to the collision with the costal part of Dinarides and Albanides. These faults are presented in Figure 6.15a for the regional tectonic of Croatia and the corresponding location of epicenters in Figure 6.14b. One can see that the seismically active areas are mostly located along the coast, only a few isolated seismic events being far from coast.

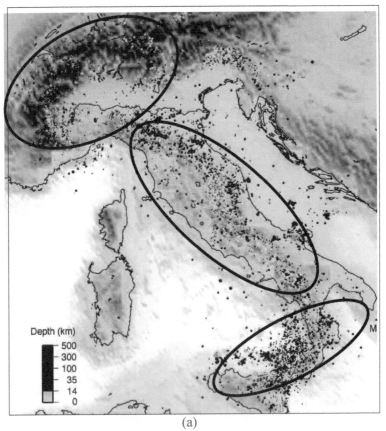

(a)

Figure 6.14 (continues)

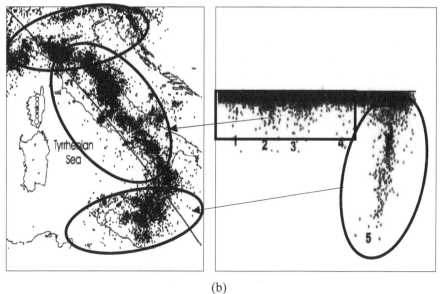

(b)

Figure 6.14 Italian earthquakes: (a) Epicenter location (after Chiarrabba et al, 2005); (b) Distribution versus depth (after Solarino and Cassinis, 2007)

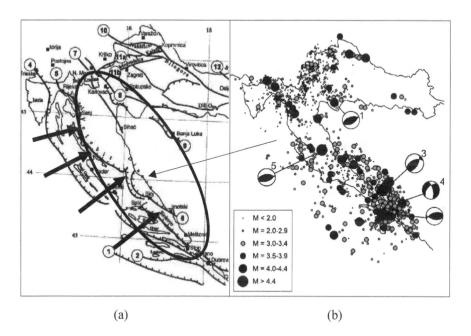

(a)   (b)

Figure 6.15 Croatian Earthquakes: (a) Map of faults; (b) Earthquakes epicenters (after Ivancic et al, 2002)

The Ural Mountains (Fig. 6.16a) are the result of the continent-to-continent collision (in a pre-Pangaea period) between the East European Plate and the Siberian Plate, producing an orogenetic belt and faults (Fig. 6.16b) (IGCP, 2006). This collision is similar to the India-Asia and Italia-Europe collisions, forming Himalayan and Alps Mountains, respectively. Generally, the magnitude of the produced earthquakes is not very large.

China territory is divided, from the seismic point of view, in two different parts (Fig. 6.17). The Western part is directly influenced by the collision of the Indian and the Eurasian plates which produced the seismogenic Tibetan plateau (see Fig. 6.13).

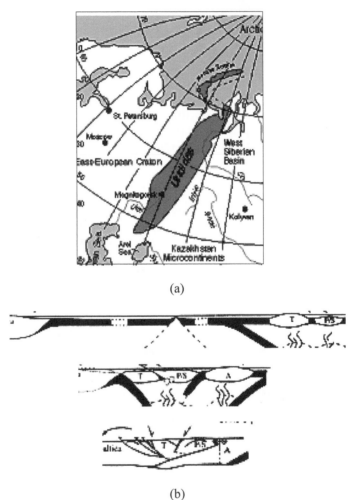

(a)

(b)

Figure 6.16 Ural Mountains: (a) Location and existing faults; (b) History of tectonic evolution during the birth of the super-continent Panagaea (modified after IGCP, 2006)

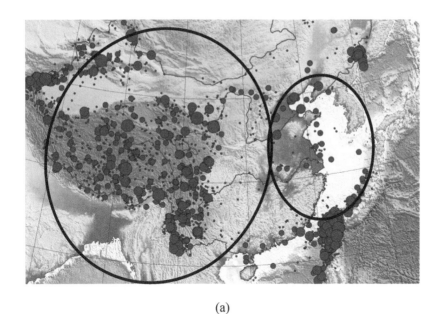

(a)

Sino-Korean block

Euro-Asia

Yangtze block

Tethys

(b)

Figure 6.17 Seismicity of China territory: (a) The two seismic areas (modified after Zhang, nd); (b) Formation of North-Eastern area in the pre-Pangaea period (after Barnes, 2003)

But the Eastern part configuration is more complex, being composed by two blocks, Sino-Korean and Yangtze blocks. Events during the pre-Pangaea cycle (150 Ma) in this region resulted in the collisions and sutures between these blocks.

The great earthquakes (i.e, the 1976 Tangshan and 2008 Sichuan earthquakes with magnitudes exceeding M 7.0) which shook these regions show that intraplate earthquakes occurring in collision zones and along the ancient faults can produce very damaging effects.

6.3.4 Earthquakes in Weak Crustal Zones

Concerning the seismic hazard, the major faults and sources of these earthquakes are not well known and the hazard assessments are more difficult than for the interplate earthquakes. Fortunately, low to moderate earthquakes occur in general in these intraplate faults. Only in some special cases, such as the soil liquefaction in densely built zones, can these earthquakes be very devastating. There are some important and specific seismic areas producing important intraplate earthquakes.

Eastern North America is the example of the earthquake produced in a rift. Although this zone of North America is far from the plate boundary, earthquakes occur frequently (Fig. 6.17). Moreover, the 1811 and 1812 New Madrid earthquakes (M 7.8 and 8.1) were two of the strongest earthquakes produced in USA. This is due to the fact that when the super-continent Pangaea was formed, a collision was produced between the North American and the African plates, giving rise to the Appalachian Mountains in the Eastern North American continent. The New Madrid earthquakes produced, due to the liquefaction phenomena, landslides, settlements, ejection of sand and water. Another seismic area is the Charleston region, where in 1886 a great earthquake (M= 7.6) shocked the Eastern US seaboard.

Another very important zone of intraplate earthquakes is the *European area*, which is divided in three distinct zones: Western, Central, and Northern areas.

The *Western area* is composed by the territories of France, the UK and Belgium.

France is a country of low-to-moderate seismic hazard but which, in the past, suffered strong earthquakes producing damage and casualties (Pequegnat et al, 2008). The earthquakes are located in four main very different tectonic regions (Fig. 6.18). The first two are the zones of the Pyrenees in the Southwest, due to the collision with the Iberian Plate and of the Alps in the Southeast, due to collision with Italian Plate. The earthquakes occurring in these zones are the result of the effect of these collisions. In the central and Western regions these seismic events have the characteristics of crust fractures.

The *UK* has a moderate rate of seismicity (earthquakes of magnitude 5 are rare), but sufficiently high to pose a potential hazard for some old buildings with a poor structure. The characteristics of these earthquakes correspond very well to the crust fracture ones, being produced by the influence of the Mid-Atlantic Ridge. The distribution of earthquakes on the territory of the UK is presented in Figure 6.19.

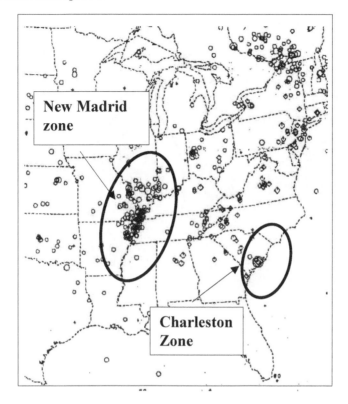

Figure 6.18 Earthquake distribution for Eastern USA (after Hinze et al, 1988)

The average recurrence is an earthquake of 4.7 every 10 years and an earthquake of 5.6 every 100 years (Ukearthquakes, 2008).

Belgium is the territory which provides examples of the high vulnerability of the stable continental interiors. It demonstrated the necessity to better understanding the seismic activity and the potential of important earthquakes in the intraplate areas. The main earthquakes in this area occur in the bordering region between Belgium, the Netherlands and Germany, often called the "Lower Rhine Embaymant" (Fig. 6.20). A part of it, the Roer (Rur) rift system appears as the most active area with several earthquakes with magnitudes larger than M 5.0. In these areas a series of grabens and rifts exist (Berg, 1994, Schafer and Siehl, 2002, BGHRC, 2003). The Rhine Valley rift system forms a NW-oriented fault between the parallel Upper Rhine and Central North Sea grabens. The earthquake epicenters from 1350 to 2004 in this region, plotted in Figure 6.21, show that many events exceed magnitude M 5.0; the strongest known seismic event occurred in 1692, being appreciated having a magnitude M 6.3 (Camelbeek, 2005). In the near region of Liege, the known activity is less important, but the earthquake of 1983, with a magnitude M 4.7 caused important damage in the city. The recent studies suggest that the magnitude could approach M 7.0.

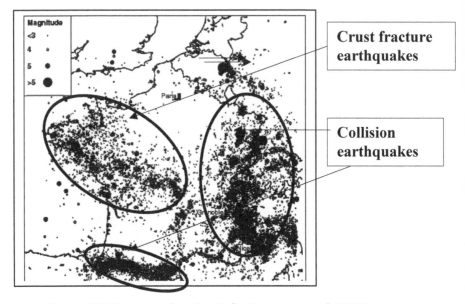

Figure 6.19 France earthquakes (after Pequegnat et al, 2008)

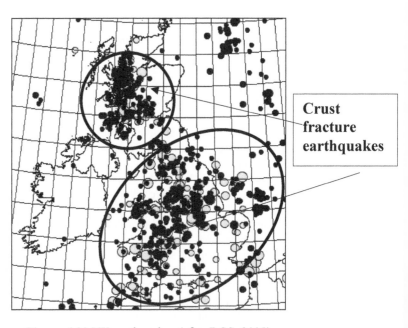

Figure 6.20 UK earthquakes (after BGS, 2008)

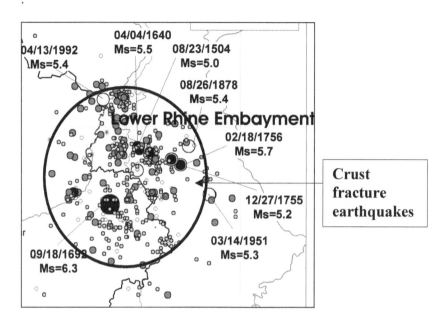

Figure 6.21 Seismotectonic map of Belgium (after Camelbeek, 2005)

Germany and Switzerland (Western zones) seismicity is influenced by the Lower Rhine and Upper Rhine grabens, respectively (Fig. 6.22). Modern earthquakes in the Rhine Valley have not exceeded magnitude M 6.0, but geological studies reveal that its border faults are capable of producing magnitudes M 6.0-6.5 earthquakes and generating ground rupture. On 1356 the most damaging intraplate earthquake known to have occurred in Central Europe devastated the city of Basel, Switzerland (RMS, 2006). Events of this magnitude (estimated between M 6.0 to 6.9) are very rare in Central Europe, because the seismic activity in this area is characterized by low levels of seismicity.

Austria, Slovakia, Hungary, Serbia, Croatia and Western Romania territories are characterized by the presence of many active faults, which produce low to moderate earthquakes (Fig. 6.23a). The best example is the Banat (Western Romania) earthquake, produced by a network of faults (Fig. 6.23b).

Northern area considers the Scandinavian countries. The intraplate areas of Scandinavia have only experienced small earthquakes. The most famous Norwegian earthquake occurred in 1904 having the magnitude M 5.4. But almost one hundred years earlier, in 1819, an even more powerful earthquake occurred, being estimate with a magnitude M 5.8 to 6.0 (Norsar 2004, Gregersen, 2006). A map of known earthquakes above M 3.0 is shown in Figure 6.23. Stress accumulation in the crust is caused by forces related to the relative movements of the tectonic plates: from the Mid-Atlantic Ridge stresses are imposed on the Eurasian Plate where Norway is located.

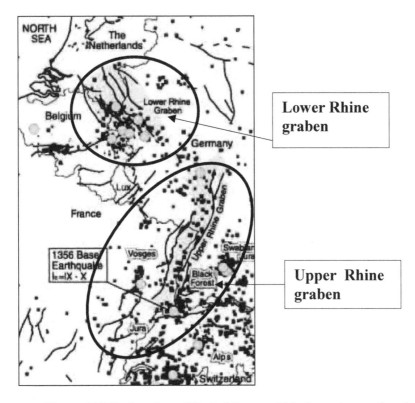

Figure 6.22 Earthquakes of Central Europe: Rhine's graben earthquakes
(after BGHRC, 2003)

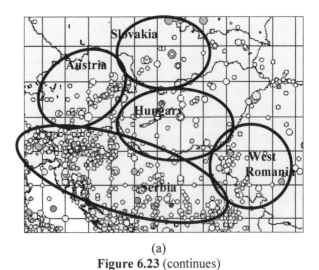

(a)
Figure 6.23 (continues)

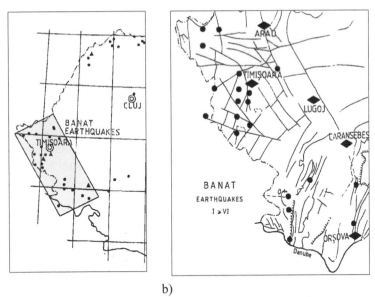

b)

Figure 6.23 Eastern Central Europe: (a) Location of epicenters (after Musson, 1999); (b) Western Romanian earthquakes (Gioncu, 1994)

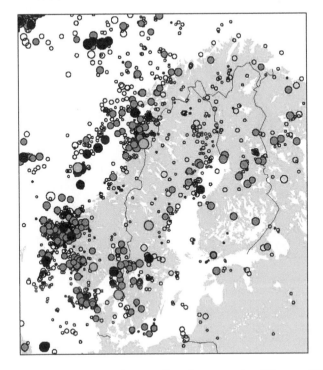

Figure 6.24 Earthquakes in Scandinavian countries (Gioncu, 2006)

Another very important zone, where intraplate earthquakes occur, is the *Australia area* (Fig. 6. 24), (Jankulovski et al, 1996). Even though it is very far from the tectonic plate boundaries, Australia is seismically active, but with moderate magnitude earthquakes. Generally, the important earthquakes frame in the magnitude M 5 to M 6. The largest earthquake which can occur in Australia is not yet known, but is expected to touch M 7.0 (Quakes, 2003). The well-known seismic events, the 1988 Tennant Creek with M 6.9 and the 1989 Newcastle with M 5.6, proved that even Australia is not immune from damaging earthquakes causing significant human and economic loss. As shown in Figure 6.25, earthquakes can occur in almost every part of Australia; it is one of the most significant examples of crustal intraplate earthquakes.

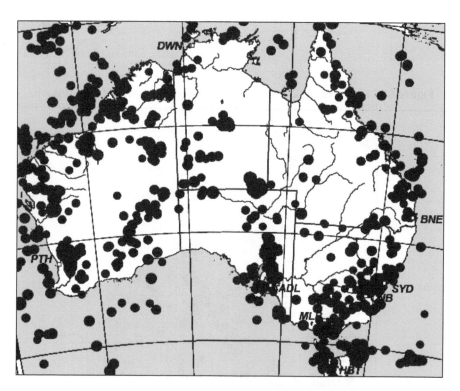

Figure 6. 25 Seismicity of Australia (after Quakes, 2003)

6.4 INTRASLAB (INSLAB) EARTHQUAKES

Scientists now know that shallow earthquakes are caused by the slippage of giant blocks of the Earth along faults, but the cause of deeper earthquakes remains a mystery (Fitzpatrick, 1996).

Contrary to crustal earthquakes, around the world there is a large number of earthquakes occurring at depth situated under the crust thickness. Figure 6.26 presents the location of large intraslab earthquakes. There are some continuous zones where intraplate earthquakes occur: Mexican and South American coasts, Western Pacific Ocean, and Eastern of Indian Ocean. At the same time, some isolated intraslab earthquakes (as in Vrancea, Sicilian, Crete, Persian and Himalayan zones) exist. The largest deep earthquake ever recorded had a magnitude M 8.3 and occurred 600 km below Bolivia in 1994.

A slab is an oceanic crustal plate that subducts under a continental plate and it is consumed by the Earth's mantle (Fig. 6.26), in a zone where the thermo-dynamic, physical and chemical processes of Earth's interior are dominant. These phenomena assure the planet's largest recycling system (Stern, 2002). Intraslab earthquakes occur in the subduction slab in a zone situated under the contact between the two plates and under the Earth's crust.

There are three categories of intraslab earthquakes (Seno and Yoshida, 2004, Boudreau, 2001, Houston, 2004, Fox, 2007): *shallow,* produced in the slab in the range of 30-70 km; *intermediate,* in the deeper range of 70 to 300 km; *deep,* when deeper than 300 km (note that these ranges differ from one investigator to another). The differences between these earthquake types consist in the influence of temperature and in the rupture process. Figure 6.27 shows a variation of the temperature with the depth of the slab and of the subducted plate. For the shallow earthquakes, the temperature does not influence the rupture characteristics, remaining a mechanical process, dominated by friction. In case of deep earthquakes, high temperature and pressure have a great influence on the characteristics of rupture. The causes of intermediate and deep earthquakes have been a very controversial problem for the last decades, being an important and "mysterious" class of earthquake. Understanding these earthquakes could be the key for unlocking the remaining secrets of plate tectonics (Reid, 2002). Shallow earthquakes represent the normal brittle failure of rocks. But the intraslab earthquakes occur when a combination of extreme heat and pressure cause the rocks to metamorphose and change into different forms, due to the various chemical reactions, changing the plate into molten lava. As they change, the rocks release water due to the *dehydratation,* which essentially lubricates the fault, causing the fault to slip (Kirby et al, 1991, Tibi et al, 2002). This explanation of the particular behavior of intraplate earthquakes is known as the Kirby's Theory

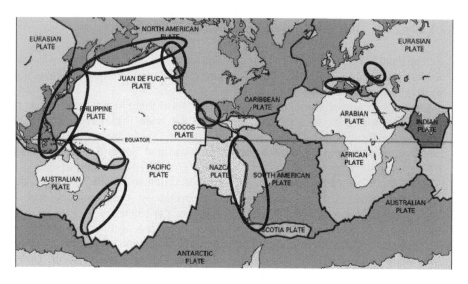

Figure 6.26 Location of large intraslab earthquakes (after USGS, nd)

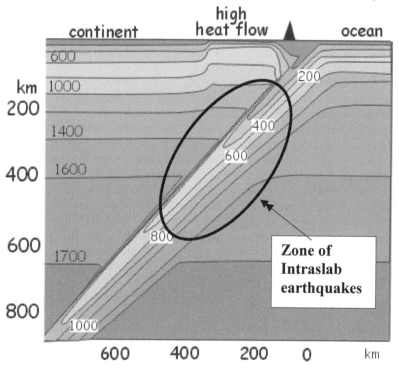

Figure 6.27 Increasing of temperature with the depth (after Dynamic Earth, 2001)

(Boudreau, 2001). In comparison with shallow ruptures, the process is very slow, the velocity being less than half of the one for shallow rupture.

The most important seismic zones, where intraslab earthquakes occur, are: Aleutian-Alaskan and Cascadia in North America; Hokkaido, Mariana and Sumatra in West Pacific; Mexico in Central America; Peru and Chile in South America. The large shallow intraslab earthquakes produced along the Circum-Pacific Ring are presented in the Figure 6.26, both in North and South American and in Western and South Pacific regions (Seno and Yoshida, 2004).

In Europe, the most known intraslab earthquakes are the Vrancea (located at the sharp bend of the South Eastern Carpathian Mountains) and the Hellenic Arc (in the South of Greece, formed as the result of the convergence between the Aegean microplate and the African Plate) (Benetatos and Kiratzi, 2004) (Fig. 6.27). *Vrancea earthquakes* occur in a site of very special features, which can be considered as being unique in the world (Wenzel et al, 2002) (Fig. 6.29a). This region is situated in the Eastern Carpathian Arc bent area, being produced by the subduction of the Black Sea microplate beneath the Intra-Alpine block (which belongs to the West Eurasian Plate) (Balan et al, 1982, Pustovitenko et al, 1994, Stiopol et al, 1994, Bala et al, 2003, Sandi, 2004, Knapp et al, 2006, Ciucu and Fulga, 2008). The particularity of this intraslab region consists in a very small and practically vertical mantle volume 30x70x130 km (Fig. 6.29b), where earthquakes

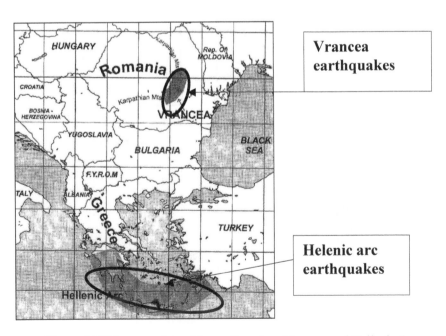

Figure 6.28 European intraslabe earthquakes: Vrancea and Hellenic Arc
(after Benetatos and Kiratzi, 2004)

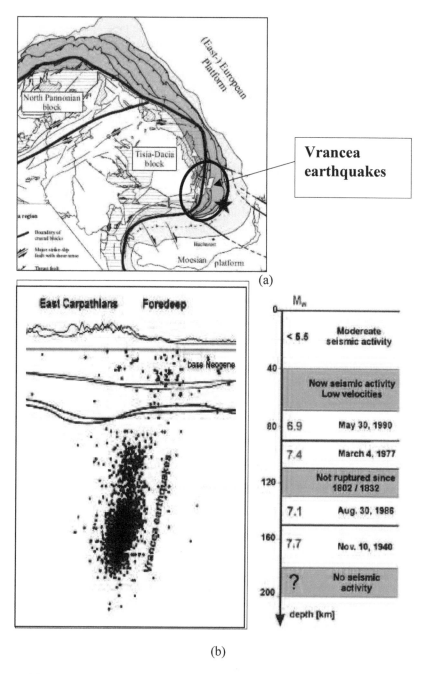

(a)

(b)

Figure 6.29 Vrancea earthquakes: (a) Geodynamic evolution of the Carpatho-Panonian region (after Wenzel et al, 2002); (b) Cross-section across the East-Carpathiaens and magnitudes related to the depths (after Ciucu and Fulga, 2008)

occur with a magnitude exceeding M 7.0 (Ismail-Zadeh et al, 2005, Cloetingh et al, 2004). In the interval of the last 600 years, 3 to 5 earthquakes per century occurred with magnitudes larger than M 7.2 (Ciucu and Fulga, 2008). The seismicity of the Vrancea zone is divided into a crustal domain (depth under 40 km) with lower magnitudes and an intermediate depth domain larger than 70 km, where all strong earthquakes are located. These two domains are separated by an almost non seismic zone ranging from 40 to 70 km in depth, showing that the earthquakes occurred in these domains belong to very different source types. It is very interesting to notice that the earthquake magnitudes increase with the depth of sources.

6.5 WORLD'S SEISMIC ZONES

Generally, the maps showing the world's epicenters (Figure 2.2) do not a make distinction between the earthquake types, showing only the epicenter locations. The advances in knowing the nature of faults allows us to establish the expected earthquake type in each seismic areas of the world and to choose, consequently, the most adequate design methodology, specific for the respective type of earthquake. In the following, these seismic areas will be described (Figs. 6.34 and 6.35) by underlining their main characteristics. The corresponding zones are numbered in function of the geographical position and the earthquake type. For North and South American continents, the seismic zones are presented in Figure 6.30.

North American seismic zones (NA). The Western coast is the most seismically active area, being influenced by the Circum-Pacific Ring. Crustal interplate subduction earthquakes, resulting from the subduction of the ancient Kula Plate under the North American Plate, characterize the Alaska zone (NA-a). In this seismic zone, the Anchorage second larger ever-recorded earthquake occurred in 1964, producing very important avalanches, landslides and tsunami. The character of this diffuse seismic zone is marked by the existence, in the continental part, of some intraplate earthquakes, dominated by the Dinali strike-slip fault (like the San Andreas Fault, but situated in a scarcely populated region). The same interplate and intraplate earthquake types characterize the Western Canadian coast, being produced by the subduction of the Kula and Juan de Fuca plates beneath the North American plate (NA-b). But the most known zone of North America is the Western coast of the USA (NA-c), dominated by the San Andreas Fault and the corresponding diffuse zone, where interplate crustal strike-slip earthquakes are the main ground motion type, due to the oblique movements between the Pacific and the North American plate. The Eastern part of North America (NA-d) is dominated by diffuse intraplate earthquakes, resulting from the continental movements during a previous cycle of the Pangaea formation. Especially Missouri, the Eastern coast of the USA and Canada are subjected to normal, reverse and thrust earthquake types.

Central American seismic zones (CA). The territory of Mexico is dominated by crustal subduction interplate and intraslab earthquakes, produced by the subduction of the Cocos plate beneath the North American plate (CA-a). In the zone of the Caribbean Isles (CA-b), some subduction interplates occur due to the interaction between the Caribbean and North American plates. The intraplate earthquakes, produced in seismic areas without very clear fault systems, dominate the Eastern zones. The central zone is practically immune from earthquakes.

South American seismic zones (SA). Generally, this seismic area is regular, without important changing along the fault. In the North part of South America, in Venezuela (SA-a), some strike-slip earthquakes occur, due to interaction between the Caribbean and the South American plate. Subduction crustal interplate and intraplate earthquakes dominate the Western seismic area (SA-b), as the result of subduction of the Nazca plate beneath the South American plate. The Eastern South American zone is immune from earthquakes.

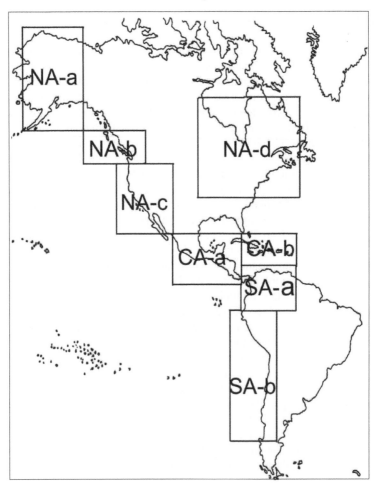

Figure 6. 30 Seismic zones in the North and South American continents

For the African, European, Asian and Australian continents, the seismic areas are presented in Figure 9.31.

African seismic zone (AF). The Northern areas (AF-a) are affected by crustal interplate earthquakes, due to the subduction of the African plate under the Mediterranean microplate. Divergent crustal earthquakes along the Red Sea and the African Rift (AF-b) dominate the Eastern African areas. The Western African zones are almost immune from earthquakes.

European seismic zones (EU). Europe has one of the most complex seismic zonifications. The first zone (EU-a), the Mediterranean one, including South of Spain and Italy, is dominated in the Southern part by crustal subduction interplate earthquakes (where the African plate subducts the Mediterranean microplate), with the exception of the South of Sicily and Crete, where some intraslab earthquakes can occur. The northern zone is located along the Pyrenees and Alps, where the Central Mediterranean and Adria microplates subduct the Eurasian plate giving rise to crustal collision interplate earthquakes. Another very important zone is the Vrancea one, characterized by intraslab earthquakes, where the Blacke Sea microplate subducts the Pannonian microplate. The most active seismic zone in the Balkans is Greece, with many intraplate earthquakes due a very active network of faults. The Anatolia zone is characterized by the presence of strike-slip earthquakes along the North Anatolian fault and collision earthquakes in the South part of Anatolia. The Central Europe (EU-b) presents many intraplate earthquakes especially along rifts in a very diffuse seismic zone. Finally, the Northern Europe (EU-c) has the same intraplate earthquakes, due to the influence of the Middle Atlantic ridge, creating diffuse seismic areas. The Eastern seismic part of Europe is dominated by the Ural zone (EU-d), where a collision between Europe and Siberia produced a chain of mountains.

Asian seismic zones (AS) The Arabian and Caucasian zones (AS-a) are dominated by collision earthquakes, produced by the contact between the Arabian and the Eurasian Plates. The North of India and the Tibet zones (AS-b) are characterized by the presence of crustal collision interplate earthquakes, due to the contact between the Indian and the Asian plates along the Himalayan Mountains. The peninsular India (AS-c) is affected by intraplate earthquakes. Another very important seismic zone is the Chinese plateau (AS-d), where intraplate earthquakes are dominating events. This diffuse zone is the result of the collision of some ancient blocks. A very large seismic zone is situated in the South Asia, where the majority of earthquakes are due to the subduction of the Indian plate under the Pacific plate (AS-e). The last zone, but a very important and complex one, is the Japanese area (AS-f), where all type of earthquakes can occur: crustal interplate, intraplate and intraslab earthquakes act in different zones of this area, being function of the existing fault type, due to the simultaneous subduction of many tectonic plates under the Eurasian Plate.

Australian seismic zone (AU) is characterized by intraplate earthquakes only, being situated far from the tectonic plate boundaries, inside of the Indo-Australian Plate.

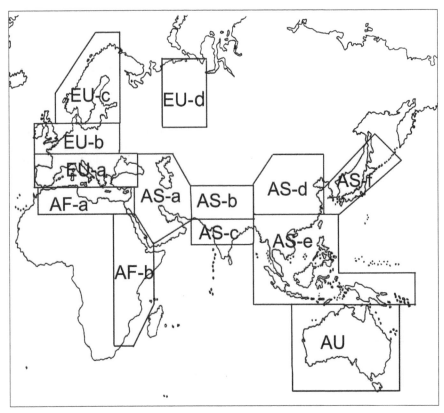

Figure 6.30 Seismic zones in Africa, Europe, Asia and Australia

6.6 REFERENCES

Anderson H., Webb, T. (1994): New Zealand seismicity: patterns revealed by the upgraded National Seismograph Network. New Zealand Journal of Geology and Geophysics, 1994, Vol. 37, 477-493

Bala, A., Radulian, M., Popescu, E.L. (2003): Earthquake distribution and their focal mechanism in correlation with the active tectonic zones of Romania. Journal of Geodynamics, Vol. 36, 129-145

Balan, S., Cristescu, V., Cornea, I. (1982): The Romanian earthquake of 4 March 1977 (in Romanian). Editura academiei, Bucharest

Barnes, G. (2003): Origins of the Japanese islands: The new "Big picture" Japan Review, Vol. 15, 3-50.

Benetatos, C.A., Kiratzi, A.A. (2004): Stochastic strong ground motion of intermediate depth earthquake: the case of the 30 May 1990 Vrancea (Romania) and the 22 January 2002 Karpathos island (Greece) earthquakes. Soil Dynamics and Earthquake Engineering, Vol. 24, 1-9

Berg, van den M.V. (1994): Neotectonics of the Roer Valley rift system. Geologie en Mijnbouw, Vol. 73. 143-156

BGHRC, Benfield Greig Hazard Research Centre (2003): A rift at the heart of Europe
http://www.benfieldhrc.org/activities/issues/issues1.pdf

BGS , British Geological Survey, (2008): Earthquakes in the UK.
http://www.earthquakes.bgs.ac.uk/earthquakes/education/uk_earthquakes.htm

Bird, P. (2003): An updated digital model of plate boundaries. Geochemistry Geophysics Geosystems, Vol. 4, No. 3, 1027-1079

Black, B. (2000): Aleutian Arc-trench. Geology 565 R, Tectonics
http://www.geology.byu.edu/faculty/rah/tectonics/students%Presentation/2000/Brian%Black/Alaskan.htm

Boudreau, D. (2001): To make a quake. ASU Research E-Magazine.
http://researchmag.asu.edu/articles/earthquake.html

Burns, P. (2005): Cordova coastal management DRAFT Plan. DGGS Review
http://www.dnr.state.ak.us/acmp/District/Plans/Cordova/comments/DGGS_Cordova_rev_6_30_05.doc

Butler, R. (2001): Dynamic earth.
http://www.see.leeds.ac.uk/structure/dynamicearth/plates/seisglobal/index.htm

Camelbeek, T. (2005): The seismology and physics of the Earth in the prevention of earthquakes: Intra-plate seismicity in NW Europe. Royal Observatory of Belgium Report.
http://bncgg/oma.be/contractforum10_05/abstract-contrib/contrib7.doc

Carminati, E., Doglioni, C., Scrocca, D. (2004): Alps vs Apennines, Special Volume of the Italian Geological Society for the IGC 32, Florence, 141-151

CGS (2007): Californian earthquake history and catalogs.
http://www.consrv.ca.gov/CGS/rghm/quakes/index.htm

Chiarabba, C., Jovane, L., DiStefano, R. (2005): A new view of Italian seismicity using 20 years of instrumental recordings. Tectonophysics, Vol. 395, 251-268

Ciucu, C., Fulga, C. (2008): Two case studies of post-seismic regimes in the Vrancea region. Romanian Reports in Physics, Vol. 60, No. 1, 173-189

Cloetingh, S., Burov, E., Matenco, L., Toussaint, G., Bertotti, G., Anddriessen, P., Wortel, M., Spakman, W. (2004): Thermo-mechanical controls on the mode of continental collision in the SE Carpathians (Romania). Earth and Planetary Science Letters, Vol. 218, 57-76

Doglioni, C., Agostini, S., Crespi, M., Innocenti, F., Manetti, P., Riguzzi, F, Savascin, Y. (2002). On the extension in Western Anatolia and the Aegian sea. Journal of the Virtual Explorer, Vol. 7, 167-181

Dynamic Earth (2001): Dynamic earth.
http://earth.leeds.ac.uk/dynamicearth/subduction/thermalbig.gif

Fitzpatrick, T. (1996): Wiens finds temperature plays role in large, deep earthquakes
http://record.wustl.edu/archive/1996/12-05-96/6453.htm

Fox, D. (2007): Study indicates unexpected earthquake danger lie beneath the Pacific Northwest. Emergency Management Division http://emd.wa.gov/5-prog/prgms/eq-tsunami/dfoxeqarticle.htm

Gangopadhyay, A., Talwani, P. (2003): Symptomatic features of intraplate earthquakes. Seismological Research Letters, Vol. 74, No. 6, 863-883

Gioncu, V. (1994): Development and design of seismic-resistant steel structures in Romania. In Behaviour of Steel Structures in Seismic Areas, STESSA 94 (eds. F.M. Mazzolani and V. Gioncu), E&FN Spon, London, 3-27

Gioncu, V. (2006): Advances in seismic codification for steel structures. Costruzzioni Metalliche, No. 6, 69-87

Gioncu, V., Mazzolani, F.M. (2002): Ductility of Seismic Resistant Steel Structures. Spon Press, London

Gioncu, V., Mazzolani, F.M. (2003): Challenges in design of steel structures subjected to exceptional earthquakes. Behaviour of Steel Structures in Seismic Areas, STESSA 2003 (ed. F.M. Mazzolani), Naples, 9-12 June 2003, Balkema, Lisse, 89-95

Gioncu, V., Mazzolani, F. M. (2006): Influence of earthquake types on the design of seismic-resistant steel structures. Part 1: Challenge for new design approaches. Part 2: Structural responses for different earthquake types. Behaviour of Steel Structures in Seismic Areas, STESSA 2006 (eds. F.M. Mazzolani and A. Wada), Yokohama, 14-17 August 2006, Taylor & Francis, London, 113-120, 121-127

Gregersen, S. (2006): Intraplate earthquakes in Scandinavia and Greenland neotectonics or postglacial uplift. Journal Geophysics Union, Vol. 10, No. 1, 25-30

Hinze, W.L., Braile, L.W., Keller, G.R., Lidiak, E.G. (1988): Models for midcontinental tectonism: An update. Revue of Geophysics, Vol. 26, 699-717

Houston, H. (2004): Earthquake seismology: Deep earthquakes. University of Washington, Seattle, MS 71

IGCP, International Geoscience Programme, Project 474 (2006): Images of the Earth's crust & upper mantle.
http://www.earthscrust.org/science/transects/urals.html

Iio, Y., Kobayashi, Y. (2002): A physical understanding of large intraplate earthquakes. Earth Planets Space, Vol. 54 1001-1004

Ismail-Zadeh, A., Mueller, B., Schubert, G. (2005): Three-dimensional numerical modeling of contemporany mantle flow and tectonic stress beneath the earthquake-prone Southeastern Carpathians based on the integrated analysis of seismic, heat flow and gravity data. Physics of the Earth and Planetary Interiors, Vol. 149, 81-98

Ivancic, I., Herak, D., Markusic, S., Sovic, I., Herak, M. (2002): Seismicity of Croatia in the period 1997-2001. Geofizica, Vol. 18-19, No. 1, 17-29

Jankulovski, E., Sinadinovski, C., McCue, K. (1996): Structural response and design spectra modelling results from some intra-plate earthquakes in Australia. 11th World Conference on Earthquake Engineering, Acapulco, 23-28 June 1996, Paper No. 1184

Kirby, S.H., Durham, W.B., Stern, L.A. (1991): Mantle phase changes and deep-earthquake faulting in subducting lithosphere. Science, Vol. 252, 216-225

Knapp, J.H., Knapp, C.C., Munteanu, L., Raileanu, V., Mocanu, V. (2006): Origin and tectonic evolution of active continental lithospheric delamination in the

Vrancea zone, Romania: Project Dracula. Surface on the Earth: Global Studies. IRIS 5-Year proposal, 97

Musson, R.M.W. (1999): Probabilistic seismic hazard maps for the North Balkan. Annali di Geofisica, Vol. 42, 1109-1124

Nature (2005): Nazca plate
http://www.nature.com/nature/journal/v436/n7052/images/43675a-i3.0.jpg

Nelson S.A (2003) Global tectonics.
http://www.tulane.edu?~sanelson/geol111/pltect.htm

Norsar (2004): Earthquakes in Norway.
http://www.norsar.no/seismology/general/earthquakesinnorway.pdf

Oceonography (2008): Origin of oceanic basins.
http://www.tulane.edu/~bianchi/Courses/Oceanography?chap%203.ppt

Pequegnat, C., Gueguen, P., Hatzfeld, D., Langlais, M. (2008): The French accelerometric network (RAP) and National Data Centre (RAP-NDC), Vol. 79, No. 1, 79-90

Persh, S.E., Houston, H. (2004): Strongly depth-dependent aftershock production in deep earthquakes. Bulletin of the Seismological Society of America, Vol. 94, No. 5, 1808-1816

Pustovitenko, B.G., Kapitanova, S.A., Gorbunova, I.V. (1994): The complex rupturing in the deep focal sources of Vrancea region. In European Seismological Commission, XXIVth General Assembly, Athens, 19-24 September 1994, 627-633

Quakes (2003): Earthquakes maps of Queensland and Australia
http://www.quakes.uq.edu.au/seis_maps/

Reid, D. (2002): UCL scientists create first earthquakes in the laboratory. Innovations Report, University College London
http://www.ucl.ac.uk/experts

RMS (2006): 1356 Basel earthquake. 650-year retrospective
http://rms.com/Publications/BaselReport_650year_retrospective.pdf

Sandi, H. (2004): Vrancea region. Geo Strategies Limited
http://www.geo-strategies.com/romania/vrancea.htm

Schafer, A., Siehl, A. (2002): Rift tectonics and syngenetic sedimentation: The Cenozoic lower Rhine Basin and relative structures. Netherlands Journal of Geosciences, Vol. 81, No. 2, 145-147

Seno, T., Yoshida, M. (2004): Where and why do large shallow intraslab earthquakes occur? Physics of the Earth and Planetary Interiors, Vol. 141, 183-206

Solarino, S., Cassinis, R. (2007): Seismicity of upper lithosphere and its relationships with the crust in the Italian region. Bollettino di Geofisica Teorica e Applicata, Vol. 48, No. 2, 99-114

Stern R.J. (2002): Subduction zones. Reviews of Geophysics, Vol. 40, No. 4, 3.1-3.42, 3.1-3.13

Stiopol, D., Radulescu, F., Nacu, V., Mateescu, D. (1994): Vertical crustal movements in the Vrancea Area (Romania), generated before and after a seismic event. In European Seismological Commission, XXIV General Assembly, Athens, 19-24 September 1994, 1178-1186

Taymaz, T., Yilmaz, Y., Dilek, Y. (2007): The geodynamics of the Aegean and Anatolia. Introduction. Geological Society, Vol. 291, 1-16

Tibi, R., Bock, G., Estabrook, C.H. (2002): Seismic body wave constraint on mechanism of intermediate-depth earthquakes. Journal of Geophysical Research, Vol. 107, No. B3, 2047, 1-23

Ukearthquakes (2008): UK earthquakes
http://www.geologyshop.co.uk/ukearthquakes.htm

US Earthquakes (2008): New Madrid Fault.
http://showme.net/~fkeller/quake/maps6.htm

USGS (nd): Major tectonic plates of the world.
http://geology.er.usgs.gov/eastern/plates.html

Valensise, G., Amato, A., Montone, P., Pantosti, D (2003): Earthquakes in Italy: past, present and future. Episodes, Vol. 26, No. 3, 245249

Vernant, P. et al (2004): Contemporary crustal deformation and plate kinematics in middle east constrained by GPS measurements in Iran and Northern Oman. Geophysics Journal International, Vol.157, 381-398

Wenzel, F., Sperner, B., Lorenz, F., Mocanu, V. (2002): Geodynamics, tomographic images and seismicity of the Vrancea region (SE-Carpathians, Romania), EGU Stephan Mueller Special Publication Series, Vol. 3 95-104

Wikipedia (nd): Earthquake
http://en. Wikipedia.org/wiki/Earthquake

Wikipedia (nd): Orogeny
http://en.wikipedia.org/wiki/Orogeny

Yates, J., Montgomery, B.C., Lowman, P.D. (1999): An integrative technique to create a digital tectonic activity map (DTAM) of Earth. International Symposium on Digital Earth (ISDE), Beijing, China

Zhang, Y. (nd): Current situation of earthquakes prediction exploration in China Mainland and its potential demands for space technology, CENC Report

Chapter 7

Earthquakes and Ground Motions

7.1 GROUND MOTIONS EVALUATION AND EARTHQUAKE ENGINEERING

7.1.1 Transfer of Engineering Seismology Knowledge to Earthquake Engineering

The advances in Seismology and the special branch of Engineering Seismology were presented in Chapters 4, 5 and 6. The proper definition of ground motions, in order to evaluate the structural response, is the first field where the contact between Engineering Seismology and Earthquake Engineering is directly established.

Earthquakes do not directly produce building collapse. Ground motions are the real cause of seismic damage. The dynamic response of buildings to ground motions is the most important cause of earthquake-induced damage in buildings. Therefore, it is very important to understand in which way the sudden movements of the source are transformed in ground motions at the building site.

The basic concepts of today's Earthquake Engineering were born almost 70 years ago, when the knowledge about seismic actions and structural response were rather poor. Today, the earthquake-resistant design has grown within the multi-disciplinary fields of Engineering Seismology and Earthquake Engineering, wherein many exciting developments are predicted in the near future. Looking to the development of the Engineering Seismology and Earthquake Engineering, it is very clear that the major efforts of researchers were directed towards the structural response analysis. As a consequence, the structural response can be predicted fairly confidently, but these achievements remain without real effects if the evaluation of seismic actions is doubtful. In fact, the prediction of ground motions is still far from a satisfactory level, due to both the complexity of the seismic phenomena and the lack of communication between seismologists and engineers. This remark can be confirmed after each important earthquake, when new and in situ lessons regarding the ground motions are learned, instead of providing in advance the missing data from seismological studies. So, the reduction of uncertainties in seismic action modeling is now the main challenge in structural seismic design. This target is possible only if the impressive progress in Seismology will be transferred into Earthquake Engineering. Any progress in this field is impossible without considering this new amount of knowledge recently cumulated in Seismology (Gioncu and Mazzolani, 2003, 2006).

Observations of damage after the earthquakes have shown that the earthquake characteristics, very different from one site to another, can have a strong influence on the structure performance. These characteristics reflect not only the source properties, but also local effects and the site earth's configuration. In order to be considered in seismic design of structures, it is very important to underline the main characteristics of these ground motion types, taking into account the source typologies. This target can be obtained by processing the recorded earthquakes, or, as a new challenge in Earthquake Engineering, by studying the rupture processes and propagation of seismic waves by numerical modeling.

7.1.2 Uncertainties in Selection of Proper Design Ground Motions

Historically, the codes were developed based on the experience of few recorded ground motions, not sufficiently close to the causative faults, due to the absence of a dense network of recording stations. Recently this situation has changed in some very urbanized seismic zones, with more near-fault records. But, unfortunately, there is still a large number of seismic zones without adequate information. In addition, due to the fault network or to the presence of some little knowledge about active faults different from the known ones, it is practically impossible to design a recording network station in such a way to get records on the sites with the maximum ground motions. So, the recorded values have a great incertitude, and only by chance, some of them are situated near the fault. But these records can be influenced by some local site conditions, so they are available only for the recording station. As a consequence, the design spectra are not, generally, applicable for other seismic sites. In these conditions, it is very difficult to detect the differences between the ground motions produced by different source types.

Therefore, the design of earthquake-resistant structures is affected by many uncertainties. One of the most important is related to the selection of the proper design ground motion. Conceptually, it should be done by considering all possible ground motions which can occur in a given site and can drive the structure to a critical response during its service life (Anderson and Bertero, 1987). The quantification of this concept creates serious difficulties due to the differences among the characteristics of different earthquake types.

During the past decades, the increasing databases of recorded earthquakes has indicated that the dynamic characteristics of the ground motion can vary significantly among recording stations which are located in the same area or in different areas. For one recorded earthquake at the different stations situated in the same area, the differences are significant especially for the stations located near the epicentral region, due to the effective distances from the epicenter and the direction of rupture. Examining the records at different sites, one can observe that they vary significantly due to several factors (fault mechanism type, path effects, local site conditions, etc.). The main differences are observed in recorded accelerations, velocities and displacements, vertical components, directivity duration, number of significant pulses and natural period of ground movements. In function of some conditions, these differences can be determinant or not in the selection of the proper design ground motions for a structure.

The recorded ground motion is influenced by the effects of some important factors (for instance the source type and local soil conditions), over which the effects of some secondary factors are superposed, changing in a certain extent the original characteristics. This is the reason why the recorded values differ so much from one site to another, making very difficult to distinguish the main characteristics of ground motions.

Due to these uncertainties, there are many adverse arguments in using only the recorded ground motions for design purposes. An alternative way to face this situation is to use the numerical computation, in which the influence of the main factors can analyzed. But these factors are so various, that a credible result is very difficult to obtain. Therefore, a rational combination between recorded data and numerical results seems to be the best way to follow for the definition of the design ground motion.

7.1.3 Factors Influencing the Ground Motions

In the last few years, the understanding of earthquake characteristics has considerably increased, due to the tragic experience from several recent earthquakes. As a consequence of the direct observations and subsequent studies, now it is possible to quantitatively predict strong motions for dangerous earthquakes, provided that the source mechanism, wave travel path and site geological conditions are correctly modeled (Huang, 1983, Kawase, 2004). Therefore, a clear representation of an earthquake can be obtained, by considering the source typology as a matrix, where the influence of some factors connected to traveling path and site conditions are overlapped (Fig. 7.1). This procedure gives the opportunity to separately analyze the influence of each basic factor in the frame of the anatomy of an earthquake.

Source. Earthquakes occur in a fault, in a network of faults or in an area of faults. In the first case, the positions of faults are known due to the seismological studies; in the second case, the fault is poorly or not known and the earthquakes can occur in a large region having uniform seismic potential. Each source has some specific characteristics, in function of the fault type and the rupture area.

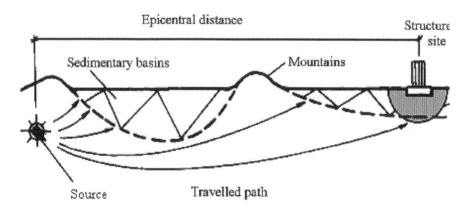

Figure 7.1 Basic factors influencing the recorded ground motions
(Gioncu and Mazzolani, 2002)

Traveled path. The propagation of seismic waves depends on the percentage of the path travel through rock or soft sediment, with the presence of mountains and valleys from the source to the site. The results of the traveled paths are the amplification, attenuation or modification, more or less, of the source basic characteristics.

Site conditions. Multi-layers with different properties and thicknesses compose the site soil profile. The alternations of these layers, the presence in the site of hills, valleys or important buildings, are very important factors, which can modify the source characteristics.

Therefore the seismicity of a site depends on some fundamental input parameters and some additional factors. In order to establish the seismic hazard for a site, the first step is the identification of the source type. The second step is the evaluation of the influence of the distance from the source and the site and the study of the local conditions. For the determination of the seismic risk, these data must be represented by means of values referring to accelerations, velocities, displacements, durations and number of pulses in the site, as a function of the source type, magnitude, duration, directivity and the local influence as attenuation or amplification, site stratification, relief, etc. These factors will be analyzed in the following sections.

7.2 SOURCE CHARACTERISTICS

7.2.1 Surface Fault Rupture

A sudden movement of rocks along a fault plane causes earthquakes. The movement or displacement is called fault rupture. When the fault rupture progresses upward, it creates a surface fault rupture. This surface can remains beneath the surface, being a *blind fault*, or reach the Earth's surface, creating a *surface faulting.*

The rupture begins at some depth called *focus* or *hypocenter* (Fig. 7.2). The rupture then spreads outward in all directions along the fault plane, but there is a dominant direction.

The characteristics of the fault rupture depend on the following factors:
- rupture surface;
- type of fault;
- depth of the fault;
- amount of the fault slip;
- age of faulting.

The earthquake magnitude depends on the length of the fault rupture. The highest magnitude earthquake, in Chile 1960 , has a length about 1300 kilometers.

The formation of a rupture surface along a fault is a very complex phenomenon, which raises many questions. There are regions where great earthquakes occur along some plate boundaries, while there are also quiet regions along the same fault. Why does it happen? Why are there segments of fault very active in some periods and quiet in others? The best example is the Mexico coast,

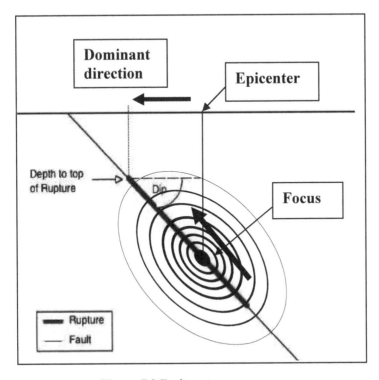

Figure 7.2 Fault rupture progress

where the Cocos plate subducts the North American plate (Lowry et al, 2001). The different segments of this plate boundary are very active in different periods. There is also a gap, named Guerrero gap, where for a long time no earthquakes were recorded. The aspects can be observed along the North Anatolian fault (see Fig. 3.4). Generally, the gaps are often considered to be the sites of the next great earthquakes. A way to explain these regional variations in seismicity along plate boundaries is to introduce the fault asperities and barriers, whose size varies from one place to another.

7.2.2 Asperities and Barriers

Asperity implies the inequality or roughness of the surface of the body, whereby some parts of it stick out beyond the rest, increasing the surface friction of any object moving over it (Wikipedia, nd). In Seismology, this term is used to describe the heterogeneity of a fault. Asperity is the small site on a fault surface of higher strength than the surrounding, where stress concentrates prior to the fault rupture.

Asperity source models are important since recent studies have clarified that the main contribution to strong ground motion comes from the asperity area. These models are defined by the fault parameters: fault rupture surface, average and maximum slips, peak slip velocity and stress drop of the overall fault and asperity area (Dalguer et al, 2004). The model to explain the seismicity variation along the fault boundaries considers that plate boundaries consist of asperities and barriers having different friction laws. These asperities are distributed in a fractal manner and each fault contains small and large asperities. The geometry of asperities is shown in Fig. 7.3.

Thus, the behavior of earthquakes is dependent on the asperity source model, characterized by dimensions, number and properties of these asperities (Seno, 2003, Dalguer et al, 2004). In case of great earthquakes, the asperity size is very large, while it is very small for minor earthquakes.

Barriers surround these asperities, with a stable sliding frictional property. The barrier surface has a higher strength than the surroundings, which is capable of stopping the rupture. In addition, the fault zones can be lubricated by the presence of pore fluid pressure (pressure within the fluid which fills voids between the solid rock particles), which removes the shear friction, invading the barrier. So, the fault surface consists of asperities, invaded barriers and un-invaded barriers. The seismic rupture can be initiated only when an asperity breaks within the invaded barriers. This is because, when an asperity breaks surrounded by invaded barriers, faulting would propagate within these barriers with almost no friction and further induces the breakage of the nearby asperities. Contrary, if the barriers surrounding the asperity are not invaded, the barriers would prohibit the propagation of fault. This model explains the differences in behavior of a fault along its length, due to the different quantity of asperities, invaded and un-invaded barriers.

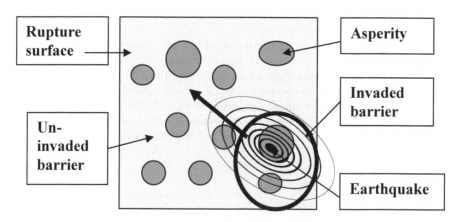

Figure 7.3 Geometry of asperities

7.2.3 Radiated Energy and Magnitude

One the most fundamental parameters for describing an earthquake is the radiated seismic energy. The total energy from an earthquake includes energy required to create new cracks in rocks, energy dissipated as heat through friction and energy elastically radiated through the Earth in the form of seismic waves. The only quantity of these energies which can be measured is the radiated one, because it shakes buildings and can be recorded by seismographs. Seismologists use a *magnitude scale* to express the seismic energy released by each earthquake, independent of the place of observation. Each earthquake has a unique amount of energy, but the magnitude values given by different seismological observatories for the same event may vary, due to the fact that seismologists use several different methods to estimate the magnitude. Therefore, there are a lot of definitions of magnitude: Richter magnitude or local magnitude, energy magnitude, moment magnitude, body-wave magnitude, etc. Therefore, one must be very careful in using the data of the earthquake magnitude, when the used scale is not mentioned. Among these magnitude types, the technical literature is very rich in correlation formulae (Bergman, 2000).

The scale of magnitude commonly used to express the seismic energy released by each earthquake is the Richter scale. The typical effects of earthquakes for various *magnitude* ranges are listed below:

Richter magnitude   Earthquake effects

a)   Micro earthquakes

Less than 3.0 M   Generally not felt, but recorded

b) Low earthquakes

3.0 – 4.5 M   Often felt, but it rarely causes damage, only for poorly
                            constructed buildings.

c) Moderate earthquakes

4.6 – 6.5 M   Major damage to poorly constructed buildings,
                            slight damage to well-constructed buildings

d) Large earthquakes

6.6 - 7.9 M   Major earthquake causing serious damage over
                               large areas in poorly constructed buildings.
                               Moderate damage to well-constructed buildings.

e) Great earthquakes

8 or greater   Great earthquake causing serious damage in areas
                               several hundred kilometers across, also for
                               well-constructed buildings.

Another measure of earthquake size is the *area of the fault* which slipped during the earthquake. Small earthquakes result in small areas. Contrary, strong earthquake occurs in large areas (Fig. 7.4). One can see that both corresponding areas and seismic magnitude for interplate earthquakes are larger than for intraplate earthquakes.

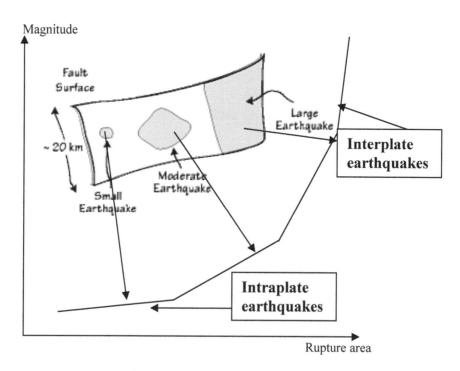

Figure 7.4 Influence of rupture area

7.2.4 Forward Directivity

After a slipping of rocks in the earthquake zone starts, due to breaking of the most exposed asperity, the rupture propagates out from this zone. The space-time history of rupture depends on many properties of the fault. For instance, the presence of asperities, invaded barriers and barriers, elastic properties of fault, etc., will hinder or encourage the propagation in a certain direction (Warren and Sheare, 2006). In addition, a branching, a curved or stepping fault will influence the rupture propagation. Therefore, the observations regarding large earthquakes show that they have primarily unilateral ruptures. One of the most representative cases for the forward directivity is the 1995 Kobe earthquake (Fig. 7.5), when the rupture directivity was exactly under the most densely populated part of the city, producing the above-mentioned great damage. Fortunately, the opposite was the situation of the Northridge earthquake, where the directivity was in the North direction, a less populated zone of the city.

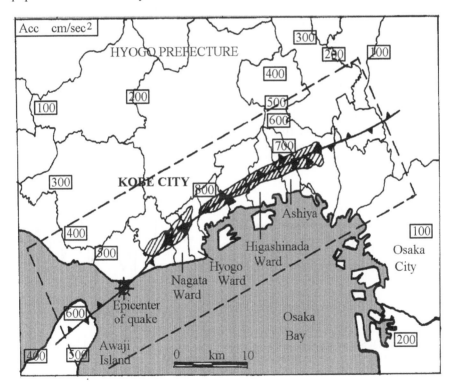

Figure 7.5 Effect of the forward directivity during the 1995 Kobe earthquake
(Gioncu and Mazzolani, 2002)

7.2.5 Rupture Duration

The duration of the rupture process depends on the amount of radiated energy, magnitude, geometry of the rupture area and speed of the rupture process (Trifunac and Novikova, 1995a,b, Lee, 2002).The duration of rupture can be described by an exponential function of the magnitude (Gioncu and Mazzolani, 2002). For low magnitude earthquakes the duration is under 0.5 seconds, for moderate earthquakes the duration is about 2 seconds and for large magnitude earthquakes the duration can reach more than 10 seconds. For great earthquakes the rupture duration is more than 15-20 seconds.

7.2.6 Pulse Types

The number of pulses depends on the number of active asperities. In many cases the propagation of the fault rupture toward a site causes most of the seismic energy to move from the rupture with a single large pulse of motion (time required for a complete cycle) (Somerville, 2000a,b). This pulse of motion represents the cumulative effect of almost all the seismic radiation from the fault.

Figure 7.6a presents the case of the 1994 Northridge earthquake, where only one asperity was active on the rupture surface. One can see that the recorded velocity has a single pulse. Contrary, in some cases where there are more asperities on the rupture surface, the velocity record has multiple pulses. Figure 7.6b shows the recorded velocity of the Kobe earthquake, where it is supposed that the earthquake was produced by three asperities.

These periods increase as far as the magnitude increases (Fig. 7.7). For low earthquakes, the pulse periods are under 0.5 seconds, while for great earthquakes, these periods can reach over 10 seconds. The pulse period depends of fault type: is larger for strike-slip earthquakes in comparison with the reverse (subduction) ones.

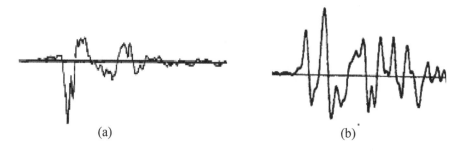

(a)   (b)

Figure 7.6 Pulses of rupture: (a) One pulse (Northridge); (b) Three pulses (Kobe)
(Gioncu and Mazzolani, 2002)

Pulse period (sec)

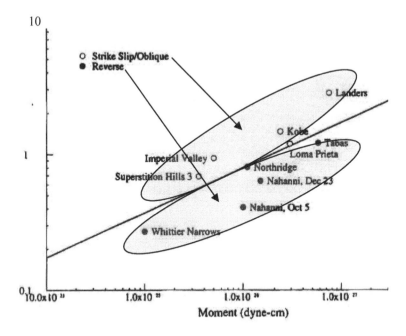

Figure 7.7 Pulse period, influence of source type (after Somerville, 1998)

7.3 PATH EFFECTS

7.3.1 Seismic Waves

When the strain accumulated in the rock exceeds its capacity limit in the asperity, the fault ruptures, rock masses are abruptly displaced and seismic waves begin to radiate from the fault. As the rupture propagates, it successively releases the strain energy stored along the activated part of the fault. Thus, each point of the rupture surface contributes, with a certain time of delay, to the total picture of seismic waves, which interfere with each other at a certain distance from the causative fault and give rise to a quite complicated wave train (Kulhanek, 1990). Unfortunately, the existing science and technology are not capable of directly observing the slip movement of faults. Therefore, the seismic waves generated by the fault movement, which travel through the Earth, are observed to indirectly clarify the seismic source rupture process.

Seismic waves are the wave of energy caused by the sudden breaking of rocks within the Earth. Seismic waves radiating from the fault break to the site are of two main types (Fig. 7.8): body (P and S) and surface (L and R) waves.

The problems of wave propagation can be understood by representing the Earth as a layered medium formed by layers of certain thickness and different

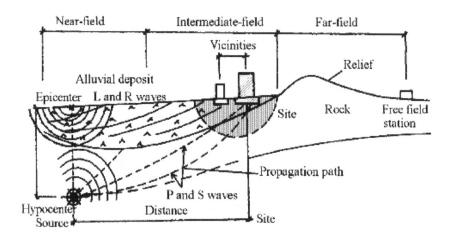

Figure 7.8 Body (P and S) and surface (L and R) seismic waves
(Gioncu and Mazzolani, 2002)

mechanical properties. Due to the discontinuities between the layers, reflection and refraction phenomena occur, making very complex the way of seismic wave to the surface site.

The *body waves* are P waves (from latin Undae Primae) and S waves (from latin Undae Secundae) (Fig 7.9a) that propagate through the Earth's crust. Richard Dixon Oldham, a British geologist and seismologist, was the first scientist who recognized the P and S-waves in the early Seismology. The first kind of body waves is the *P wave*. They push and pull the rock and move through just like the sound waves push and pull the air, by compressions and dilatations. The rock particles, affected by a P-wave, oscillate backward and forward in the same direction as the wave propagates. They are the ones that travel fastest, having velocities 5 to 7 km/sec, and thereby are usually felt first. This wave has very high frequency. The second type of body wave is the *S wave*, which moves rock up and down, or side-to-side, the Earth vibrating perpendicularly to the direction of the wave travel. This wave is much slower than the P waves, with a velocity of 3 to 4 km/sec, and therefore, arrives after the primary wave, having a reduced frequency in comparison with the first wave. Due to the difference in velocities, a time delay between P-wave and S-wave arrival exists, and this delay can serve for locating the earthquake epicenter.

The second type of waves are the *surface waves*, which travel along the free Earth's surface (Fig. 7.9 b): L waves (from the British mathematician Love, who worked out the mathematical model for this kind of wave in 1911) and R waves (from Lord Rayleigh, who mathematically predicted the existence of this kind of wave in 1885). The surface waves propagate also along the discontinuities in media of the Earth's interior. Other modes of wave propagation exist (for example reflected and refracted waves, effects of conversion and dispersion, etc.), but they are of comparatively minor importance. Surface waves carry the greatest amount of

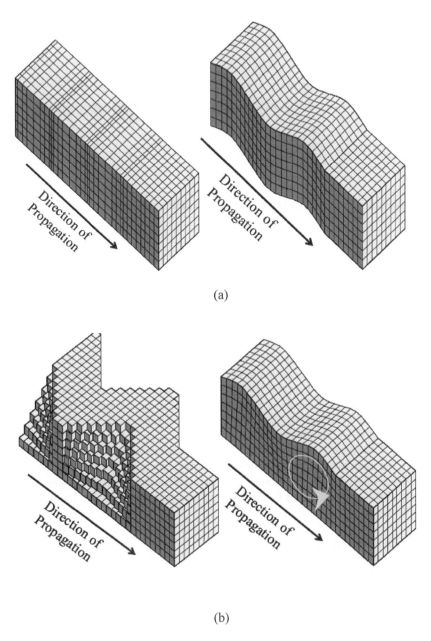

(a)

(b)

Figure 7.9 Seismic wave types: (a) Body waves;
(b) Surfaces waves (FEMA, Session 5, 2005)

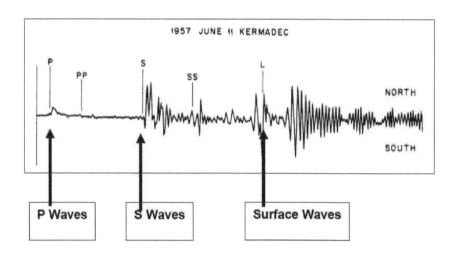

Figure 7.10 Typical recorded accelerogram (FEMA, Session 5, 2005)

energy from shallow shock and are usually the primary cause of destruction which results from earthquakes affecting densely populated areas (Kulhanek, 1990).

The first kind of surface wave is the *L wave*, which moves the ground from side-to-side, having a velocity of 2 to 4.4 km/sec. The second kind of surface wave is the *R wave*, which is a combination of P and S waves and moves the ground up and down and side-to-side in the same direction in which the wave is moving. The velocity of the surface waves is practically the same as the one of the L waves, which, generally, arrives to the site at the same time, with a gentle time in comparison with the first type.

Figure 7.10 illustrates some of the basic properties of recorded body and surface waves. The first is the P wave, followed by the S wave after some time. After a short time, one can see the increasing of ground motion amplitudes due to the arriving surface waves, L and R. Therefore, the surface waves are the largest signal in a seismogram.

7.3.2 Path Effects for Crustal Sources

The seismic waves lose energy as they propagate through the Earth along the travel path. The rate, at which the earthquake ground motion decreases with the distance, is a function of the source, seismic wave types, regional geology and inherent characteristics of the earthquake. These major factors affect the severity of the ground shaking at the site. The attenuation is very different, depending on the source type.

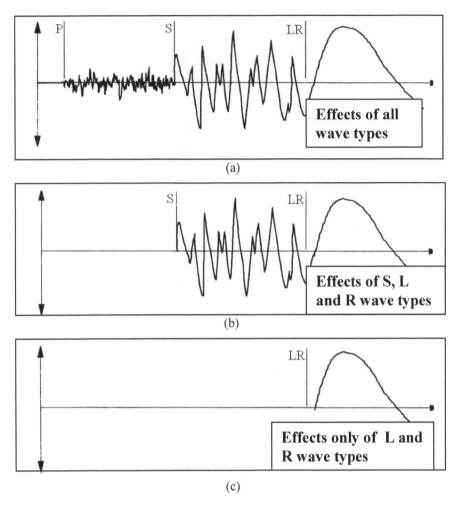

Figure 7.11 Accelerogram types in function of the distance from the epicenter:
(a) Near-source area; (b) Intermediate-field area; (c) Far-field area
(after Filiatrault, 1996)

In case of crustal sources, the source-site distance plays a leading role, considering that the attenuation is very different for the different wave types. Attenuation is very high for the P-waves, moderate for S-waves, and reduced for surface waves. Figure 7.11 presents the recorded acceleration types for near-source sites, where all the wave types are present, for intermediate-source sites, where the P-wave disappears due to the very important attenuation, and for the far-source sites, where the surface waves only remain (Filiatrault, 1996).

Therefore, in function of source-site distance the following *site classification* may be considered (Fig. 7.12):

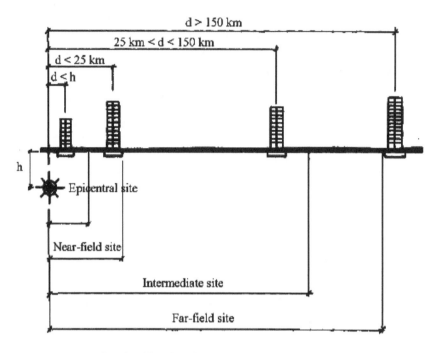

Figure 7.12 Site classifications in function of the distance from the epicenter
(Gioncu and Mazzolani, 2002)

- *Epicentral site*, including the area around the epicenter, generally with a radius equal to the source depth;
- *Near-source site* (near-field site), including an area within a distance of 25-30 km from the epicenter;
- *Intermediate-source site* (intermediate-field site), including an area within a distance of about 150 km from the epicenter;
- *Far-source site* (far-field site), including an area with distance more than 150 km.

In the following, the main effects of the path traveling are presented.

Traveled soil types. The path effects on ground motions depend on the percentage of the path travel through rock or through soft sediments (see Figure 7.1). Deviations from uniform horizontally layered crust model occur along the path of wave propagation from the fault to site. These deviations are produced due to the topography of basement rock, the path being a collection of sedimentary basins with alluviums, separated by irregular basement rock, forming mountains and geological and topographical irregularities. Using a map showing these distributions of rock, soft rock and alluvium, it is possible to characterize the transmission path for the fault site (Trifunac and Novikova, 1995a). The simplest procedure is proposed by Erdik (1995), who suggest to consider the propagation-path effects as a sum of different soil conditions depending on the source distance and to multiply the corresponding value by an amplification coefficient. The

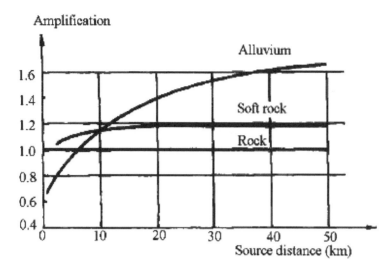

Figure 7.13 Amplification due to traveled soil conditions
(Gioncu and Mazzolani, 2002, after Erdik, 1995)

traveled soil can be classified in three classes: hard rock, soft rock and alluvium. Figure 7.13 indicates the amplification coefficients of peak-ground acceleration on alluvium and soft rock traveled soil with respect to the hard rock soil. No amplification results in the rock, a reduced amplification for the soft rock. Figure 7.13 indicates that attenuation occurs for alluvium soil at source distances, while for long distance amplification may be noticed. Amplification is observed on all distances in the soft rock.

Attenuations, due to the propagation-path effects; horizontal and vertical ground motions are presented in Figure 7.14 (Gioncu and Mazzolani, 2002, Ambraseys, 1995) for European strong data. One can see that the focal depth has a very important effect for the near-source sites, while small differences characterize the intermediate and far-source sites. Reduced attenuations are observed for near-source sites and an important reduction is observed for intermediate and far-source sites.

Earthquake duration. Another very important effect of the propagation-path is the increasing *duration.* One can see that the records in near-source sites are shorter in duration than the ones recorded in far-source sites (Fig. 7.15a). The duration of ground motion is the sum of the duration of the rupture process at the source and the increasing in duration due to propagation-path effects (Fig. 7.15b) (Trifunac and Novikiva, 1995, Lee, 2002). One can see that the increasing of duration is function of the soil type; the maximum increasing is obtained for alluvium soils. At the same time, the effect of soil type is maximum in the near-source areas and the difference disappears at far-source distance. The duration increases also in function of magnitude (Fig. 7.15c): for large magnitudes, the duration increasing is higher than for low magnitude.

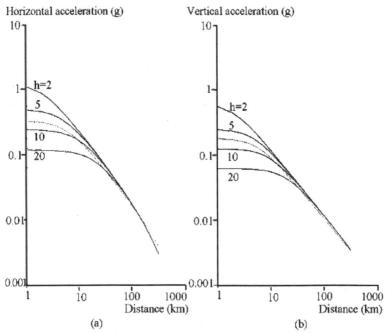

Figure 7.14 Attenuation of peak ground acceleration: (a) Horizontal attenuation; (b) Vertical attenuation (Gioncu and Mazzolani, 2002, after Ambraseys, 1995)

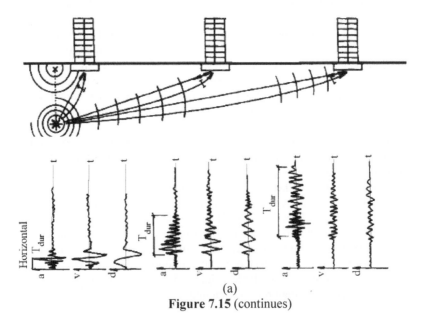

(a)

Figure 7.15 (continues)

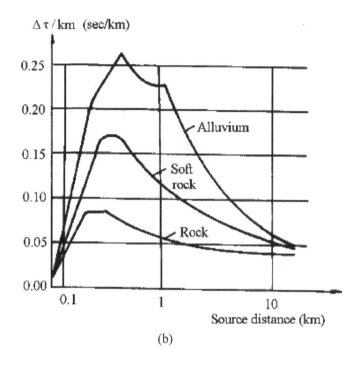

(b)

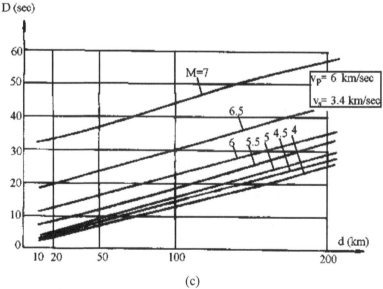

(c)

Figure 7.15 Duration increasing: (a) Function of epicentral distance; (b) Function
of traveled soil conditions; (c) Function of magnitude
(Gioncu and Mazzolani, 2002, afterTrifunac and Brandy, 1975)

7.3.3 Path Effects for Deep Sources

The amplitude of surfaces waves is the largest at or near the surface and rapidly decreases with the depth. Consequently, the shallow crustal earthquakes generate large surface waves, but with the increasing of the focal depth, the surface waves become smaller and smaller. Therefore, if surface waves usually dominate the seismogram for crustal earthquakes, they often become insignificant for the deeper source (intraslab earthquakes), the body waves being dominating (Kulhanek, 1990). These aspects concerning the differences in seismic wave attenuation provide the designer with a powerful tool for a reliable discrimination of crustal earthquakes against deep sources. Unfortunately, there are not many data about the influence of the propagation path. Generally, it is accepted that P-waves are not influenced by the path characteristics, having a very high frequency, very different from the traveled path. Contrary, this traveled path influences the S-waves, having higher frequencies (Elnashai and Papazouglou, 1997).

7.4 GROUND RESPONSE TO EARTHQUAKES

7.4.1 Significance of Local Soil Conditions

Observations of damage during numerous earthquakes have shown that local site conditions can have a strong influence on the performance of structures during earthquakes. Significant damage and loss of life has been directly related to the effect of local site conditions in recent earthquakes, such as the 1964 Niigata, 1977 Bucharest, 1985 Mexico City, 1988 Armenia, 1989 Loma Prieta, 1990 Iran, 1994 Northridge, 1995 Kobe, 1999 Kocaeli and 2009 L'Aquila earthquakes. Therefore, during the last decades the role of the local site effects has been widely discussed. Local site effects can be responsible for micro-seismic variations, which can be more important than the propagation-path influence. The ignorance or inappropriate consideration of site effects, which can be still encountered in practice, is, therefore, a professional negligence (Studer and Koller, 1995, Gioncu and Mazzolani, 2002).

The local site characterization is a very complicated multi-disciplinary problem. An attempt to classify the local site effects must consider the local layered deposits, soil type and shear wave velocity of each layer, topographic surface irregularities, alluvial valley and liquefaction.

7.4.2 Local Horizontally Layered Deposits

Local horizontally layered deposist can be identified by characterizing the multi-layers with different mechanical properties and thickness, corresponding to some soil categories. Figure 7.16a shows the soil layers for Eastern Bucharest (Lungu et al, 1997), which determine an important pulse with long periods in accelerogram due to the presence of deposits of lacustral layers (Fig. 7.16b) (Gioncu and Mazzolani, 2002, Ifrim et al, 1986). The presence of these pulses in ground

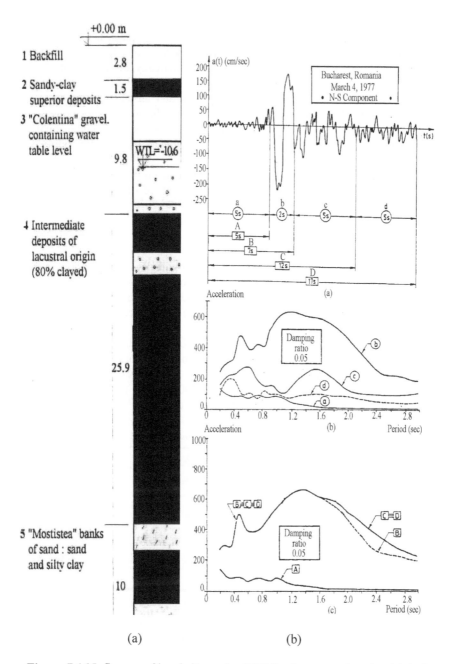

Figure 7.16 Influence of local site on the 1977 Bucharest earthquake: (a) Soil profile in recorded accelerogram (with the permission of Lungu et al, 1997); (c)) Acceleration pulse Gioncu and Mazzolani,, 2002, after Ifrim et al, 1986)

motions depends on the earthquake magnitude. For magnitudes larger than M 7, pulses are present in ground motions with natural periods of 1 to 2 seconds, while for magnitudes smaller than M 7, these pulses practically disappear (see Figure 4.9)

To consider the local site profile only by recorded ground motions is a very difficult task. Therefore, numerical methods were introduced in design, considering one dimensional site response. For a refined solution, two or three dimensional site analysis may be performed by means of finite elements but this methodology is rarely used in the design practice.

The one-dimensional method has now been introduced in the Building Code of Japan (Midorikawa et al, 2000, Otani, 2004). The verification procedure for developing the seismic design spectra includes (Fig. 7.17):

(i)   The basic design spectra defined at the engineering bedrock;
(ii)   The evaluation of the site response from the geotechnical data of surface soil layers.

The verification procedure considers the soil as a multi-degree-of freedom system, composed by the soil layers and transforms it into a single-degree-of freedom for which the spectrum is determined. Using this analytical model, the amplification factor for surface soil layers is performed and the design acceleration response at ground surface is determined (Fig.7.18).

The computer program SHAKE (in the new versions SHAKE-91 and SHAKE 2000) is available for the evaluation of the response of a horizontally layered soil system, by using an equivalent soil traveled by seismic waves (Schnabel et al, 1972, Idriss and Sun, 1993, SHAKE 2000, 2007). This program provides an approximation of the dynamic response of the soil deposit.

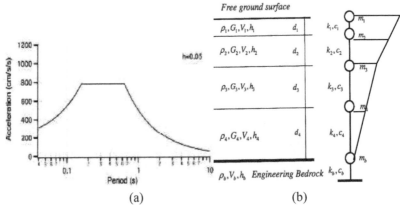

(a)   (b)

Figure 7.17 Analytical models: (a) Basic spectrum at engineering bedrock; (b) Soil model (Gioncu, 2006, after Otani, 2004)

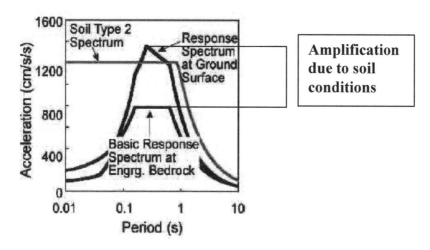

Figure 7.18 Amplification of basic spectrum: Design acceleration response at
ground surface (Gioncu, 2006, after Midorikawa et al, 2000)

7.4.3   Soil Type

The site classification scheme is based on the following parameters:
-    Type of deposit (i.e. different types of rocks and soils);
-    Average shear wave velocity;
-    Depth of layers.

There are different types of soil classifications, but the most close to the design
requirements is due to Borcherd (1994) and Rodriguez-Marek et al (2000). The
classification of soil conditions contains five soil classes:

- Class A refers to hard rock, igneous rocks, conglomerates, sandstones,
characterized by sheare wave velocity greater than 1500 m/sec and natural
period smaller than 0.1 sec;

- Class B considers gravelly soil and soft to firm rock, characterized by shear
wave velocity greater than 750 m/sec, natural period smaller than 0.2 sec and
soil depth of minimum 10 m;

- Class C refers to stiff clays and sandy soils and clays, medium stiff to hard
clays and silt clays, with shear wave velocity greater than 350 m/sec, natural
period smaller than 0.4 to 0.8 sec, and soil depth between 6 and 60 m;

- Class D considers soft clay soils, with shear velocity greater than 200 m/sec
and soil depth between 60 to 200 m;

- Class E refers to special soft soils (liquefiable soils, highly organic clays,
very high plasticity clays and soft soils), for which special geotechnical
investigations are recommended.

A similar classification, but with some differences in required velocities, is
presented in EUROCODE 8 (Sabetta and Bommer, 2002).

In the last decades there was a clear gap between seismologists and geotechnical engineers regarding the effects of soft soils, considering the nonlinear behavior of such soil type (Bard, 1995).

The presence of soft sediments produces a decrease of site dominant frequencies. For low and moderate earthquakes, the modifications refer only to the increase of the natural period. Contrary, for large earthquakes, this modification is accompanied also by the increase of site acceleration (Trifunac, 1990). Opposite effects, expressed by the decrease of amplification factors for peak acceleration exceeding 0.4g, are referred by Bard (1995), Liam and Matsunga (1995), Dobry et al (2000), Rodriguez-Marek et al (2001) (Fig. 7.19). For low rock accelerations on the order of 0.05g to 0.20g, corresponding to Mexico City and Loma Prieta earthquakes, the soft soil accelerations are 1.5 to 4.0 times greater than the rock accelerations. The amplification factor decreases as the rock acceleration increases, this phenomenon being directly related to the nonlinear behavior of soil (Faccioli, 1996).

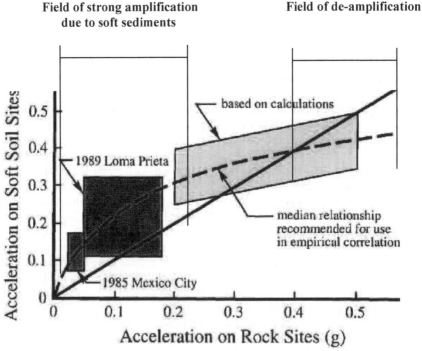

Figure 7.19 Maximum rock accelerations versus soil site accelerations (after Idriss, 1990)

A tendency to a de-amplification occurs for rocks with acceleration greater than 0.4g. More recent studies of Anastasiadis and Klimis (2002) for C and D site types have shown that the de-amplification occurs for 0.55g and 0.25g, respectively.

Therefore, the nonlinear inelastic soil behavior, even in the small acceleration range (for instance, in low to moderate seismic regions), introduces a complexity in evaluating the response of soil layers during the earthquakes. The coupling between earthquake characteristics and local site conditions may play an important role. It is essential to take into account in earthquake analysis the geotechnical aspects for hazard mitigation (Ansal, 1995).

7.4.4   Topographic Surfaces Irregularities

The effects of topographic amplification of seismic response have been observed in numerous earthquakes, where damage has been concentrated near cliff and ridge crests, due to the interference of waves causing very complex patterns of frequency-dependent amplifications. A characteristic example of increased earthquake damage close to the crest of a step-like topography is illustrated in Figure 7.20. In this case, during the 1980 earthquake in Italy, the damage of Ipirna, an Italian village, sitting at the top of a hill, was concentrated close to the crest of a steep slope, whereas it was insignificant in the direction away from the crest (Athanasopoulos et al, 1998). The main types of topographical surface irregularities are (Faccioli and al, 2002): slopes and cliffs, valleys and hills and ridges. Amplification of seismic waves due to topographical irregularities may be ascribed to different causes:

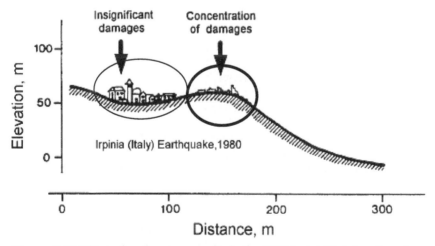

Figure 7.20 Effect of surface topography in the 1980 Irpina (Italy) earthquake
(after Athanasopoulos et al, 1998)

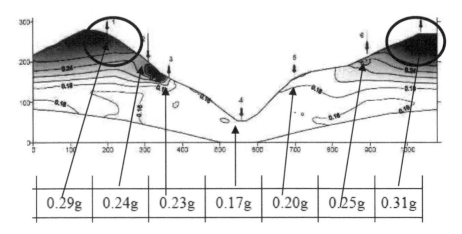

| 0.29g | 0.24g | 0.23g | 0.17g | 0.20g | 0.25g | 0.31g |

Figure 7.21 Spectral accelerations at characteristic points of the valley in a near-source site (after Klimis and Anastasiadis, 2002)

- *Focusing phenomenon,* due to the incidence of waves on a locally convex or concave surface profile;
- *Dynamic phenomenon,* producing a resonant motion of a whole hill or mountain, if the waves incident at the base meet some values;
- *Presence of local deposit* of unconsolidated soil for a limited extent.

Paolucci and Rimoldi (2002) have studied the case of the presence of a hill (Civita di Bagnoregio) in the Italian geological environment. The results show very important amplifications of ground motion on the hill, which are double those of the surrounding area.

The valley effect is studied by Klimis and Anastasiadis (2002). Two valley types with different geometry and mean slopes are theoretically analyzed (Fig. 7.21). The amplification at near crest free surface is about 1.4 times for milder slope inclination and higher, about 1.85 times, for steeper slope. Reduced amplifications are obtained for the intermediate-source sites, and the amplification effect disappears for the far-source sites.

7.4.5   Aluvial Basins

This aspect began to present interest after the 1976 Tangshan, 1985 Mexico City and 2009 L'Aquila earthquakes, when the alluvial basin produced an important amplification of the source accelerations. The finite lateral extent of soil surface layers generates surface waves at the edge and two-dimensional resonance patterns in the lateral direction and increases the amplitude, natural period and duration of ground motions. The influence of the variation in the basin dimensions and the effects of soil type and input rock motion characteristics are investigated by Rassem et al (1997). The differences in the maximum amplitude of accelerations for deep and shallow basins in the Beijing area are presented in Figure 7.22.

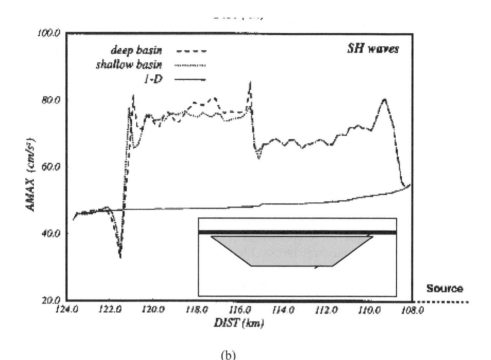

(b)

Figure 7.22 Effects of alluvial valley: Increasing of accelerations (after Panza et al, 1999)

The thick low velocity deposits (alluvium sands and clays poorly consolidated with high water content) are responsible for large increment of the acceleration values inside the basin, explaining the anomalous high micro-seismic intensity caused by the 1976 Tangshan earthquake (with magnitude M 7.8).

7.4.6   Liquefaction

The destructive power of liquefaction was abruptly brought to the specialists' attention by the two large earthquakes that occurred in 1964, the Great Alaskan (M 9.2) and Niigata (M 7.5). The liquefaction during the Alaskan earthquake produced large damage in highway and railway bridges. During the Niigata earthquake, the liquefaction phenomenon reduced the bearing capacity beneath many buildings, causing settlement and tipping (Youd, 2003). The seismic liquefaction of soils was noted in numerous other earthquakes: 1811-1812 New Madrid (producing the greatest seismic effects in the USA), 1989 Loma Prieta (USA), 1995 Kobe (Japan) and 1999 Kocaeli (Turkey).

In common usage, liquefaction produces the loss of strength in saturated, cohesionless soils due to the build-up of pore water pressures during dynamic loading. It results from the tendency of soils to decrease in volume when subjected to shearing stresses. Liquefaction occurs when the soil beneath the surface actually behaves like a fluid under strong shaking (Brandes, 2003, Yeats and Gath, 2004). Liquefaction refers to a process resulting in a loss of shear strength of soil due to the excess of pore water pressure and to the cyclical shearing. The soil particles initially have large voids between them. Due to shaking, the particles are relatively displaced against each other and tend to more tightly pack, decreasing the void volume. The water, which occupies the voids, being incompressible, comes under increasing pressure and migrates upward to the surface, where the pressure is relieved (Fig. 7.23) (Dobry, 1989).

As far as the structural behavior is concerned, the liquefaction during the earthquakes produces the buoyant rise of buried structures such as tanks, ground settlement and inclination of buildings (see also Figure 3.6) and failure of retaining walls.

The failure of ground by large displacements due to the liquefaction can be vertical, horizontal or a combination. During the failure, the surface layer commonly breaks into large blocks (Fig. 7.24), which move back and forth and up and down in the form of ground waves, as they progressively migrate horizontally (Youd, 2003). Ground oscillation occurs on the flat terrain in response to inertial forces within the broken blocks.

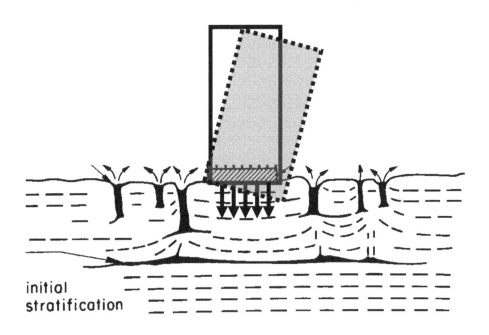

Figure 7.23 Effects of liquefaction: Migration of water toward the surface and inclination of buildings (modified after Dobry, 1989)

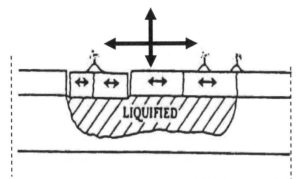

Figure 7.24 Mode of lateral permanent ground displacement (after Scawthorn, 1997)

The liquefaction effect is more significant for the strong earthquakes with long duration (large number of cycles). The minimum values of earthquake magnitude and peak acceleration required to generate liquefaction in function of soil characteristics are the following (Youd, 2003):

Magnitude	Acceleration for stiff sites	Acceleration for soft sites
M < 5.2	< 0.4g	< 0.1g
5.2 < M < 6.4	< 0.1g	< 0.05g
6.4 < M < 7.6	< 0. 05g	<0.025g
M > 7.6	< 0.025g	<0.025g

It is very clear that for low and moderate magnitudes, the influence of site type is very important. Contrary, for strong earthquakes, the influence of site type is insignificant. This means that the site type plays a very important role in the case of intraplate earthquakes.

7.4.7 Six-degree-of-freedom Ground Motions

The general motion of a body is uniquely specified by three components of displacement and three components of rotation. Nevertheless, it is standard to observe only translational motions and the study of rotations in the context of earthquakes has little attention, partly because rotational effects generated by earthquakes are considered to be small and partly because no instruments exist for directly measuring the absolute rotation (Igel et al, 2005).

The *translational motions* consider two horizontal components in perpendicular direction, (NS and EW or arbitrary) and one vertical component. If there are no problems regarding the direction for the vertical motions, for the horizontal ones it is not sure that the two selected perpendicular directions provide the maximum effects produced by an earthquake. Therefore, using the recorded values, it is very important to know the direction which yields the largest value. This value is measured by varying the orientation of an elastic oscillator in different directions (Erkay and Karaesmen, 1996, 1998, 2002). The ground accelerations are determined into two perpendicular directions, (1) and (2), corresponding to the recorded acceleration directions. By composing these values, one can plot a set of oriented values. The obtained three-dimensional representative diagrams are presented in Figure 7.25, which allows determining the upper value of oscillator response. The example of this methodology refers to the records of the 1985 Vina del Mar (Chile) earthquake (Vasquez, 1996). One can see that the spectrum is strongly influenced by the recorded direction (in the range of 0.6 to 1.0 periods), the largest acceleration being obtained for an orientation of 15 degrees from the recorded direction. Contrary, for degrees between 45 and 75, the largest accelerations occur in the range of 0.2 to 0.4 sec periods.

The *rotational component* of ground motions has been basically ignored in the past decades, compared to the substantial research in the field of translation components. Even the torsional oscillations of the Earth were predicted in the 19th Century, by examining the rotation of some objects on the ground surface (i.e. the rotation effects at different heights of some obelisks, Teisseyre, 2006). During the last decades, numerous observations pointed to rotation effects at the ground surface, so a new direction of research can be framed into the so-called *Rotational Seismology* (Lee et al, 2007). In 2007, the First International Workshop on Rotational Seismology and Engineering Application was organized in California-USA (USGS, 2007). Three main directions of research can be noted:

(i)   Development of fundamental theory.

(ii)   Providing additional information to Earthquake Engineering.

(iii)   Developing some special instruments to measure these Earth's rotations.

From the Structural Engineering point of view, the interesting aspects are related to the effect at free surface, rotation and tilt, produced by Love and Rayleigh waves. It seems that in the near-source of an earthquake the tilt effects are not always negligible compared to the effects of translation motions. Therefore, in these areas it is necessary to measure and to consider in the design all the six components of the motion. Contrary, in far-field areas these effects can be neglected (Graizer, 2005).

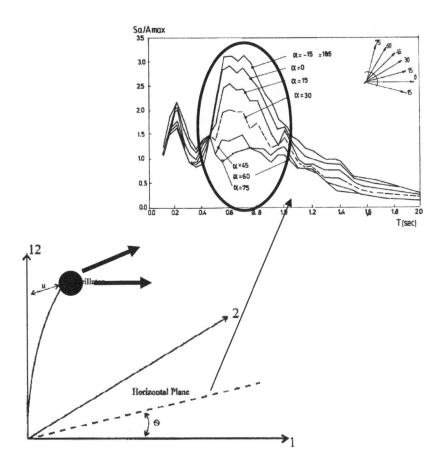

Figure 7.25 Parametric spectra for 1985 Vina del Mar Chile earthquake
(modified after Vasquez, 1996)

7.4.8 Spatial Variation of Ground Motions

The term spatial variation of ground motions denotes the differences in amplitude
and phase of seismic motions recorded over extended areas. The spatial variation
of ground motions has an important effect on the response of buildings and
lifelines, especially in long structures such as bridges, pipelines, communication
transport systems, etc. (Zerva and Zervas, 2002).

Spatial variation in ground motions arises from the following causes (Der
Kiureghian and Keshishian, 1996) (Fig. 7.26):

- *Incoherence effect*: loss of coherency of seismic waves due to scattering in heterogeneous medium of ground and superposition of the waves arriving from different point of the source, which is extended on a surface;
- *Wave-passage effect*: differences in the arrival time of waves at different surface points;
- *Attenuation effect*: gradual decay of wave amplitude with the distance due to geometric spreading and energy dissipation in the ground medium;
- *Site-response effect*: spatially varying local soil profiles and the manner in which they influence both amplitude and frequency content of the ground motion underneath each surface point.

These variations in ground motions produce differential motions of building foundations. The estimation of the effects of these motions on the structural response suggests that such effects are usually small. However, in the near-source areas, where the amplitudes of ground motions are large, and in the soft soils, the effects can be significant.

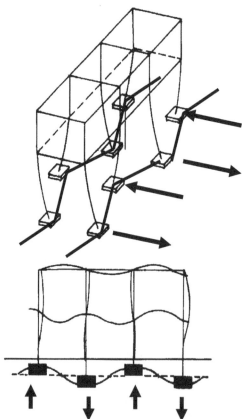

Figure 7.26 Wave effects: Torsional excitation of columns by passage of Love waves and Rocking excitation by passage of Rayleigh waves (Gioncu and Mazzolani, 2002)

Figure 7.26 schematically illustrates the column performance at the first story of a building during the passage of Love and Rayleigh waves, considering that the waves propagate along the longitudinal axis of the buildings. The first case is very important for long structures, while the second for high structures (Trifunac, 2006).

7.5 PECULIAR FEATURES OF NEAR-SOURCE GROUND MOTIONS

7.5.1 New Challenge in Seismic Design: Near-source Ground Motions

Due to the source position near to Earth's surface, the main severe values for the crustal interplate earthquake are the recorded strong ground motions around the epicenter and very strong attenuation for far-source areas. Up to now a large number of earthquakes have been recorded, but only a few of them are recorded within the distance of 15 km from this epicenter. All these records show some very important characteristics (i.e. forward rupture directivity and fling step), significantly different from those recorded away from the seismic source. Therefore, a new challenging research topic in Engineering Seismology and Earthquake Engineering is the characterization of *near-source earthquakes* (named also near-fault or near-field earthquakes). These earthquakes may shake major urban centers (see 1994 Northridge, 1995 Kobe, 1999 Kocaeli and Chi-Chi earthquakes) and, therefore, pose delicate problems and challenges for determining their main characteristics, considering that global urbanization (see Section 1.2) means that many urban centers are situated in the near-source area.

Therefore, very urgent challenges to improve the structural design methodologies for near-source earthquakes are required.

Examining the recorded ground motions during Northridge earthquake, it is very clear that the main characteristic of the recorded values for the near-source earthquakes is the velocity (Fig. 7.27). It is recognized that the definition of all the aspects related to velocity pulse problems for near-source earthquakes is very complex and very difficult to setup. Ground motions close to a ruptured fault can be significantly different from those further away from the seismic source. The near-source area is typically assumed to be within 15-30 km from a rupture fault. Within this near-source area, ground motions are significantly influenced by the rupture mechanism, the direction of rupture propagation relative to the site and possible permanent ground displacements resulting from the fault slip. Their effects are termed herein as *rupture directivity pulse* and *fling step*. These two characteristics of near-source ground motions result in *long period* and *large velocity pulse* in the velocity history and large step pulse in the displacement time history. So, the directivity pulse is a concentrated pulse like ground motion generated by constructive interference of S-waves traveling ahead of the rupture

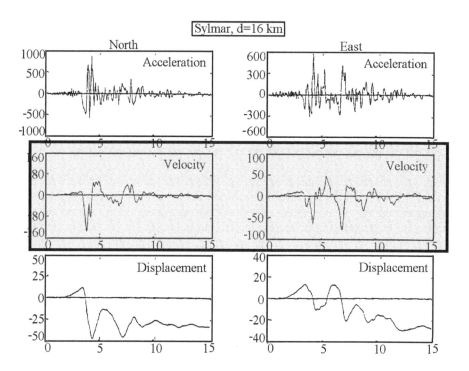

Figure 7.27 Characteristic records for near-source earthquakes: 1994 Northridge earthquakes (Gioncu and Mazzolani, 2002)

front, while the fling step is the permanent displacement of the ground near the rupture fault. The estimation of ground motions close to an active fault should account for these characteristics of near-source ground motions.

An earthquake is produced by a shear dislocation which begins at a given point on a fault (asperity) and spreads with a velocity being almost as large as the shear wave velocity. This phenomenon is known as *rupture directivity pulse*. The fault rupture tends to propagate slightly slower than the shear wave generated by slip and the rupture reinforces the shear waves as far as it propagates. If the rupture propagates towards a site (forward direction), then the energy arrives over a much shorter time interval than the duration of the entire rupture, whereas if the rupture propagates away from the site (backward direction), the energy arrives over a time interval which roughly matches the duration of the entire rupture. This feature follows the same principle as the Doppler Effect. Not all near-source earthquakes present this effect, because two conditions must be satisfied: the rupture front propagates towards the site and the direction of slip on the fault is aligned with the site. The propagation of fault rupture toward a site at very high velocity causes most of the seismic energy from the rupture to arrive in a single large long period pulse of motion (Somerville, 1997). This pulse of motion represents the cumulative effect of almost all of the seismic radiation from the fault. Figure 7.28 shows the directivity effect of ground velocity in strike-slip earthquake from two near-source recordings of the 1992 Landers earthquake (M 7.3). This recorded earthquake is the most representative one among the large number of records for the near-source effects, where one can see the differences in records in the forward and backward rupture directivity.

Due to the forward directivity, the Lucerne record shows a large brief velocity pulse of motion, while the Joshua Tree record, situated in the backward directivity region, consists of a long duration without significant velocity pulses and low amplitude record. Although this example refers to a strike-slip earthquake, the conditions required for forward directivity are also met in thrust earthquakes (resulting from subduction or collision faults). The alignment of both the rupture direction and the slip direction up the fault plane produces rupture directivity effects at sites located around the surface exposure of the fault. Consequently, it is a general case when all sites located near the surface exposure of crustal faults experience forward rupture directivity for an earthquake occurring on these faults.

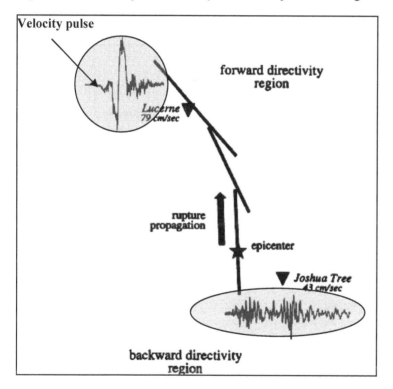

Figure 7.28 Directivity effect during 1992 Landers (USA) earthquake
(after Somerville, 1997)

7.5.2 Main Characteristics of Near-source Ground Motions

The near-source earthquakes have some very specific features, which preserve the characteristics of the source, due to the short source-site distance.

Rupture directivity. Figure 7.29 shows the effects of the rupture directivity for the 1999 Chi-Chi earthquake, which is the result of slipping of a thrust fault. One can see that the fault normal velocities of ground motions are larger than the

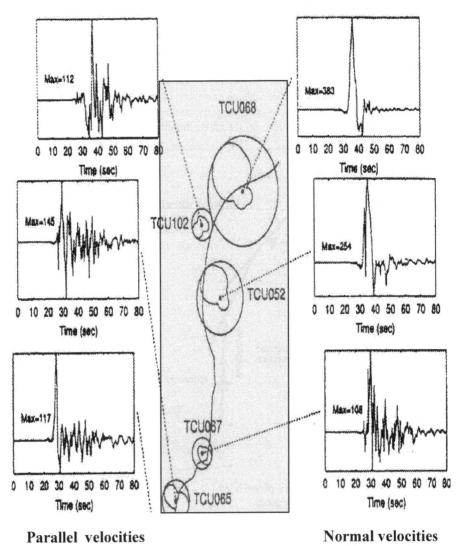

Parallel velocities   **Normal velocities**

Figure 7.29 Velocity time histories of maximum velocity direction for the Chi-Chi (Taiwan) earthquake (after Huang and Chen, 2000)

parallel ones. Contrary, the ground motions from the 1999 Kocaeli (Turkey) earthquake, recorded at Yarimca, show the near-source effects, which are characteristic for strike-slip earthquakes: ground velocities with almost the same values in all directions.

The largest dynamic ground velocity is coincident in time with the ground displacement and occurs over a time interval of several seconds. It is, therefore, necessary to treat the dynamic components of the seismic load as coincident loads (Somerville, 2002). Comparing the ground motions, resulting during these two different faulting mechanisms, the effects of rupture directivity are similar, with the exception of the fling slip displacements, which are not present in the case of thrust earthquakes. This indicates that, for the engineering purposes, it is not necessary to distinguish between different fault types in characterizing the near-source rupture directivity effects (Somerville et al, 1996). The key ground motion parameters for near-source earthquakes are: amplification of horizontal and vertical components, peak ground velocity, pulse period and number of significant pulses (Bray and Rodriguez-Marek, 2000).

Horizontal and vertical components amplification. The increase of the horizontal accelerations with the site distance from the fault, the period larger than 0.6 sec (for the earthquake magnitude M 7.0) are shown in Figure 7.30 (Somerville, 1997). One can observe that the influence of near-source increase with the ground motion period and it is reduced at the distance greater than 30 km. For large magnitudes, large periods and near source, important increasing can be noticed, over 50%.

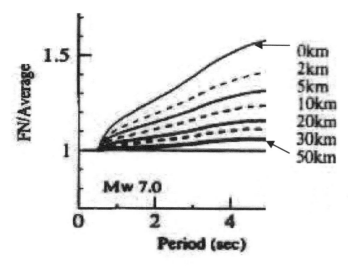

Figure 7.30 Amplification factors of ground motions due to near-source effect (after Somerville, 1997)

Vertical components. Another interesting phenomenon observed from the near-source earthquakes is the increasing of vertical accelerations. It is generally acknowledged that vertical motions are smaller than the horizontal ones. Recent observations during the last important earthquakes have demonstrated that at the near-source areas the vertical accelerations are sometimes greater than the horizontal ones, due to the *effect of last ball* (Fig. 7.31a) and *direct propagation* of P-waves (Fig. 7.31b) Many strong-motion recordings obtained in the near-source areas have exceeded unity and the widely used 2/3 rule-of-thumb (Fig. 7.31c) (Gioncu and Mazzolani, 2002, Elnashai et al, 1998). The ratio of the vertical to horizontal spectra is shown in Figure 7.32a as a function of distance and periods (Gioncu and Mazzolani, 2002, Bozorgnia et al, 1996) and distance and magnitude in Figure 7.32b (Gioncu and Mazzolani, 2002, Elnashai and Papazoglu, 1997). The increase in ratio at small distance from the fault in near-source areas can be dramatic (Gioncu and Mazzolani, 2002, Carydis, 2005).

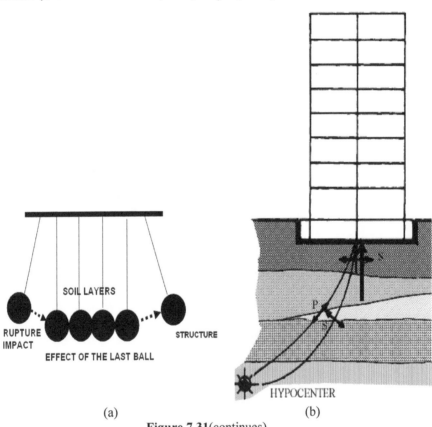

(a)   (b)

Figure 7.31(continues)

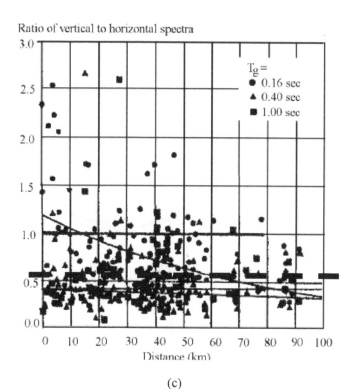

(c)

Figure 7.31 Vertical component: (a) Effect of last ball; (b) Influence of P waves
(Gioncu and Mazzolani, 2002); (c) Vertical-to-horizontal ratios, the Northridge
earthquakes (Gioncu and Mazzolani, 2002, after Hudson et al, 1996)

Peak ground velocity. Looking to Figures 7.27 and 7.29, one can note that the
velocity records are the most significant for characterizing the ground motions in
the near-source areas. It is a damage potential indicator, due to the high influence
on the material properties (strain rate) (Gioncu and Mazzolani, 2002). Near-source
records obtained during the 1979 Imperial Valley strike-slip earthquake contain
severe velocities. Figure 7.33a shows the location of the recording stations, while
Figures 7.33b-e present the recorded velocities of the fault-normal components in
the four stations: two can be considered as influenced by forward directivity, one is
in a neutral position and one is situated in backward directivity conditions. One can
see that the maximum-recorded velocity in the forward directivity is about 120
m/sec, while it is about 40 cm/sec only in the backward directivity (Mollaioli et al,
2006). During the 1989 Loma Prieta strike-slip earthquake (M 6.9), the recorded
velocity was about 179 cm/sec (Lexington Dam station); for the 1994 Northridge
thrust earthquake (M 6.7), the maximum recorded velocity was 174 cm/sec
(Rinaldi station) and for the 1995 Kobe strike-slip earthquake (M 6.9), a velocity
of 174 cm/sec (Takatori station) was recorded (Fu and Menun, 2004).

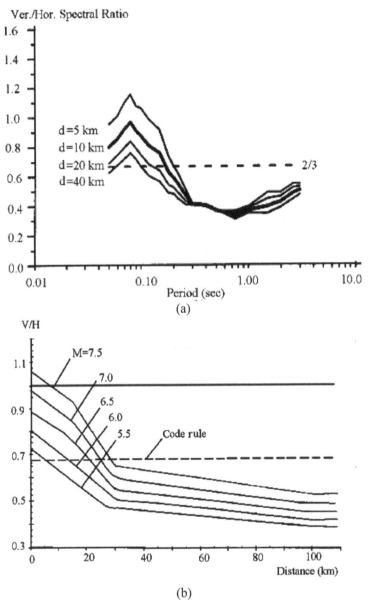

(b)

Figure 7.32 Ratio of vertical to horizontal spectra: (a) Influence of periods and distance; (b) Influence of magnitude (Gioncu and Mazzolani, 2002, after Bozorgnia et al, 1996, Elnashai and Papazoglou, 1997)

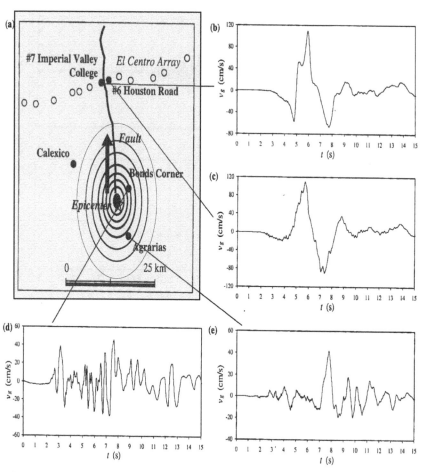

Figure 7.33 Near-source records of the 1979 Imperial Valley earthquake (after Mollaioli, 2006)

The largest values of peak ground velocity which could be generated in the near-source areas are estimated around 300 cm/sec (Bommer and Alarcon, 2006). The majority of the velocity records for the Chi-Chi (Taiwan) thrust earthquake (with a very dense number of recording stations, which allow having 420 records) framed into this limit (Fig. 7.34) but the recorded value of 380 cm/sec from the TCU068 station exceeded this limit (RMS, 2000). The presumed maximum value for peak ground velocity can reach even 600 cm/sec, but only for interplate earthquakes (Fig. 7.35). For intraplate earthquakes, the values of peak ground velocity are reduced, generally under 50 cm/sec

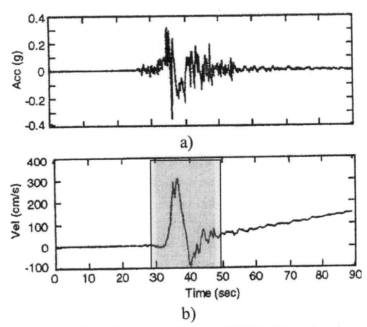

Figure 7.34 Time-history records from 1999 Chi-Chi earthquake
(Gioncu and Mazzolani, 2006)

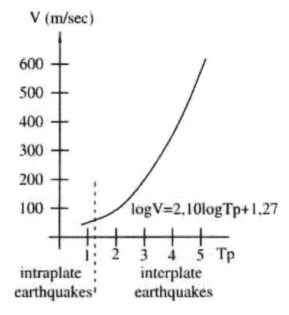

Figure 7.35 Ground velocity in near-source (Gyorgyi et al, 2006)

Velocity pulse period. Looking to Figure 7.35, which shows the velocities recorded on the near-source areas, on can see that the main feature of ground motions is the long-pulse pattern of velocity records. Generally, these pulses are around 1-2 sec, but some exceptional very long values are recorded in some cases: 1978 Tabas (Iran), 5.26 sec, 1989 Loma Prieta (Gilroy station), 4.24 sec, 1992 Landers, 8.23 sec, 1999 Kocaeli (Gebze station), 6.47 sec, 1999 Chi-Chi (TCU129 station), 7.41 sec (Menun and Fu, 2002a,b, Fu and Menun, 2004). The pulse period dependence on the magnitude (Fig. 7.36) is observed (Gyorgyi et al, 2006, Stewart et al, 2001, Rodriguez-Marek, 2000). The influence of the pulse period on the acceleration and velocity spectra is presented in Figure 7.37. One can see that short pulse periods and low magnitudes create important amplifications for short structure periods and low amplification for large magnitudes and pulse periods. Contrary, velocities increase with the increasing of magnitudes and pulse periods (Somerville, 2003). This remark is very important, showing that for long pulse periods (this is the case of many near-source crustal interplate earthquakes), the velocity and not the acceleration is the main factor affecting the structure behavior.

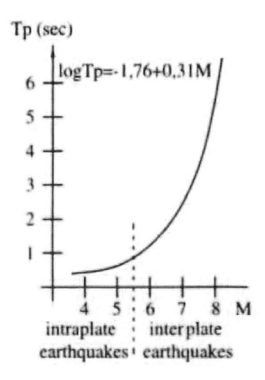

Figure 7.36 Pulse periods-magnitudes (Gyorgyi et al, 2006)

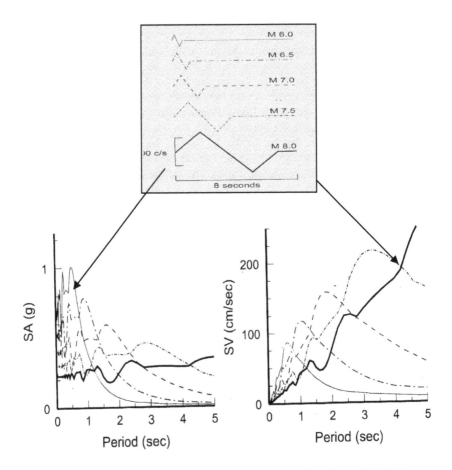

Figure 7.37 Influence of pulse period on the acceleration and velocity
(after Somerville, 2003)

Number of significant pulses. The number of significant pulses is defined as the number of half-cycle velocity pulses crossing the axis, whose amplitudes are at least 50% of the peak velocity of ground motion (Fig. 7.38). Examining the same recorded earthquakes (see Figs. 7.27 to 7.29 and 7.33), one can observe that the main characteristic of near-source earthquakes is the reduced number of significant velocity pulses satisfying these criteria and that there are cases with one, two, three or maximum four pulses. The number of significant pulses in fault normal component of 48 near-fault records (including Californian, Japanese and Turkey earthquakes) is presented in Figure 7.39 (Rodriguez-Marek, 2000, Stewart et al, 2001).

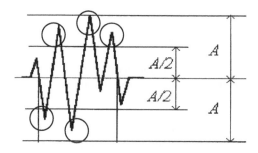

Figure 7.38 Definition of significant pulse (after Aptikaev and Erteleva, 2005)

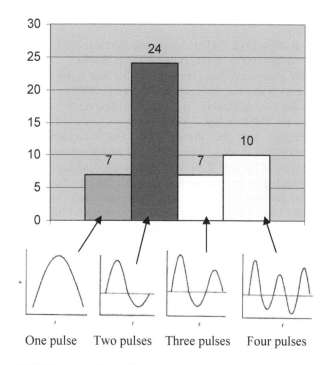

One pulse   Two pulses   Three pulses   Four pulses

Figure 7.39 Number of significant pulses (after Rodriquez-Marek, 2000)

The majority of records contain two significant pulses (i.e., one full cycle of pulse type ground motion). Somerville (1998) suggests that the number of half pulses in the velocity time history might be associated to the number of asperities in a fault. This is a difficult phenomenon to be estimated a priori. There are no models currently available for predicting the number of significant pulses in the velocity time history. For the majority of cases, the number will vary from 1 to 3, but the consideration of one or maximum two half-cycle pulses is a good value to be used in seismic evaluation (Stewart et al, 2001).

7.5.3 Analytical Models for Velocity Pulses

To generate an analytical model for velocity pulses, one must specify the values of several parameters which control the ground motion in the above-mentioned near-source areas (peak ground velocity, pulse period and number of pulses). The analytical models use the so-called *wavelet* analysis to extract the largest velocity pulse from a given ground motion (Baker, 2007). The analytical expression of the time variation of velocity must be established so that it fits the recorded variation (Fig. 7.40) (Fu and Menun, 2002).

Some proposals for analytical model of wavelet are the following:

- *Polygonal variation* (Fig. 7.41a) is proposed by Alavi and Krawinkler (2000) for symmetrical waves and by Mateescu and Gioncu (2000) for symmetrical and asymmetrical waves;

-*Sinusoidal variation* (Fig. 7.41b) (Sasani and Bertero, 2000);

-*Polynomial variation* (Fig. 7.41c) (Gyorgyi et al, 2006);

-*Four different trigonometrical functions* for the four pulse types (Makris and Chang, 1998, Agrawal and He, 2002, He and Agrawal, 2008);

-*Five-parameter mathematical expression* (Mavroeidis and Papageorgiou, 2002, Menun and Fu, 2002a,b, Fig. 7.42).

Some conclusions can be drawn from the comparison among those representative analytical models. The polygonal and sinusoidal function are the simplest, but much different from the earthquake motions. The polynomial variation is constructed to be symmetrical with reference to the maximum velocity pulse. The model adopted by Makris and Chang does not consider the alterable amplitudes of velocities. The Mavroeidis and Papageorgiou (2002) and Menun and Fu (2002 a,b) and Fu and Menun (2002) proposals may consider this variation of pulse amplitudes.

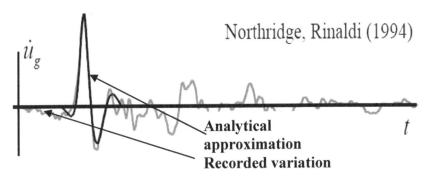

Figure 7.40 Analytical and recorded velocities (after Fu and Menun, 2002)

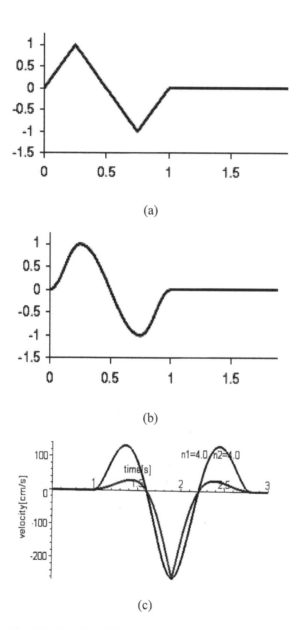

(a)

(b)

(c)

Figure 7.41 Simplified pulses: (a) Polygonal variation (Mateescu and Gioncu, 2000); (b) Sinusoidal variation (after Sasani and Bertero, 2000); (c) Polynomial variation (Gyorgyi et al, 2006)

$$\ddot{u}_m(t;\theta) = V_p \exp\left[-n_1\left(\frac{3}{4}T_p - t + t_0\right)\right]\sin\left[\frac{2\pi}{T_p}(t-t_0)\right] \qquad t_0 < t \le t_0 + \frac{3}{4}T_p$$

$$= V_p \exp\left[-n_2\left(t - t_0 - \frac{3}{4}T_p\right)\right]\sin\left[\frac{2\pi}{T_p}(t-t_0)\right] \qquad t_0 + \frac{3}{4}T_p < t \le t_0 + 2T_p$$

$$= 0 \qquad\qquad\qquad\qquad\qquad\qquad\qquad\qquad\qquad\text{otherwise}$$

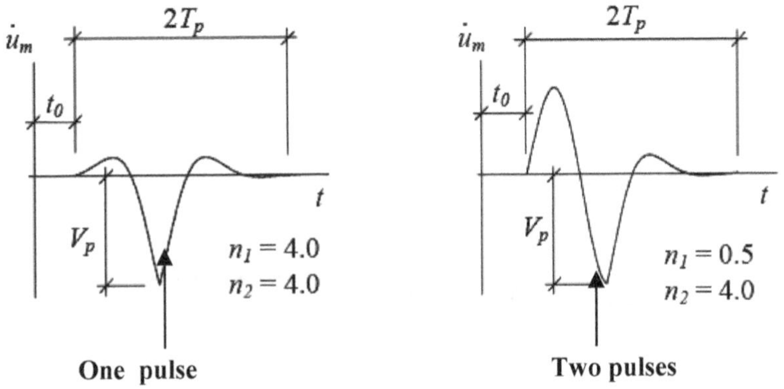

Figure 7.42 Menun's proposal (after Menun and Fu, 2002 a,b)

Another analytical model of ground motion pulses is the one presented by Agrawal and He (2002), He and Agrawal (2008). It is very important to notice that this proposal can generate various forms of pulses, by appropriately selecting pulse parameters.

7.5.4 Numerical Simulations: 3-D Computational Methods in Engineering Seismology

During the last decades, two earthquakes have been a reminder about what can be still learned about the fault rupture effect and the resulting ground motions. The 1994 Northridge earthquake produced the destruction which is possible when the fault rupture arises below a populated area. Fortunately, most of the seismic energy propagated away from the center of population. The recorded ground motions in the most affected area are very scarce, showing the difficulty of achieving effective network stations. In additions, the singular very high values recorded at the Tarzana station seem to be produced by local site conditions. One year later, during the 1995 Kobe earthquake, the rupture propagated toward the city, showing that even a moderate earthquake can cause substantial damage. Records from these two events significantly increased the knowledge about the ground motions close to source. But this information was not enough to underline the differences between the two earthquake types, knowing that the Northridge earthquake was produced by a thrust fault source, while the Kobe earthquake by a strike-slip fault source.

The solution for this complex problem is based on the use of computational methods in seismology in order to model the fault rupture and the propagation of seismic waves. Seismology and Engineering Seismology have been strongly affected by the set-up of high-performance computing tools (Bielak and Ghattas, 1999), especially related to the activity for searching for petroleum. During the first step, only some supercomputers were used. Now, the developing software for parallel supercomputers allows an increase in the amount of obtained data. Simulation involving hundreds of thousands to millions degree of freedom requires hundreds of megabytes to gigabytes of memory and billions of floating point operations. Luckily, only the parallel computer provides a suitable environment for solving such problems by distributing both the storage and computing among many processors (Bao et al, 1996). Therefore, only by using the computing centers equipped with a large number of supercomputers (such as Caltech Center, Advanced Computing Research, CACR, at the California Institute of Technology, Pasadena) are able to solve this huge computational problem.

Using the 256 processors of Caltech Center, the simulation of dynamic fault ruptures was performed by Aagaard (2000). The discretization of the considered volume of soil (Fig. 7.43) is using a hexahedral (six-sided) and tetrahedral (four-sided) finite-element mesh with special elements for fault plane, which impose a dislocation in finite-element model.

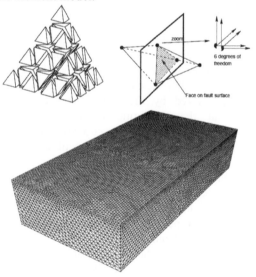

Figure 7.43 FE model for simulating the wave propagation:
Finite element, element for fault plane and soil mesh
(after Aagaard et al, 2001a)

The main characteristics of soil (velocity and density) are also considered in the analysis. In order to have an image of the dimensions of this computational problem, it must be mentioned that, during the modeling of the fault rupture, the discretization contains 9 million elements, 1.7 million nodes and 5.1 million degrees of freedom. Using this numerical approach, the Northridge (USA), Chi-Chi (Taiwan) and Denali (Alaska) earthquakes were modeled (Aagaard et al, 2001c, Aagaard et al, 2004, Anderson et al, 2003). The numerical modelling was used also to analyze the characteristics of different source types.

Another methodology for the simulation by computer of strong ground motions near the fault was presented by Miyatake (2000). This method is based on fourth-order 3-D finite-difference simulation method with staggered grids, which allows determining the spatial pattern of ground velocity vectors near the fault.

In the future, the ground motion numerical simulations will likely play an increasingly important role in the Engineering Seismology and Earthquake Engineering.

7.5.5 Characterization of Near-source Ground Motions with Finite Element Simulation

Due to the different influencing factors affecting the recorded ground motions by alteration of values caused by the source features, it is very difficult in practice to determine these values and to state the differences between the primary characteristics of the different source types. Additionally, the scarce coverage of recording stations limits the ability to capture ground motions close to fault ruptures, where the influence of these factors is reduced. Therefore, when in the future the ground motion numerical simulation will be able to eliminate the parasite influences, it will likely play an increasingly important role in the structural design, helping to understand the nature of these ground motions, including the variability caused by the changes in the seismic parameters.

Using the facilities offered by the supercomputers Centre for Advanced Computing Research at the California Institute of Technology and the implementation of parallel processing, Aagaard and his team (2000 to 2004) performed a systematic analysis of the influence of the source parameters on earthquakes, such as:

- Source type: thrust (including subduction and collision faults) and strike-slip faults;
- Source parameters: fault dip and rupture surface dimensions (for thrust faults), rupture speed, peak slip rate, hypocenter location, distribution of slip, average slip, fault depth, etc.

The numerical simulations refer to the characteristics of long period near source ground motions for thrust (blind and surface rupture types) fault, with M 6.6-7.0 magnitudes and strike-slip fault, with M 7.0-7.1 magnitudes.

Blind thrust faults (Fig. 7.44). The first analyzed problem is the effect of forward directivity (Aagaard, 2000). The examined Earth block is a region of 60x60 km down to a depth of 24 km. The rupture surface is burying one with the dimensions of 28 km long and 18 km wide and 8.6 km below the surface and a slope of 23 degrees. This rupture surface is close to the Elysian fault, which is assumed to have produced the 1994 Northridge earthquake. Figure 7.44 shows the evolution of horizontal velocity at the site surface and Figure 7.45 presents the evolution of vertical velocity. The results of the numerical simulation of the rupture evolution along the rupture surface (during 6 seconds) and the influence of this rupture on ground motions are presented in Figure 7.46. One can see that, even if the fault is blind, at the end of ground motions some residual ground deformations remain at the affected surface. The duration of ground motions was about 22 seconds.

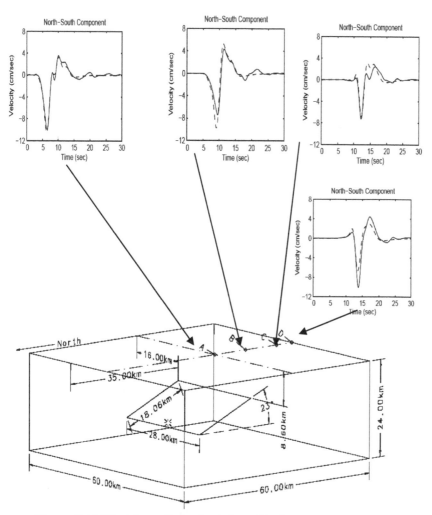

Figure 7.44 Blind thrust fault: Evolution of horizontal velocity at surface
(after Aagaard, 2000)

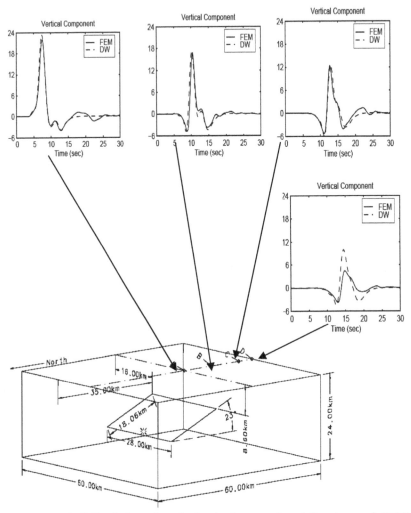

Figure 7.45 Evolution of vertical velocity at surface (after Aagaard, 2000)

The maximum horizontal and vertical ground motions along the directivity line across the top of the fault provide a good indication of the severity of these motions near to the surface rupture. There are some important problems which must be underlining.

The first examined problem is the *directivity effect*. The magnitude of the velocity vector showing the forward directivity is given in Figs 7.44 and 7.45, at different surface sites. The velocity variation corresponds to pulse one with a half-semi cycle. Site A lies directly above the top of the fault and site D lies at the block boundary, at 25 km from the top of the fault One can observe that the vertical components of velocity are higher than the horizontal ones; the maximum values resulting in the site directly located above the top of the fault.

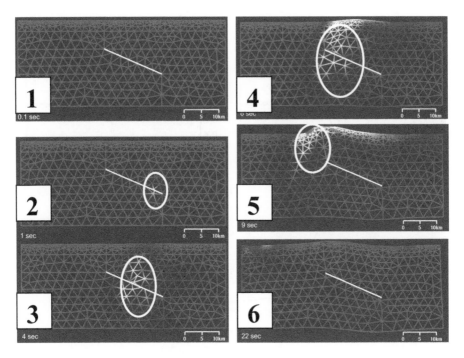

Figure 7.46 Evolution of rupture along the rupture surface (after Aagaard, 2003)

The attenuation of these velocities is very strong at site D, being only a quarter of the maximum value. Contrary, the horizontal velocity components remain very high, showing a reduced attenuation with the distance from the source.

The second examined problem is related to the *differences between normal and parallel ground motions* (Aagaard, 2000). For the same Earth block the velocities at two sites S1 and S2 are examined (Fig. 7.47). Site S1 lies 10 km North of the North end of the fault surface and site S2 lies 10 km East of the fault center. One can see that only the velocities of site S1, situated on forward directivity, have important pulse velocity values. The site S2 is practically unaffected by the ground motion. The maximum horizontal velocity obtained at 5 km from the top of the fault is about 140 cm/sec.

The third analyzed problem refers to the *influence of dip angles* (Aagaard et al, 2001c). The results show that the maximum horizontal value is obtained for a dip angle of around 60 degrees and the horizontal velocities are higher than the vertical ones in the field of 90 to 60 degrees. Contrary, the vertical velocities have the highest values for 45-50 degrees, being higher than the horizontal ones in this field (Fig. 7.48). These results confirm the great values of the vertical components recorded in some seismic stations during the Northridge and Taiwan earthquakes.

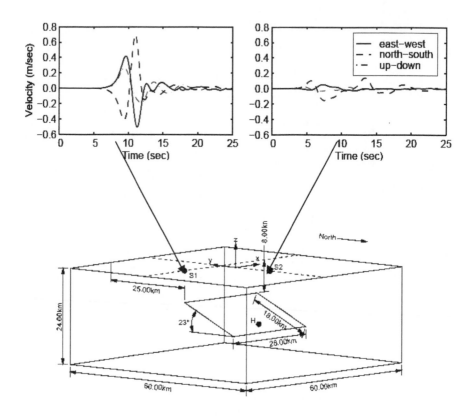

Figure 7.47 Differences between normal and parallel ground motions
(after Aagaard, 2000)

Strike-slip faults. The geometry of the strike-slip fault (Fig. 7.49) corresponds to the 1992 Landers earthquake (M 7.2). The analyzed Earth block is 100 km long, 40 km wide and 32 km deep. The vertical rupture surface corresponds to 60 km long and 15 km wide. The velocities on the ground surface give a clear picture of the effect of directivity on the ground motions. Figure 7.50 shows the numerical simulation of the ground motion from the beginning to the end of the earthquake. One can see that the main characteristic of the ground motion is the important lateral oscillations. Therefore, very important lateral ground motions perpendicular to the fault occur, larger than the ground motions along the fault. The total duration of earthquake was about 50 sec

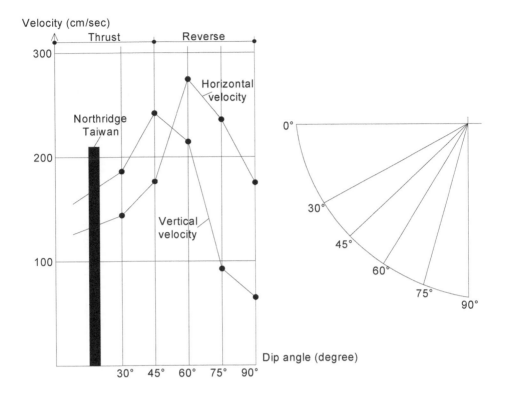

Figure 7.48 Influence of dip angles

In order to examine the velocity in different positions, two sites are considered: S1 lies 10 km North to the end of the fault and site S2 lies 10 km East to the fault center. The main shear wave at S1 has velocity components perpendicular to the fault plane and increases as the rupture propagates, the normal velocity having very reduced values. Contrary, at the site S2, the normal and perpendicular velocity components have comparable values. At site S1 the peak velocity is 2.6 times greater than the peak velocity at site S2, due to the directivity effect. One can see that the velocity pattern has 3-4 semi-cycles and the pulse period is around 4 seconds. The maximum obtained velocity results of about 290 cm/sec at a site located 6.7 km South and 0.5 km East of the North end of the fault. This value confirms the large recorded velocities during Landers and Kobe earthquakes.

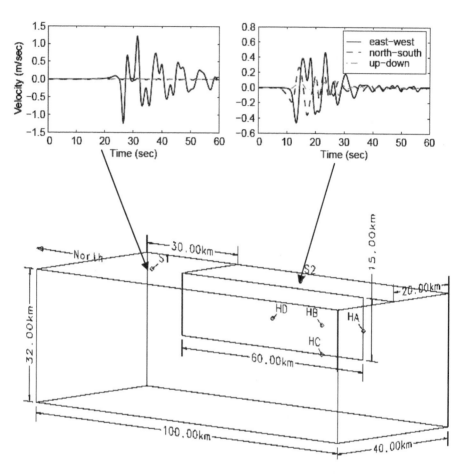

Figure 7.49 Strike-slip earthquakes (after Aagaard, 2000)

Comparison between these two earthquake types. The numerical simulation offers the opportunity to compare the two earthquake types and to underline the main differences, comparison which was practically impossible from the recorded values, due to parasite influences. The analysis of the obtained results can lead one to conclude that some important differences exist, which must be introduced in the design specifications.

  - For thrust earthquakes the dominant velocities are the vertical ones, while for the strike-slip earthquakes the main velocities occur in plan normal to the rupture surface.
  - The duration of strike-earthquakes is larger than the one of the thrust earthquakes.

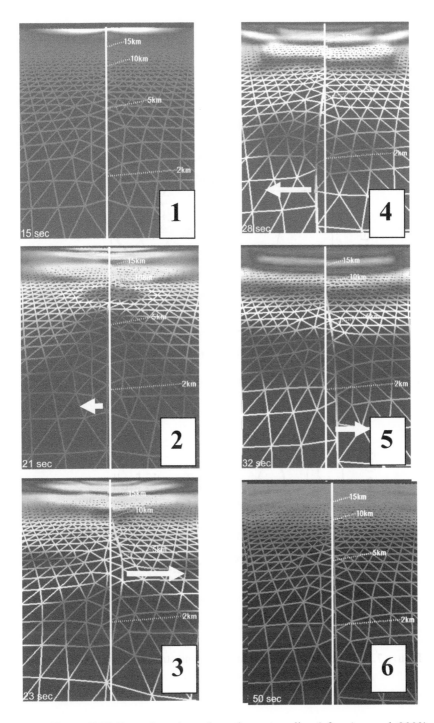

Figure 7.50 Ground motions along the rupture line (after Aagaard, 2003)

- The pulse velocity has only 1-2 half-cycles for thrust earthquakes and 3-4 half-cycles for strike-slip earthquakes.
- Resulting radiated energies (Fig. 7.51) are larger for strike-slip earthquakes in comparison with the thrust earthquakes.

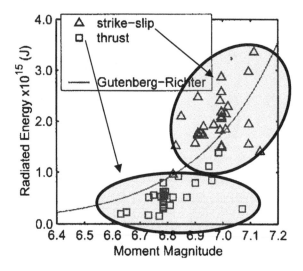

Figure 7.51 Radiated energy for thrust and strike-slip earthquakes (after Aagaard, 2000)

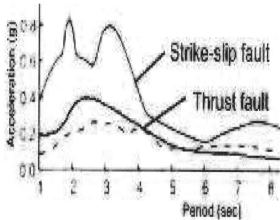

Figure 7.52 Acceleration spectra for thrust and strike-slip earthquakes (after Aagaard, 2000)

- Amplification of horizontal accelerations (Fig. 7.52) is 0.4 for thrust earthquake, at 2.6 sec period, and 0.82 for strike-slip, being maximum at around 3 seconds. In both cases, the amplification is more severe in site S1 than in site S2, due to the effect of directivity.

Considering all these aspects, a very important conclusion results: in comparison with the thrust earthquakes, the strike-slip earthquakes are the most damaging, explaining the dramatic effects of the strike-slip 2004 Kobe earthquake, in comparison with the effects of the thrust 2004 Northridge earthquake.

7.6 GROUND MOTIONS PECULIARITIES OF CRUSTAL INTRAPLATE EARTHQUAKES

7.6.1 Intraplate versus Interplate Earthquakes

The intraplate earthquakes occur in the so-named *Stable continental regions*, away from the boundaries between tectonic plates. The occurrence of these earthquakes apparently violates the plate tectonic model, which considers that the earthquakes occur only at these boundaries. Unfortunately, the design for earthquake resistance has only recently been required by codes in regions of low-to-moderate seismicity, contrary to the areas of high seismicity. The research works on the seismic response of structures are mainly concentrated on the cases of recent earthquakes that have occurred in zones with important seismicity, where the seismic activity is very frequent and can strongly affect human lives. Therefore, such research is not very interested to study the moderate seismicity effects.

There are many zones in the world where the earthquakes do occur, but either rather infrequently or with a moderate intensity, so that they tend to be perceived more as "accidental" than regular (Pinto, 2000). In Europe, the crustal intraplate earthquakes are the most frequent and this finding is very important, due to their particularities. The most frequent ground motions have magnitudes from 4 to 5 (Fig. 7.53a), framing into the category of *low-to-moderate earthquakes*. The majority of the records are from 0 to 25 km, 60 percent of them being in the range from 4 to 14 km (Fig. 7.53b). Intermediate earthquakes occur in Romania, Greece and Italy only (Gioncu and Mazzolani, 2002).

Figure 7.54 reports the histograms of seismic intensities of all the seismic events in Germany, Switzerland and Greece, updated to 1985 (Pinto, 2000). All these countries are situated in diffuse seismic zones, but Germany is far from the collision seismic zone of the Alps, Switzerland is near to this zone, and Greece is situated in a special diffuse zone, having various fault types. The number of events is very small for Germany, greater for Switzerland, but reduced in comparison with Greece. The maximum earthquakes' intensity in Germany and Switzerland is about VI to VII, while for Greece the intensities reach IX. Therefore, the comparison with Greece clearly suggests that in the former two countries the seismic hazard should be treated in a different way than in Greece.

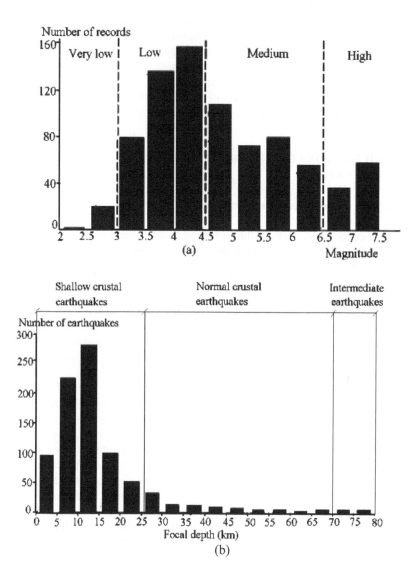

Figure 7.53 Distribution of European earthquakes: (a) In terms of magnitude; (b) In terms of focal depth (Gioncu and Mazzolani, 2002)

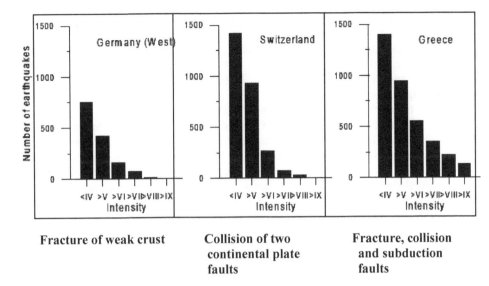

Fracture of weak crust   Collision of two   Fracture, collision
                                continental plate   and subduction
                                faults   faults

Figure 7.54 Histograms for intensity in three European countries: Germany,
Switzerland and Greece (after Pinto, 2000)

In the last few years, the understanding of earthquakes that occur in low to
moderate zones has improved, as Eastern North America, Central and Northern
Europe, South of China and Eastern Australia, made progress in the establishment
of the main characteristics of these earthquake types (Mooney et al, 2004, Schulte
and Mooney, 2005, 2007). Consequently, the results derived from these analyses
show that the recorded intraplate earthquakes were generally different from the
interplate-recorded earthquakes. Unfortunately, the information referring the
characteristics of these intraplate earthquakes remains until now very reduced in
comparison with that existing for interplate earthquakes.

Although only about 5% of the global seismic energy is released in continental
interiors by intraplate earthquakes, some of the most damaging earthquakes
occurred there (see Tangshan and Sishuan-China and Bhuj-India), due to the near-
source effects and the local site conditions in the urbanized areas. The great limit in
the structural design for moderate earthquakes is that the results obtained for the
other earthquake types, mainly for interplate earthquakes, are normally extended
also to these earthquakes. In the last decades a significant progress has been
reached in the understanding of the seismic risk for intraplate earthquakes,
especially based on studies of earthquakes that have occurred in Central United
States and Northern Europe (Gioncu, 2008).

7.6.2 Low-to-moderate Near-source Ground Motions of Intraplate
Earthquakes

First of all, one must mention that the following characteristics of intraplate
earthquakes refer only to the ones produced by rifts. When the intraplate

earthquakes occur along some ancient faults, the main characteristics correspond to the interplate collision faults.

The catalogue for crustal earthquakes, with magnitude larger than M 4.5, produced in stable continental regions, contains 1373 events from 495 to 2003 (Schulte and Mooney, 2005, 2007). These earthquakes are classified in the following categories:

- Interior rifts, if located within 20 km of identified rifts.
- Rifted continental margins, if located within 20 km of these margins.
- Non-rifted crust, if located further than 50 km of any of the above.

If one compares this classification with the one presented in Section 5.3, the interior rifts correspond to the normal faults produced in zones of crust weakness, non-rifted crust to reversal-thrust faults, while the rifted continental margins to faults produced in subducted and collided zones. From the analysis of these data, one can find that 27% are associated to the rifted crust, 25% to rifted continental margins, 36% to non-rifted crust and 12% remain uncertain. So, over half of all events are associated with the two rifted types. From 1373 earthquakes presented in the database, only 118 earthquakes have magnitude larger than M 6 and 14 exceed magnitude M 7 (all in rifted crust). Examining the map of intraplate earthquakes, one can observe that they are not evenly distributed and several zones of concentrated seismicity seem to exist. So, 12 regions are responsible for 74% of all the events and 98% of the seismic energy is released in these zones. An analysis of the main intraplate earthquake characteristics is performed by Gangopadhyay and Talwani (2003) and Stepp (1996).

Situated in crustal zones, the ground motions must have the features of near-source earthquakes, but with some modified characteristics. Figure 7.55 shows the case of Banat (Romania) earthquake recorded on 11 December 1991 (Gioncu, 2000). The records refer to the Timisoara site, situated 40 km from the epicenter, and to the Banloc site which is the place of the epicentre. In the epicentral Banloc area, one can see that their records have the characteristics of near-source ground motions (velocity pulse) (Fig. 7.55), but having very short period (0.2 to 0.4 sec), low accelerations and short duration. The records of the Timisoara site have the characteristics of intermediate–source area, due to the attenuation effect. The same characteristics of intraplate earthquakes were observed for other earthquakes. So, the intraplate ground motions have the features of near-source earthquakes, but with reduced periods, accelerations and shorter duration.

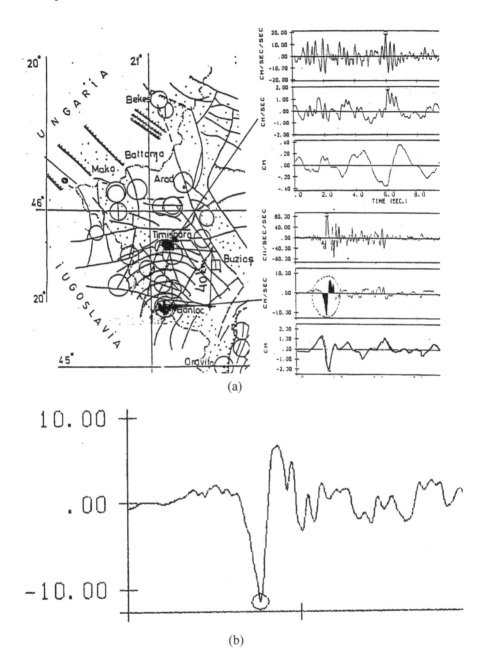

(a)

(b)

Figure 7.55 1991 Banat (Romania) intraplate earthquake: (a) Records at Timisoara and Banloc sites; (b) Velocity pulse (Gioncu, 2000)

The principal characteristics of low-to-moderate seismic regions, compared to high seismicity areas, can be summarized as follows (Nordenson and Bell, 2000, Pezeshk, 2004):

- There are many uncertainties in delineating the seismic source zones, the geometry and location being not well known, due to lack of fault traces on the ground surface. The historic seismicity record, at the same site, is too short comparing to the recurrence periods of interplate earthquakes, where the frequency of occurrence is higher. Therefore, seismic sources and hazard are less readily defined.

- Large magnitudes are very rare events. The probability for large intraplate earthquakes is 2% in 50 years, in comparison with interplate earthquakes where the probability of large earthquakes is 10% in 50 years. This is mainly due to the reduced rupture areas in comparison with the interplate rupture areas. The territory of the USA is the best example to compare the occurrence of these two earthquake types. While the Eastern part is dominated by intraplate earthquakes, in the Western part the interplate earthquakes are the most frequent. Figure 7.56a shows the return periods for peak accelerations for some cities of Eastern and Western USA (USGS, 2008, Nordenson and Bell, 2000). One can see very important differences between

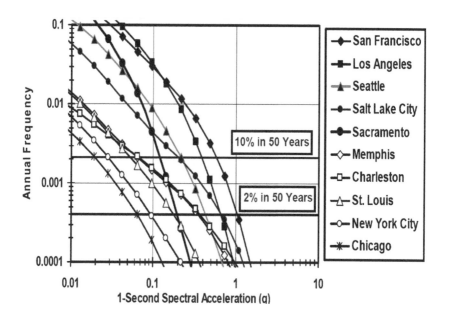

(a)

Figure 7.56 (continues)

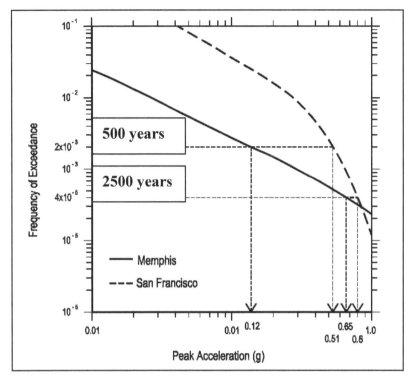

Return Period (yrs)	PA (g)		
	Memphis (MEM)	San Francisco (SF)	MEM/SF
500	0.12	0.51	0.24
2500	0.65	0.80	0.81

(b)

Figure 7.56 Return periods for interplate and intraplate earthquakes (USA):
(a) For some selected cities (USGS , 2008); (b) Memphis (intraplate) and San
Francisco (interplate) return periods (Gioncu, 2006)

the hazard curves for these selected cities. Figure 7.56b selects these curves for two
cities: Memphis in the Eastern USA and San Francisco in the Western USA. The
differences in accelerations are very high (MEM/SF = 24%), but this difference
decreases for a return period of 2500 years (MEM/SF = 81%). One can notice that,
if a return period of 500 years gives rise to low or moderate earthquakes, but in

some exceptional cases the magnitude can be higher (Pezeshk, 2004, Gioncu, 2006).

- The same observations result by examining the European earthquakes recorded in the so-called stable regions. As documented in the historical records (Ambraseys and Bommer, 1991, Gioncu and Mazzolani, 2002), these regions usually experienced earthquakes with magnitude under M 6 and only in a relatively small number of cases a magnitude M 6.5. The majority of earthquakes have magnitude under 4.5. But this observation does not exclude the possibility that some more important earthquakes occurred in the preceding millennia (Woo, 1996). This remark is also valuable for the future.

- The rare occurrence of severe seismic loading at a site situated in "Stable continental regions" suggests that the philosophy for the design of buildings must be different from the one used in areas where interplate earthquakes occur.

- Buried thrust or rift fault types, caused by compression or tension of the crust, produce the great majority of intraplate earthquakes. Only in very rare cases do these earthquakes result from a strike-slip fault. Local factors such as the strength of crustal rock, stress concentration, stress drop and level of stress play a dominant role in the earthquake process (Jankulovski et al, 1996).

- The duration of intraplate earthquakes are shorter than the one of interplate earthquakes. Intraplate records show that the strongest ground motions last just for a few seconds. The motions start with one or two relatively large amplitude cycles, followed by 5 or 6 cycles of reduced amplitude, then rapidly decay towards the end.

- The ground motions for intraplate earthquake are characterized by higher frequency content (which corresponds to short natural vibration periods) and very short pulses (Chandler et al, 1992). Therefore, the shapes of interplate and intraplate earthquakes are very different (Fig. 7.57).

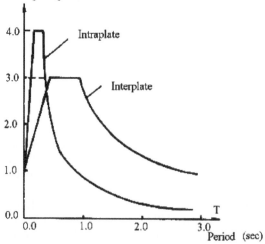

Figure 7.57 Normalized spectra for interplate and intraplate earthquakes
(Gioncu and Mazzolani, 2002)

- Attenuation of ground motions is different; the full areas of intraplate earthquakes have been quite large, even if the magnitude is moderate.
- Higher contrast of shaking on soft soils versus hard rock, showing the high influence of site soil type. Especially the liquefaction of non-cohesive soils seems to be very important for this earthquake type.
- Intraplate earthquakes with relatively small magnitude can produce large peak accelerations and large ground velocities only in the near-source region (see 1396 Basel, Switzerland, and 1811-1812 New Madrid, and 1994 Northridge, USA, earthquakes). This is due to the reduced depth, approximately 80% being less than 10 km deep.
 - For source-to-site distance less 10 km, the predicted levels of ground motion are very strong, depending on the focal depth of the earthquake (see Figure 7.18). Considering the threshold of 0.2g for damaging motions, it can be observed that only those motions with focal depths less than 10 km from the surface would be expected to generate potential damage on limited surface around the epicenter.
 - In the near-source zone, the intraplate earthquakes also depend on the source type and the rupture propagation. There are observation evidences that thrust faults produce higher ground motion than normal faults for comparable hypocentral depths and equivalent magnitudes. In the same conditions, the magnitude of thrust faults can be more than three times the one corresponding to normal faults (Duan and Oglesby, 2005).

Due to these characteristics, only the near-source of intraplate earthquakes can be very damaging. Unfortunately, the lack of intraplate records at short distance from the epicenter encouraged the use of interplate earthquake records, even if it is known that there are important differences between these two near-source earthquake types. This is a high priority of professionals to create a network of seismic stations also in stable continental zone, where some earthquakes have been experienced, especially along the existing rifts which are already known.

7.7 GROUND MOTIONS PECULIARITIES OF INTRASLAB SOURCES

Earthquakes delineate the subducting lithosphere down to nearly 700 km depth below the Earth's crust. The classification of intraslab faults is presented in Section 6.4. Unfortunately, there is not information about the earthquakes produced by these faults at the same level as for the other earthquake types.

For shallow intraslab earthquakes (depth until 70 km), the events are produced by shear dislocations as a primary mechanism, which is therefore not very different from the one of subduction earthquakes. Therefore, the same rules as for these earthquakes can be used, with the only exception for the magnitude, which could be larger.

The characteristics of earthquakes at depth below 70 km are the subject controversial discussions, by provoking the interest of seismologists. For intermediate intraslab earthquakes (70 km to 300 km), the events include both mechanical and thermal processes (the transformation and dehydration of the crust components) (Kirby et al, 1991), but the last are moderate. For these earthquakes, the interests manifested by seismologists are more concentrated in the phenomenology of this earthquake type, rather than in the engineering aspects. The data for intermediate intraslab earthquake characteristics mainly come from the Western part of Canada (British Columbia, Cascadia zone), Northern part of Japan and Carpathian zone (Vrancea).

For deep earthquakes (below to 300 km), where the thermal effects are dominant, the knowledge about the characteristics of ground motions are practically missing in the technical literature. It is not clear how these earthquakes are generated. So, deep earthquakes remain mysterious! Deep earthquakes will not give up their secrets easily (Houston, 2007).

However, some general conclusions regarding the ground motions produced by intraslab may be drawn.

-The rate of occurrence of earthquakes deeper than 70 km is several times lower than those of crustal earthquakes (Houston, 2007). Eighteen percent of the earthquakes with the magnitude greater than M 5.8 have depth greater than 100 km and only 6 percent occur at depths between 400 and 700 km. Only about 2 percent occur within the seismically quiet interval from 300 to 500 km (Houston, 2007).

- In contrast with the crustal earthquakes, for which the location of sources is scattered over a large surface, the deep earthquakes are located in very small surfaces, showing the repeatability of rupture in the same fault surface (Houston, 2007).

-Contrary to crustal earthquakes, deep earthquakes are notable for producing very few aftershocks. But large deep earthquakes can trigger, after some time, other large deep earthquakes at great distance. The time delays range from 2 to 140 minutes and the spatial separations range from 70 to 320 km (Houston, 2007).

-Generally, all the recorded ground motions have shown that the magnitudes are stronger than for crustal events of the same size. All professionals accept this as a matter of fact. So, for the South-western British Columbia, where crustal and intraslab earthquakes frequently occur at the same places, the magnitude recurrence for both earthquake types shows that for magnitude M 6, the predicted rate of intraslab earthquakes is three to ten times the rate of crustal earthquakes. For a recurrence of 475 years, the expected magnitudes are M 5.5 for crustal earthquakes and 7 M for intraslab earthquakes (Adams and Halchuk, 2000). Therefore, the expected magnitudes of intraslab earthquakes are greater than for both subduction and intraplate earthquakes.

-As the fault breaks so deeply, the radiated seismic wave energy spreads over a much larger area than in crustal earthquakes and very large areas experience significant shaking.

-Therefore, the near-source effects, very important for the crustal earthquakes, are not significant in the case of intraplate earthquakes.

-It is very interesting to notice that the initial rupture time is shorter for the deep earthquakes than for the shallow ones (Houston, 2007). However, the duration of ground motions frames in the category of long duration ground motions, due to the increased distance from the source to the site, traveling through different upper mantle rocks and crust soil layers.

-Accordingly, the number of important cycles is larger than for crustal earthquakes, an observation involving special requirements for structures in the areas affected by this earthquake type.

-Contrary to the crustal earthquakes, where the seismic-radiated energy is dominated by surfaces waves (L and R), for the deep earthquakes the radiated energy is mainly produced by the body waves (P and S).

-Due to the fact that the velocities are strongly attenuated by the large distance of the source-site, they have no great influence on the characteristics of ground motions, contrary to the great role they have for the crustal events. So, the accelerations remain the most important characteristic for this kind of ground motions.

-The local soil conditions may have a leader role in the pattern of these ground motions. The bad soil conditions, as in the Eastern part of Bucharest, influenced the recorded accelerations, introducing a long time vibration pulse.

-By comparing the North and South American earthquakes, Saragoni et al (2004) show that the intraslab earthquakes are more damaging than the crustal ones. The same conclusion results for Europe, if the crustal earthquakes from Western and Central zones are compared with the intraslab Vrancea (Romania) earthquakes.

7.8 REFERENCES

Aagaard, B.T. (2000): Finite-element simulation of earthquakes. Ph. D. Thesis, California Institute of Technology, Pasadena, California

Aagaard, B.T. (2003): High quality earthquake animations, US Geological Survey, Pasadena, California
http://pasadena.wr.usgs.gov/office/baagaard/research/animations/animations.html

Aagaard, B.T., Hall, J.F., Heaton, T.H. (2000): Simulation of near-source ground motions with dynamic failure. ASCE Structure Congress, Philadelphia, May 2000

Aagaard, B.T., Hall, J.F., Heaton, T.H. (2000): Sensitivity study near-source ground motion. 12th World Conference on Earthquake Engineering, Auckland,

Aagaard, B.T., Hall, J.F., Heaton, T.H. (2001a): Characterization of near-source ground motions with earthquake simulations. Earthquake Spectra, Vol. 17, No. 2, 177-207

Aagaard, B.,T., Heaton T.H., Hall, J.F. (2001b): Dynamic earthquake ruptures in the presence of lithostatic normal stresses: Implications for friction models and heat production. Bulletin of Seismological Society of America, Vol. 91, No. 6, 1761-195

Aagaard, B.T., Hall, J.F., Heaton, T.H. (2001c): Effect of fault dip and slip rake on ground motions: Why Chi-Chi was a relatively mild M7.6 earthquake. Pacific Earthquake Engineering Research Center, University of California, Berkeley Report. (See also Bulletin of Seismological Society of America, Vol. 94, No. 1, 155-170)

Aagaard, B.T., Heaton, T.H. (2004): Near-source ground motions from simulation of sustained intersonic and supersonic fault ruptures. Bulletin of Seismological Society of America, Vol. 94, Vol. 6, 2064-2078

Aagaard, B.T., Anderson, G., Hudnut, K.W.(2004): Dynamic rupture modeling of the transition from thrust to strike-slip motion in 2002 Denali fault, Alaska, earthquake. Bulletin of Seimological Society of America

Adams, J., Halchuk, S. (2000): Knowledge of in-slab earthquakes needed to improve seismic hazard estimates for southwestern British Columbia. Intraslab Earthquakes in the Cascadia Subduction System: Science and Hazard Workshop, Victoria, September 2000, 1-6

Agrawal, A.K., He, W.L. (2002): A closed-form approximation of near-fault ground motion pulses for flexible structures. 15th Engineering Mechanics Conference, EM 2002, New York, 2–5 June 2002

Alavi, B., Krawinkler, H. (2000): Consideration of near-fault ground motion effects in seismic design. 12th World Conference on Earthquake Engineering, Auckland, 30 January–4 February 2000, Paper No. 2665

Ambraseys, N.N. (1995): The prediction of earthquake peak ground acceleration in Europe. Earthquake Engineering and Structural Dynamics, Vol. 24, 467-490

Ambraseys, N.N., Bommer, J.J. (1991): Database of European strong motions records. European Earthquake Engineering, No. 2, 18-37

Anastasiadis, A.J., Klimis, N.S. (2002): Effect of soil non-linearities and site characteristics on evaluation of site coefficients. 12th European Conference on Earthquake Engineering, London, 9-13 September 2002, Paper no. 498

Anderson, J.C., Bertero, V.V. (1987): Uncertainties in establishing design earthquakes. Journal of Structural Engineering, Vol. 113, No. 8, 1709-1724

Anderson, G., Aagaard, B., Hudmit, K. (2003): Fault interactions and large complex earthquakes In Los Angeles area. Science 302 (5652), 1946-194

Ansal, A.M. (1995): Effects of geotechnical factors and behaviour of soil layers during earthquake. State-of-the-art lecture. 10th Conference on Earthquake Engineering, (ed. G. Duma), Vienna, 28 August - 2 September 1994, Balkema, Rotterdam, 467-476

Aptikaev, F.F., Erteleva, O.O. (2005): Some problems of the synthetic accelerograms generation. Earthquake Engineering in the 21st Century, Skopje – Ohrid, 27August–1 September, 2005

Athanasopoulos, G.A., Pelekis, P.C., Leonidou, E.A. (1998): Effects of surface topography and soil conditions on the seismic ground response-including liquefaction-in the Egion (Greece) 15/6/1995 earthquake. 11th European Conference on Earthquake Engineering, Paris, 6-11 September 1998

Baker, J.W. (2007): Quantitative classification of near-fault ground motions using wavelet analysis. Bulletin of the Seismological Society of America, Vol. 97, No. 3, 1486-1501

Bao, H., Bielak, J., Ghattas, O., Kallivokas, L.F., O'Hallaron, D.R., Shewchuk, J.R., Xu, J. (1996): Earthquake ground motion modelling on parallel computers. Conference on High Performance Networking and Computing. Proceedings of ACM/IEEE Conference on Supercomputing, Pittsburgh, Pennsylvania

Bard, P.Y. (1995): Effects of surface geology on ground motion: Recent results and remaining issues. 10th European Conference on Earthquake Engineering (ed. G. Duma), Vienna, 28 August-2 September 1994, Balkema, Rotterdam, 305-323

Bergman, E. (2000): Seismic scaling relations.
http://seismo.um.ac.ir/education/Seismic%20Scaling%20Relations.htm

Bielak, J., Ghattas, O. (1999): Computational challenges in seismology. 1st ACESWorkshop (ed. P.Mora), Brisbane and Noosa, 31 January-5 February 1999

Bommer, J.J., Alarcon, J.E. (2006): The prediction and use of peak ground velocity. Journal of Earthquake Engineering, Vol. 10, No. 1, 1-31

Borcherdt, R. (1994): Estimation of site-dependent response spectra for design (methodology and justification, Earthquake Spectra, Vol. 10, No. 4, 617-653

Bozorgnia, Y., Niazi, M., Campbell, K. W. (1996): Relation between vertical and horizontal response spectra for Northridge earthquake. 11th World Conference on Earthquake Engineering, Acapulco, 23-28 June 1996, Paper No. 893

Brandes, H.G. (2003): Geotechnical and foundation aspects. Earthquake Engineering Handbook (eds. W.F. Chen and C. Scawthorn), Chapter 7, CRC Press, Boca Raton

Bray, J.D., Rodriguez-Marek, A. (2000): Near-fault seismic site effects. Effects of Near-Field Earthquake Shaking, US-Japan Workshop, San Francisco, 20-21 March 2000, 39-46

Carydis, P.G. (2005): The effect of the vertical earthquake motion in near field. Earthquake Engineering in the 21st Century, Skopje-Ohrid, 27August–1 September 2005

Chandler, A.M., Hutchinson, G.L., Wilson, J.L. (1992): The use of interplate derived spectra in intraplate seismic regions. 10th World Conference on Earthquake Engineering, Madrid, 19-24 July 1992, Balkema, Rotterdam, 5823-5827

Dalguer, L.A., Miyake, H., Irikura, K. (2004): Characterization of dynamic asperity source models for simulating strong ground motion. 13th World Conference on Earthquake Engineering, Vancouver, 1-6 August 2004, Paper No. 3286

Der Kiureghian, A., Keshishian, P. (1996): Effect of site response on spatial variability of ground motion. 11th World Conference on Earthquake Engineering, Acapulco, 23-26 June 1996, Paper No. 708

Dobry, R. (1989): Some basic aspects of soil liquefaction during earthquakes. Earthquake hazards and the Design of Constructed Facilities in the Eastern United States (eds. K.H. Jacob, C.J. Turkstra), Annales of the New York Academy of Science, Vol. 558, 172-195

Dobry, R., Borcherdt, R.D., Crouse, C.B., Idriss, I.M., Joyner, W.B., Martin, G.R., Power, M.S., Rinne, E.E., Seed, R.B. (2000): New site coefficients and Site classification system used in recent building seismic code provision. Earthquake Spectra, Vol. 16, No. 1, 41- 67

Duan, B., Oglesby, D.D. (2005): The dynamics of trust and normal faults over multiple earthquake cycles: Effects of dipping fault geometry. Bulletin of Seismological Society of America, Vol. 95, No. 5, 1623-1636

Elnashai, A.S., Papazouglou, A.J. (1997): Procedure and spectra for analysis of RC structures subjected to strong vertical earthquake loads. Journal of Earthquake Engineering, Vol. 1, No. 1, 121-155

Elnashai, A.S., Bommer, J.J., Martinez-Pereira, A. (1998): Engineering implications of strong-motion records from recent earthquakes. 11th European Conference on Earthquake Engineering, Paris, 6-11 September 1998

Erdik, M. (1995): Developments on empirical assessment in the effects of surface geology on strong ground motion. In 10th European Conference on Earthquake Engineering, Wienna, 8 August-2 September 1994, Balkema, Rotterdam, Vol. 4, 2593-2598

Erkay, C., Karaesmen, E. (1996): Site effects and considerations for seismic code renewals. 11th World Conference on Earthquake Engineering, Acapulco, 23-28 June 1996, Paper No. 1516

Erkay, C., Karaesmen, E., Karaesmen, Er. (1998): A study on spatial variability of seismic motion through planar spectrum concept. 11th European Conference on Earthquake Engineering, Paris, 6-11 September 1998, T1-88

Erkay, C., Karaesmen, En., Karaesmen, Er. (2002): Significance of spatial variation of seismic motion and consequences affecting design procedures. 12th European Conference on Earthquake Engineering, London, 9–13 September 2002, Paper No. 130

Faccioli, E. (1996): On the use of engineering seismology tools in ground shaking scenarios. 11th World Conference on Earthquake Engineering, Acapulco, 23-28 June 1996

Faccioli, E., Vanini, M., Frassine, L. (2002): Complex site effects in earthquake ground motion, including topography. 12th European Conference on Earthquake Engineering, London, 9-13 September 2002, Paper No. 844

FEMA, Session 5 (2005): Earthquake hazard and emergency management. Characteristics of earthquakes: magnitude, intensity and energy. Courses http://training.fema.gov/EarthquakeEM/.../Session2005/...pdf

Filiatrault, A. (1996): Elements de Genie Parasismique et le Calcul Dynamique des Structures. Editions de l'Ecole Polytechnique de Montreal

Fu, Q., Menun, C. (2004): Seismic-environment-based simulation of near-fault ground motions. 13th World Conference on Earthquake Engineering, Vancouver, 1-6 August 2004, Paper No. 322

Gangopadhyay, A., Talwani, P. (2003): Symptomatic features of intraplate earthquakes. Seismological Research Letters, Vol. 74, No. 6, 863-883

Gioncu, V. (2000): Design criteria for seismic resistant steel structures, in Seismic Resistant Steel Structures (eds. F.M. Mazzolani and V. Gioncu), CISM Courses, Udine, Springer Wien, 19-99

Gioncu, V. (2006): Advances in seismic codification for steel structures. Costruzioni Metalliche, No. 6, 69-87

Gioncu, V. (2008): Structural problems in the low-to-moderate seismic regions. Part 1: Seismic risk and hazard. Part 2: Building's vulnerability. Seismic Risk, Liege 11-12 September 2008, 249-260, 261-272

Gioncu, V., Mazzolani, F.M. (2002): Ductility of Seismic Resistant Steel Structures. Spon Press, London

Gioncu, V., Mazzolani, F.M. (2003): Challenges in design of steel structures subjected to exceptional earthquakes, in Behaviour of Steel Structures in Seismic Areas, STESSA 2003 (ed. F.M. Mazzolani), Naples, 9-12 June 2003, Balkema, Lisse, 89-95

Gioncu, V., Mazzolani, F.M. (2006): Influence of earthquake types on the design of seismic-resistant steel structures. Part 1: Challenge for new design approaches. Part 2 : Structural responses for different earthquake types, in Behaviour of Steel Structures in Seismic Areas, STESSA 2006 (eds. F.M. Mazzolani and A. Wada), Yokohama, 14-17 August 2006, Taylor & Francis, London, 113-120, 121-127

Graizer, V.M. (2005): Effect of tilt on the strong motion data processing. Soil Dynamics and Earthquake Engineering, Vol. 25, 197-204

Gyorgyi, J., Gioncu, V., Mosoarca, M. (2006): Behaviour of steel MRFs subjected to near-fault ground motions, in Behaviour of Steel Structures in Seismic Areas, STESSA 2006 (eds. F.M. Mazzolani and A. Wada), Yokohama, 14-17 August 2006, Taylor & Francis, London, 129-136

He, W.L., Agrawal, A.K. (2008): Analytical model of ground motion pulses for the design and assessment of seismic protective systems. Journal of Structural Engineering, Vol. 134, No .7, 1177-1188

Houston, H. (2009): Deep earthquakes. In Earthquake Seismology: (ed. H. Kanamori), Elsevier

Huang, M.J (1983): Investigation of local geology effects on strong earthquake ground motions. California Institute of Technology Report, EERL 83-03

Huang, C.T., Chen, S.S (2000): Near-field characteristics and engineering implications of the 1999 Chi-Chi earthquake. Earthquake Engineering and Engineering Seismology, Vol. 2, No. 1, 23-41

Hudson, R.L., Skyers, B.D., Lew, M. (1996): Vertical strong motion characteristics of the Northridge earthquake. 11th World Conference on Earthquake Engineering, Acapulco, 23-28 June 1996, Paper No 728

Hwang, H. (2000): Comments on design earthquake specified in the 1997 NEHRP provisions. 12th World Conference on Earthquake Engineering, Auckland, 30 January-4 February 2000, Paper No. 0657

Idriss, I.M. (1990): Response of soft soil sites during earthquakes. Proceedings of the Symposium to Honor professor H.B. Seed, Berkeley, May 1990, 273-289

Idriss, I.M., Sun, J.I. (1993): User's manual for SHAKE91: a computer program for conducting equivalent linear seismic response analyses of horizontally layered soil deposits. Report of Center for Geotechnical Modeling, University of California

Ifrim, M., Macavei, F., Demetriu, S., Vlad, I. (1986): Analysis of degradation process in structures during the earthquakes. 8th European Conference on Earthquake Engineering, Lisbon, 65/8-72/8

Igel, H., Cochard, A., Schreiber, U., Velikoseltsev, A., Flaws, A. (2005): Rotational motions induced by earthquakes: theory and observations. Ludwig-Maximilians-Universitat Munchen Report

Jankulovski, E., Sinadinovski, C., McCue, K. (1996): Structural response and design spectra modelling results from some intra-plate earthquakes in Australia.

11th World Conference on Earthquake Engineering, Acapulco, 23-28 June 1996, Paper No. 1184

Kawase, H. (2004): Strong motion prediction considering the effects of surface geology: introduction and overview. 13th World Conference on Earthquake Engineering, Vancouver, 1-6 August 2004, Paper No. 5054

Kirby, S.H., Durham, W.B., Stern, L.A. (1991): Mantle phase changes and deep-earthquake faulting in subducting lithosphere. Science, Vol. 252, 216-225

Klimis, N.S., Anastasiadis, A.J. (2002): Comparative evaluation on topography effects via code recommendations and 2-D numerical analysis. 12th European Conference on Earthquake Engineering, London, 9-13 September 2002, paper No. 745

Kulhanek, O. (1990): Anatomy of Seismograms. Elsevier, Amsterdam

Liam, F.W.D., Matsunga, S. (1995): The effects of site conditions on ground motions. 10th European Conference on Earthquake Engineering (ed. G. Duma), Vienna, 28 August–2 September 1994, Balkema, Rotterdam, 2607-2611

Lee, V.W. (2002): Empirical scaling of strong earthquake ground motion. Duration of strong motion. Journal of Earthquake Technology. Paper No. 426, Vol. 39, No. 4, 255-271

Lee, H.K., Celebi, M., Todorovska, M. (2007): An introduction to Rotational Seismology and engineering applications. http://pubs.usgs.gov/of/2007/1114/abstracts/Abstract_LeeWHK-etal.pdf

Lowry, A.R., Larson, K.M., Kostoglodov, V., Bilham, R. (2001): Transient fault slip in Guerrero, southern Mexico. Geophysical Research Letters, Vol. 28, No. 19, 3753-3756

Lungu, D., Mazzolani, F.M., Savidis, S. (1997): Design of Structures in Seismic Zones, EUROCODE 8, Worked Examples, Tempus Phare Project, 01198 Report

Makris, N., Chang, S.P. (1998): Effect of damping mechanisms on the response of seismically isolated structures. PEER Report 1998/06, University of California, Berkeley

Mallaioli, F., Bruno, S., Decanini, L.D., Panza, G.F. (2006): Characterization of the dynamic response of structures to damaging pulse-type near-fault ground motions. Meccanica, Vol. 41, 23-46

Mateescu, G., Gioncu, V. (2000): Member response to strong pulse seismic loading. Behaviour of Steel Structures in Seismic Areas, STESSA 2000 (eds. F.M. Mazzolani and R. Tremblay), Montreal, 21-24 August 2000, Balkema, Rotterdam, 55-62

Mavroeidis, G.P., Papageorgiou, A.P. (2002): Near-source strong ground motion: Characteristics and design issues. 7th U.S. National Conference on Earthquake Engineering, 7NCEE, Boston, 21-25 July 2002

Menun, C., Fu, Q. (2002a): An analytical model for near-fault motions and the response of SDOF systems.7th National Conference on earthquake Engineering, Boston, 21-25 July 2002

Menun, C., Fu, Q. (2002b): An analytical model for near-fault ground motions and the response of MDOF systems. 12th European Conference on Earthquake Engineering, London, 9-13 September 2002, Paper No. 647

Midorikava, M., Hiraishi, H., Okawa, I., Ilba, M., Teshigawara, M., Isoda, H. (2000): Development of seismic performance evaluation procedures in building code of Japan. 12th World Conference on Earthquake Engineering, Auckland, 30 January-4 February 2000, Paper No. 2215

Miyatake, T. (2000): Computer simulation of strong ground motion near a fault using dynamic fault rupture modelling: Spatial distribution of the peak ground velocity vectors. Pure and Applied Geophysics, Vol. 157, 2063-2081

Mooney, W.D., Schulte, S., Detweiler, S.T., (2004): Intraplate seismicity and the discrimination of nuclear test events using an updated global earthquake catalogue. 26th Seismic Research Review – Trend in Nuclear Explosion Monitoring, 439-448

Nordenson, G.J.P., Bell, G.R. (2000): Seismic design requirements for regions of moderate seismicity. Earthquake Spectra, Vol. 16, No. 1, 205-225

Otani, S. (2004): Japanese seismic design of high-rise reinforced concrete buildings: An example of performance-based design code and state of practice. 13th World Conference on Earthquake Engineering, Vancouver, 1-6 August 2004, Paper No. 5010

Panza, G.F., Vaccari, F., Romanelli, F. (1999): IGCP Project 414: Realistic modelling of seismic input for megacities and large urban areas. Episodes, Vol. 22, No. 1, 26-32

Paolucci, R., Rimoldi, A. (2002): Seismic amplification for 3D steep topographic irregularities. 12th European Conference on Earthquake Engineering, London, 9-13 September 2002, paper No. 087

Pezeshk, S. (2004): Seismic code issues in central United States. 13th World Conference on Earthquake Engineering, Vancouver, 1-6 August 2004, paper No. 1843

Pinto, P.E. (2000): Design for low/moderate seismic risk. 12th World Earthquake Engineering, Aukland, 30 January-4 February, 2000, paper No. 2830

Rassem, M., Ghobarah, A., Heidebrecht, A.C. (1997): Engineering perspective for the seismic site response of alluvial valleys. Earthquake Engineering and Structural Dynamics, Vol. 26, 477-493

RMS, Risk Management Solutions (2000): Chi-Chi Taiwan Earthquake, Event Report

Rodriguez-Marek, A. (2000): Near-fault seismic site response. Ph. D. Dissertation, Department of Civil Engineering, University of California, Berkeley

Rodriguez-Marek, A., Bray, J.D., Abrahamson, N.A. (2000): A geotechnical seismic site response evaluation procedure. 12th World Conference on Earthquake Engineering, Auckland, 30 January–4 February 2000, paper No. 1590

Rodriguez-Marek, A., Brady, J.D., Abrahamson, N. (2001): Ground motion. Chapter 3. Commentary. www.bsscoline.org/NEHRP2003/comments/C3.pdf

Sabetta, F., Bommer, J. (2002): Modification of the spectral shapes and subsoil conditions in EUROCODE 8, 12th European Conference on Earthquake Engineering, London, 9-13 September 2002, Paper No. 518

Saragoni, R.G., Astroza, M., Ruiz, S. (2004): Comparative study of subduction earthquake ground motion of North, Central and South America. 13th World

Conference on Earthquake Engineering, Vancouver, 1-6 August 2004, Paper No. 104

Sasani, M., Bertero, V.V. (2000): Importance of severe pulse-type ground motions in performance-based engineering: Historical and critical review. 12th World Conference on Earthquake Engineering, Auckland, 30 January-4 February 2000, Paper No. 1302

Scawthorne, Ch. (1997): Earthquake engineering. Handbook of Structural Engineering (ed. W.F. Chen), CRC Press, New York, 5.1-5.83

Schnabel, P.B., Lysmer, J., Seed, H.B (1972): SHAKE: a computer program for earthquake response analysis of horizontally layered sites.   University of California, Berkeley, UBC/EERC-72/12 Report

Schulte, S.M., Mooney, W.D. (2005): An updated global earthquake catalogue for stable continental regions: reassessing the correlation with ancient rifts. Geophysical Journal International, Vol. 161, No. 3, 707-721

Schulte, S.M., Mooney, W.D. (2007): An updated earthquake catalogue for stable continental regions. Interplate earthquakes (495-2002). http://earthquake.usgs.gov?research/data/scr_catalog.php

Seno, T. (2003): Fractal asperities, invasion of barriers, and interpolate earthquakes. Earth Planets Space, Vol. 55, 649-665

SHAKE2000 (2007): a computer program for the 1-D analysis of geotechnical earthquake engineering problems.

Somerville, P. (1997): Engineering characteristics of near fault ground motions. SMIP97 Seminar on Utilization of Strong-Motion Data, Los Angeles, 8 May 1997, 9-28

Somerville, P. (1998): Development of an improved representation of near fault ground motions. SMIP98 Seminar on Utilization of Strong Motion Data, Oakland, 15 September 1998, 1-20

Somerville, P. (2000a): Characterization of near-fault ground motions. Effects of Near-field Earthquake Shaking. US-Japan Workshop, San Francisco, 20-21 March 2000, 21-29

Somerville, P. (2000b): Observation of earthquake source and ground motion scaling at the macro level. http://www.tokyo.rist.or.jp/ACES_WS2/extended_abstract/PDF

Somerville, P. (2002): Characterizing near fault ground motion for the design and evaluation of bridges. 3rd National Seismic Conference & Workshop on Bridges and Highways, Portland, Oregon, 29 April-1 May 2002

Somerville, P. (2003): Magnitude scaling of the near fault rupture directivity pulse. Physics of Earth and Planetary Interiors, Vol. 137, No. 1, 21-212

Somerville, P., Smith, N.P., Abrahamson, N.A. (1996): Accounting for near-fault rupture directivity effects in the development of design ground motions. 11th World Conference on Earthquake Engineering, Acapulco, 23-28 June 1996, paper No. 711

Stepp, J.C. (1996): Characteristics of seismicity in stable continental regions important for seismic hazard assessment. In 11th World Conference on Earthquake Engineering, Acapulco, 23-28 June1996, CD-ROM 2159

Stewart, J.P., Chiou, S.J., Bray, J.D., Graves, R.W., Somerville, P.G., Abrahamson, N.A. (2001): Ground motion evaluation procedures for performance-based design. PEER Report 2001/09, University of California, Berkeley

Studer, J. et al, (1997): Erdbeden in der Westturkei von 17 august 1999, Schweizer Ingenieur and Architekt, No 43, 938-944

Studer, J., Koller, M.G. (1995): Design earthquake: The importance of engineering judgement. 10th European Conference on Earthquake Engineering (ed. G. Duma), Vienna, 28 August-2 September 1994, Balkema, Rotterdam, Vol. 1, 167-176

Teisseyre, R. (2006): Seismic effects and rotation waves. Polish Academy, Institute of Geophysics, M-29 (395)

Trifunac, M.D. (1990): How to model amplification of strong earthquake motions by local soil and geological site conditions. Earthquake Engineering and Structural Dynamics, Vol. 19, 833-846

Trifunac, M.D. (1997): Relative earthquake motion of building foundations. Journal of Structural Engineering, Vol. 123, No. 4, 414-422

Trifunac, M.D. (2006): Measurement of rotations- condition sine qua non- for comprehensive interpretation of strong motion. http://www.rotational-seismology.org/library/downloads/IWGoRS_Charte_v3_03 Nov06.pdf

Trifunac M.D., Brandy, A.G. (1975): A study on the duration of strong earthquake ground motion. Bulletin of the Seismological Society of America, Vol. 65, 581-626

Trifunac, M.D., Novikova, E.I. (1995a): State of the art review on strong motion duration. 10th European Conference on Earthquake Engineering, Vienna, 28 August-2 September 1994, Balkema, Rotterdam, Vol .1, 131-140

Trifunac, M.D., Novikova, E.I. (1995b): Duration of earthquake fault motion in California, Earthquake Engineering and Structural Dynamics, Vol. 24, 781-799

USGS (2007): Rotational Seismology and Engineering Applications. Menlo Park, California, 18-19 September 2007

USGS, (2008): Introduction to SDPRG and changes proposed for 2009 NEHRP provisions. FEMA report Ch. Kircher, N. Luco, A. Whittaker, September 2008

Vasquez, J. (1996): Spectral superposition under two-directional excitation. Acapulco, 23-28 June 1996, Paper No. 614

Warren, L.M., Sheare, P.M. (2006): Systematic determination of earthquake rupture directivity and fault planes from analysis of long-period P-wave spectra. Geophysics Journal, Vol. 164, 46-62

Wikipedia (nd): Asperity htpp://en.wikipedia.org/wiki/Asperity

Woo, G. (1996): Northwest European seismic hazard: The search for a regional perspective. In 11th World Conference on Earthquake Engineering, Acapulco, 23-28 June 1996, Paper No. 2156

Yeats, R.S., Gath, E.M. (2004): The role of geology in seismic hazard mitigation. Earthquake Engineering. From Engineering Seismology to Performance-Based Engineering (eds. Y. Bozorgnia and V.V. Bertero), CRC Press, Boca Raton, 3.1-3.23

Youd, T.L. (2003): Liquefaction mechanisms and induced ground failure. International Handbook of Earthquake & Engineering Seismology (eds. W.H.K. Lee et al), Academy Press, Amsterdam, 1159-1173

Zerva, A., Zervas, V. (2002): Spatial variation of seismic ground motions: An overview. Applied Mechanical Review, Vol. 55, No. 3, 271-297

Chapter 8

Ground Motions and Structures

8.1 STRUCTURE INFLUENCE ON GROUND MOTIONS

8.1.1 Building, the New Subject in the Seismic Approach

For a long time, the realistic evaluation of the structural response of buildings during an earthquake was considered inaccessible for practical design. Although the design procedures already started to be proposed in the middle of the 20th Century, it is only during the last decades that improved and intensified research works have revealed how to effectively evaluate the actual response of structures.

In addition to the previous chapters, where only the complex ensemble of source, travel path and site soil are considered, this chapter is dealing with a new subject, the building as a whole (Fig. 8.1).

During an earthquake, seismic waves arise from sudden movements in a rupture zone in the Earth's crust. Waves of different types and velocities travel different paths before reaching a building's site, where the local ground is subjected to various motions. The ground moves rapidly back and forth in all directions, usually mainly horizontally, but also vertically. Then the foundations of the building are forced to follow these movements. The upper part of the building remains in delay in respect to the foundation moving, due to its mass. This causes strong vibrations of the structure

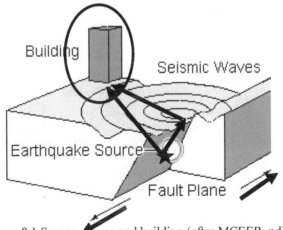

Figure 8.1 Source, waves and building (after MCEER, nd)

with resonance phenomena between the structure and the ground, and thus large inertial forces arise. This frequently results in plastic deformation of the structure and substantial damage with local failures, leading, in extreme cases, to the overall collapse (Bachmann, 2003).

8.1.2 Effects of Building's and City's Weight on Ground Motions

The free-field ground motions are usually considered as the source of the structure motions. However, there is the evidence that the actual ground motions which shock the structures depend on the building presence, being substantially modified due to the building's weight.

Building's weight. For a proper structural design, it is very important to have records in free-field conditions, measured at a convenient distance from the building and at its base, in order to evaluate the effect of the structure weight. Generally, this demand cannot be answered by the actual network of the instrumented sites, so only the theoretical results can be considered. Due to the building weight, the soil is overloaded, becoming stiffer and, therefore, producing a change in its behavior. In some cases, especially for stiff soils, the weight effect is to reduce the acceleration peak (Pitilakis, 1995, Gioncu and Mazzolani, 2002) (Fig. 8.2). But for soft soils, it is possible to observe a relative increasing in acceleration peaks due to the presence of a building.

City–soil effect. The analysis of ground motion effects on structures generally disregards the influence of the surface structures on the free-field motions in densely urbanized areas. The effect of the building density increasing is studied by Semblat et al (2000, 2002a,b, 2004), using the boundary element model (BEM) for alluvial basins with soft soil (Fig. 8.3a). The idea is that a part of the vibrating energy of buildings is released into soil through waves produced by buildings, significantly modifying the

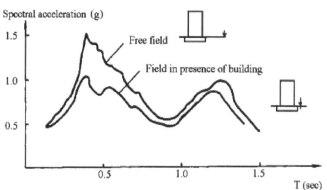

Figure 8.2 Influence of building's weight
(Gioncu and Mazzolani, 2002, after Pitilakis, 1995)

free-field ground motions. This approach is confirmed by many observations during some important earthquakes (Gueguen et al, 2002a,b). Numerical analyses consider the alluvial basin located in the center of Nice (France), where city–site effects are upposed to be significant. Five different building densities are considered: 1, 3, 7, 15 and 30 identical structures along the 2 km width of the basin (Fig. 8.3b). The influence of building density is shown in Figure 8.3c, for the first frequency value of buildings (0.5 HZ). One can see that the amplification level of free-field ground motions is significantly modified by the presence of buildings on site. There are three important zones: until 500m and beyond 1500m the influence is insignificant, but in the area between 500 and 1500m the differences are very important. For largest building density, the amplification of ground motions is very high (three times), when compared to the free-field case. It is possible to explain the damage during the 1995 Kobe earthquake due to the city-site effect, because of the presence of very soft soil and large density of buildings. The conclusion of these researches is that the modifications of the free-field amplification due to the city–site effect lead to introduce a specific factor, the *urban field amplification,* to be introduced in the design code, which is very important especially in case of soft soils.

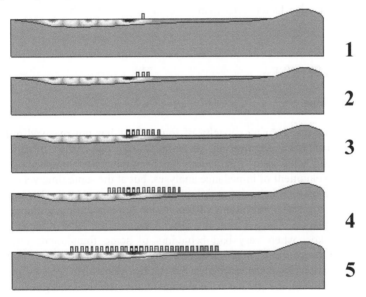

(a)
Figure 8.3 (continues)

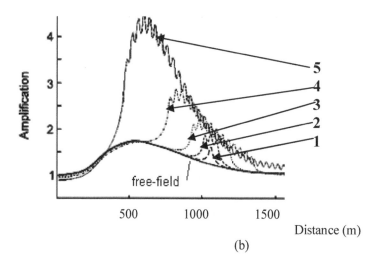

Figure 8.3 City-site effect for Nice (France); (a) Various building density;
(b) Amplification factor (after Semblat et al, 2000, 2002a, b)

8.2 FOUNDATION RESPONSE

8.2.1 Basic Considerations

The evaluation of the seismic bearing capacity of foundations has not received much attention from the earthquake engineering community. On one hand, it can be recognized that the cyclic behavior of foundations has been extensively studied in the last decade, with the development of impedance analyses, but at the same time very few researches have initiated on their behavior at failure due to the seismic loads. The major reason probably lies on the few observations of foundation failures during earthquakes (Pecker, 1996). This situation changed after the 1985 Mexico City, 1999 Kocaeli and 1999 Chi-Chi earthquakes, when very bad behaviors of some foundations due to settlement, tilting and overturning were observed. Significant improvement in the seismic design of foundations has been reached in the last few years after these events.

Once the ground motion on the site has been evaluated, the structural designer needs to proceed with the selection of the foundation system and its sizing. For this purpose, he appeals to the *Foundation Engineering,* which is an amalgam of experience, judgment, technical knowledge, theory and practice, being one of a most challenging task for the structural engineering. In the last period, due to the requirements developed in the frame of Earthquake Engineering, the problems are complicated by the necessity to use the knowledge of the dynamic soil mechanics, soil-foundation-structure interaction and structural dynamics. The bad behavior of many foundations during the above-mentioned earthquakes confirms the complexity of the phenomena produced during the ground motions. The *Soil Dynamics* is a new branch of Soil Mechanics, which deals with the behavior of soil and foundations under dynamic loads. Earthquake ground motions constitute a class of these dynamic loads which usually challenges the engineers in their design of different kinds of foundations.

One main problem of the seismic response of the foundation is related to the change of soil properties for seismic loads. Even for static conditions, the correct evaluation of the soil strength is a challenge for the geotechnical engineers. Additional factors must be considered for seismic conditions, depending on the soil type, which can be characterized as cohesive or cohesionless (Pecker, 1996):

-   the rate of loading may significantly affect the value of the statically determined strength. The modification is not very high for cohesionless soils, but can by very important also for cohesive soils, increasing the static strength of 30% to 60%;
-   the repetition of alternate cycles of loading may cause a degradation of the soil and a subsequent decrease in its strength. Once again the cohesionless soils are insensitive to this effect and the cohesive soils experience degradation when they are strained beyond a given strain threshold which is soil type dependent;
-   the saturated cohesionless soils experience an increase in their pore water pressure due to the cyclic loading, which may lead to a liquefaction condition, unless the drainage conditions allow for a rapid dissipation of this pressure.

The seismic behavior of foundations depends on their conformation. The following main foundation types are used in seismic areas:

-   *spread shallow individual* or *continuous foundations*, used for the structures insensible to differential settlements;
-   *mat foundations* (plate foundations), applicable for structures sensible to the differential settlements;
-   *pile foundations*, used in the case of soft soils and structures sensible to differential settlements.

In general case, the seismic forces have six components acting on the foundations:

-   vertical forces, which can be very important if the seismic vertical components are high (such in case of near-source earthquakes). Only in the far-field areas can these components be neglected, since their magnitude is small with respect to the static one;
-   two shear forces, introducing an inclination of the resultant force. These forces can be very important in the zones where seismic horizontal components are dominant;
-   two overturning moments, introducing an eccentricity of the resultant force and, in some cases, the uplifting of the structure;
-   torsional moment, very important in case of asymmetric structures.

The forces acting on the foundation are different for static and dynamic loads. If the static loads are constant in time, the dynamic ones vary with the time. The excessive static loads may generate a general failure; the dynamic loads induce permanent irreversible displacements. So, the limit seismic load can be defined as the one producing excessive permanent displacements, which impede the proper functioning of the structure (Pecker, 1996).

The seismic behavior of these kinds of foundations under both static and seismic forces will be discussed in the following sections.

8.2.2 Spread Shallow Foundations

Spread shallow foundations are presented in Figure 8.4; they are used when the soil conditions are adequate and the acting forces are not too strong. In seismic areas it is recommended to inter-connect the footing elements, in order to restrain any relative movement of the foundations.

There are two methods to evaluate the dynamic response of spread shallow foundations:

Impedance method, which considers the opposition to the motion of the foundations subjected to the dynamic forces acting on the structure. The structural impedance of the foundation is the ratio between the force introduced by the structure and the resulting velocity at the foundation level. This method is mainly used in the approach which considers the interaction soil-foundations-structure, the inelastic deformation of soil being ignored. This method is mainly based on the expansion of computers and numerical analysis and it comes from the elastic half-space approach.

The main problem is the determination of the soil parameters: stiffness for different displacement types. The displacements, which are considered for a single foundation, are the vertical, the rocking and the horizontal (Carrubba et al, 2000):

(i)   In case of vertical displacements, the foundation is constrained to settle uniformly, by increasing the vertical load in the mass center of the basement (Fig. 8.5a).

(ii)   In case of rocking displacements, the applied system of forces being in the mass center, the foundation is constrained to rotate, without uplifting, settling or swaying (Fig. 8.5b).

(iii)   In case of swaying displacements, the applied system of forces being in the mass center, the foundation is constrained to translate, without rotating, settling or uplifting (Fig. 8.5c).

For each displacement, the corresponding stiffness is determined. The method allows

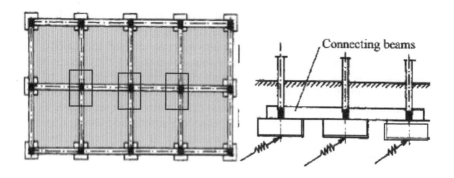

Figure 8.4 Spread shallow foundations (Gioncu and Mazzolani, 2002)

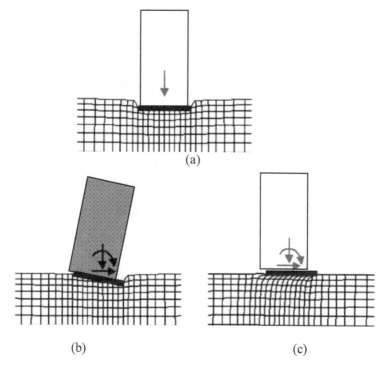

Figure 8.5 Displacements of foundations: (a) Vertical displacements; (b) Rocking displacements; (c) Swaying displacements (after Carrubba et al, 2000)

analyzing the soil-foundation-structure system in elastic linear field only. In all these analyses the foundation is considered as infinitely rigid, so its response to dynamic loading arises solely from the displacement of the supporting ground.

The known impedance solutions do not take into account the soil inelasticity, but consider the non-linear behavior of the ensemble foundation-soil by the decreasing of soil stiffness with the strain level. So, the method examines the behavior of the foundation-soil ensemble from the beginning of loading until the non-linear behavior, assuming the ultimate load as a limitation of the soil deformation at an acceptable level.

Bearing capacity method. Contrary to the Impedance Method, the Bearing Capacity Method determines only the ultimate load corresponding to the failure mechanism. Figure 8.6 illustrates the various failure modes of shallow foundations which may occur during seismic events (Pappin, 2002):

- *Horizontal sliding failure* may occur when the lateral load exceeds the lateral capacity of the foundation as shown in Figure 8.6a. Generally, the lateral capacity has a minimum value when the vertical applied load is also at a minimum. It is important to note that is refers to seismic loads only.

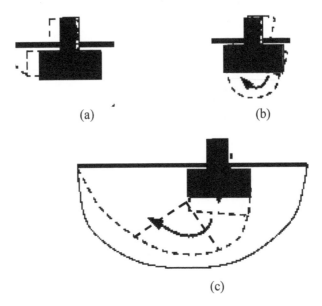

(a)   (b)

(c)

Figure 8.6 Failure modes of shallow foundations: (a) Sliding; (b) Overturning; (c)Shear failure (after Pappin, 2002)

- *Overturning*, illustrated in Figure 8.6b, occurs as a result of the overturning loads arising from the seismic ground motion. Here the building tilts in such a way that the P-delta effect is sufficient to keep the building moving after the initial movement due to seismic loads.

- *Shear failure* may occur when the shear stresses along the sliding surface exceed the shear strength of the soil, producing the formation of a slip mechanism (Fig. 8.6c). It is the more general form of the sliding case and again can arise from a combination of the maximum lateral load with the minimum vertical load, but may also occur due to the combination of maximum lateral load and maximum vertical load. Liquefaction is a major problem for this failure mode, especially for low seismicity levels.

The most significant among these failure modes for the ultimate bearing capacity is the dynamic shear failure due to seismic loads. Figure 8.6c shows the shear failure kinematical mechanism, composed by three rigid zones in limit equilibrium: the first one below the foundation being the active zone, the second one in the form of fan, being an intermediate zone, and the last one, the passive zone. The dimensions of these zones depend on the angle of internal friction. Considering that the effect of cyclic loads produces the reduction of this angle, the shape of failure mode is different in comparison with the static loading. Due to the reducing of the passive zone, the corresponding bearing capacity decreases drastically. Some numerical examples, using pseudo-static approach, performed by Choudhury and Subba Rao (2006), show a reduction of more than 50%, especially if the foundation is embedded in slope ground.

8. 2.3 Mat Foundations

The mat foundation is a continuous footing, used when the soil conditions at the site are poor, in order to distribute the heavy loads transmitted by columns and walls across the entire building area, ensuring a relative uniform load transfer and lowering the contact pressure compared to the conventional spread foundations (Fig. 8.7). For the seismic approach, the mat foundation can be assumed to be a rigid body. Due to this assumption, the main structural problem is the evaluation of the ultimate load due to the foundation rocking or overturning on the deformable base, when the supporting soil is soft and weak.

The problem is becoming of increasing engineering interest after the 1995 Kobe earthquake and especially both the 1999 Kocaeli (Adapazari) and Chi-Chi (Taiwan) earthquakes (Gazetas et al, 2003). Large settlements, permanent tilting and complete overturning of numerous buildings, which otherwise retained their integrity, captured the attention of the world geotechnical and earthquake engineering community. The liquefaction of shallow sediment and soil layers was evident in the ground surface. Significant tilting and overturning were observed only in slender buildings, with aspect ratio $H/B > 2$, free from other buildings on one of their sides. Small or no rotations were detected in wider buildings.

The behavior of mat foundations is influenced by the condition of contact between

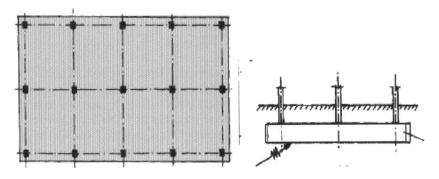

Figure 8.7 Mat foundations (Gioncu and Mazzolani, 2002)

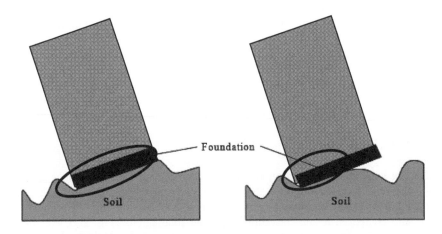

Figure 8.8 Condition of foundation-soil contact (after Savidis et al, 2000)

the foundation and the soil beneath. Tensile contact stresses are basically excluded and a partial or full uplift becomes possible (Fig. 8.8). Therefore, the so-called condition of partial contact is allowed for mat foundations (Savidis et al, 2000).

8.2.4 Pile Foundations

Pile foundation is used when firm soil is at a considerable distance below the natural level of the site, in order to transfer building loads down to a suitable bearing layer, when the soil immediately below a construction is unsuitable for direct bearing of footing (Fig. 8.9). Piles support loads either by bearing directly on rock or suitable soil and/or by developing friction along their very long length.

The passage of seismic waves through soft soil during strong earthquakes may cause significant strains developed in the soil and piles, causing, in some cases, the failure of piles. The main examples refer to the 1964 Niigata, 1978 Off-Miyagi, 1989 Loma Prieta and 1995 Kobe earthquakes, where much damage was suffered by the pile foundations. Examining the causes of this kind of damage, producing the failure of piles, one can conclude that it is the result of three main reasons:
-   bad behavior of the pile to cap connections;
-   reduced bearing capacity due to the discontinuity of layers, producing high stress concentration in the discontinuity zone;
-   liquefaction of soil under buildings, producing a loss of lateral support to the piles and consequent free buckling of them.

Seismic behavior of pile-cap connection. The seismic performance of piles is directly related to the earthquake response of the pile-cap connections. Potential inelastic damage occurs at the interface between pile heads and pile cap as evidenced in recent earthquakes. Some seismic design conceptions consider the possibility of the formation of plastic hinges at the pile–cap contact, but it presents serious risks, because the repair of damaged piles in high–rise building systems is impracticable, because of the elevated cost and the difficulty associated with ground excavations (Teguh et al, 2006). It should be noted that, from the reason presented above, it is desirable to design the piles to remain undamaged. So, the design concept should aim at dissipating seismic energy just by the structure above the foundations. The pile foundations should be provided with sufficient strength to ensure, as far as possible, that they remain in the elastic range. Therefore, special attention must be paid to the pile-cap connection details.

Layer discontinuity effect, due to passage of seismic waves through two different layer soils, generating bending moment in the piles. The research results show that bending moments due to this effect, named *kinematic moments,* tend to be amplified in the vicinity of interfaces between stiff and soft soil layers (Pappin, 1996, Mylonakis, 2002). This effect can produce the damage of piles, due to the fact that these moments can exceed the moments resulting in the pile-cap connection. The same observation as for the pile-cap connections is valuable in general: the pile must remain in elastic field, because it is impossible to repair this kind of damage.

Liquefaction effect. Collapse of pile foundation in liquefiable areas has been observed in the majority of the recent strong earthquakes. Two phenomena are considered to be the cause of these collapses (Bhattacharya et al, 2003):

- *The soil pushes the pile,* so the failure is due to lateral spreading of soil crust, the soil liquefies producing large lateral displacements and the soil crust drags the pile with them, causing bending failure due to soil lateral pushing. The deformation of the ground surface adjacent to the pile foundation is indeed indicative of this collapse mechanism.

- *The pile pushes the soil,* considering that failure is due to the pile buckling. This is the case of slender precast piles. During earthquake-induced liquefaction, the soil surrounding the pile loses its effective strength and it can no longer offer sufficient support to the pile, which now acts as an unsupported column prone to axial instability.

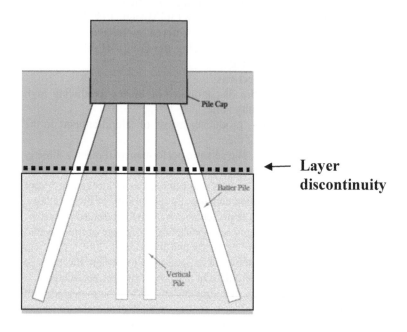

Figure 8.9 Pile foundations (after Bhattacharya et al, 2003)

Therefore, the buckling of piles must be considered.

The pile collapse mechanism is presented in Figure 8.10 (Bhattacharya, 2003). This collapse mechanism seems to be very attractive for the dissipation of seismic energy. But, as it was presented before, this form of structural collapse must be avoided, due to the difficulties of consolidating the damaged piles.

8.3 STRUCTURAL RESPONSE

8.3.1 Behavior of Structures during Ground Motions

The main effect of the ground motions on a structure is the dynamic nature of the earthquake loading. As a consequence of time variability, the ground motions are characterized by the time history of the three ground motion parameters at the level of foundation, acceleration, velocity and displacement. For the structure subjected to such ground motions, these actions will propagate through the structure as waves, causing large oscillations (Fig. 8.11). Therefore, the structural response also varies with time, involving dynamical movements. The structure performs a series of *forced oscillations*

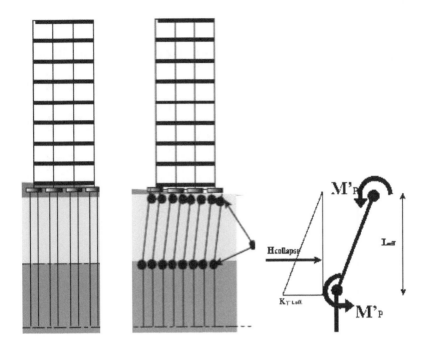

Figure 8.10 Collapse mechanisms of foundations (after Bhattacharya, 2003)

during the earthquake, having a very complex chaotical movement, characterized by peaks of acceleration, velocity and displacement, produced at different times. After the end of the seismic action, the structure continues to move under form of *free oscillations*, which depend on its level of damping. For strong damping, the movements stop very quickly, while for weak damping structure continues to move a long time after the end of the seismic action. Generally, the maximum values of movements occur during the forced oscillations, but for short seismic actions (such as pulse loadings), the maximum values can be reached during free oscillations

The structure movements are characterized by *vibration modes* (Fig. 8.12), being a superposing of these modes in function of the participation factor. The vibration modes are horizontal, vertical and torsional. For horizontal modes, generally the most important for seismic design, the number of vibration modes depends on the number of masses. But, in the majority of cases, the first three modes are the most important for the structural analysis. Which mode is determinant for the structural response depends on the ground motion and the structure characteristics.

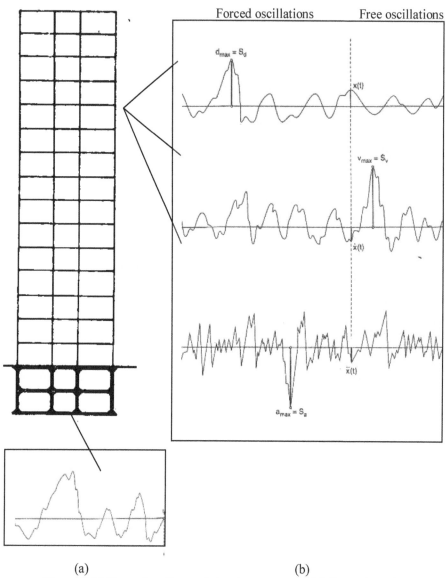

(a)   (b)
Figure 8.11 Ground motion and structure response: (a) Ground motions;
(b) Structural response: displacements, velocities and accelerations (after Ifrim, 1984)

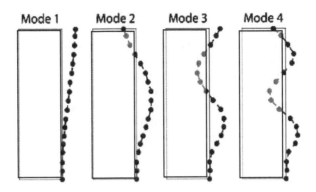

Figure 8.12 Vibration modes (after Kohler et al. 2007)

Looking to ground motions and structural response, one can see that the latter is much larger that the input movements. The reason of this amplification is due to the phenomenon of *resonance*, which is maximum when one of the frequencies of the ground motion oscillations is equal to the one of the natural frequency of the structure. In this situation, there is a very important amplification of the structural response, as a function of damping effects.

For instance, in some cases, especially for interplate earthquakes, when the main frequency of ground motions corresponds to the frequency of the first vibration mode of the structure, the resonance increases the amplitude of this mode. But in some cases, mainly for interplate earthquakes, when the main ground motion frequency corresponds to the frequencies of the second or the third structural mode, only these modes will be amplified by resonance (Fig. 8.13) (Gioncu and Mazzolani, 2006). This aspect gives rise to a very complex problem, which cannot be ignored during a proper seismic design process. It is a very important observation, because in case of resonance with superior mode of vibrations, the most affected part is the top of the structure, while in case of resonance with the first mode, the main damage is concentrated at the first structure levels (Gyorgyi et al, 2006). The main resonance with the superior modes is characteristic for intraplate crustal earthquakes, producing ground motions with very short natural periods of vibration, while for interplate crustal and intraslab earthquakes, with medium and large ground motion natural periods, the main resonance occurs with the first mode.

Fortunately, due to the very chaotic nature of ground motions, these resonance and amplification phenomena take place for a short time only, the ground motion acting as a damper for the structure motions during the remainder time. Unfortunately, in exchange, during the earthquakes, there are multiple resonance phenomena, when the superior structural vibration modes interact with the corresponding vibration modes.

The best example is the recorded accelerogram during the 1977 Vrancea earthquake. The main duration of 17 seconds was divided in four segments of time: the first of 5 seconds, the second of 2 seconds, the third of 5 seconds and the last of 5 seconds (Fig. 8.14a). The first segment corresponds to the beginning of the earthquake with moderate ground motions (arrival of P and S waves). The second segment, with the duration of only 2 seconds, marks the arrival of the surface waves, which occur under form of acceleration pulse with long period due to very bad soil conditions. The last two segments represent the ground motion as the consequence of the main shock. Considering each interval as an independent earthquake, Figure 8.14b shows the domination of the second segment corresponding to pulse acceleration.

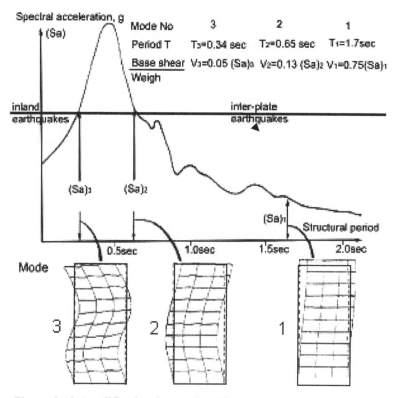

Figure 8.13 Amplification factors for different structure natural periods
(Gioncu and Mazzolani, 2006)

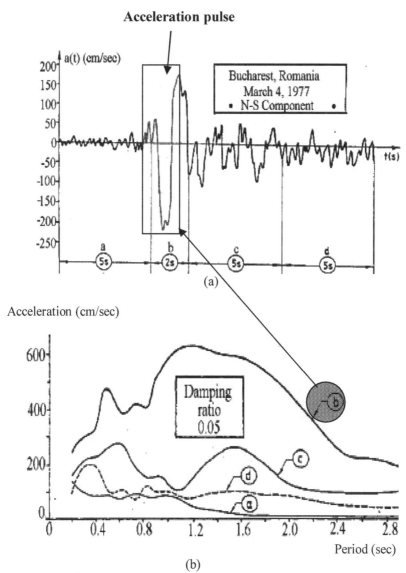

Figure 8.14 Influence of resonance in structural response: (a) The four segments of accelerogram; (b) Influence of each segment on the amplification
(Gioncu and Mazzolani, 2002, after Ifrim et al, 1986)

So, it is very clear that the ground motion is dominated by the very short segment of 2 seconds, the other segments having just a modest contribution.

Another conception regarding the structural response, valuable especially for near-source ground motions characterized by pulse velocity time history and acting on tall and regular buildings, is the so-called *wave-propagation approach*, developed by Iwan (1996, 1997). During an earthquake, a fault initiates seismic waves which are transmitted outward in all directions through the Earth's layers. The waves, after reaching the surface, are reflected back into the Earth when the surface is free. When there is a structure on the surface, however, the waves continue to propagate into the structure, causing the structure to vibrate. In other words, the vibration of a structure during an earthquake is caused by the propagation of seismic waves into the structure itself (Safak, 1999). Theoretically, the vibration and wave-propagation approaches represent two alternative solutions for the same problem. Vibration concept, such as the modal representation of structure movements, has been the standard approach in code provisions. Wave-propagation concept is useful when the structure can be modeled as a continuous medium. This concept, used for directly determining the interstory drifts, avoids the complex analysis of modal resonance. For structures subjected to near-source ground motions, the coherent pulse propagates through the structure as a *single harmonically oscillating wave*, causing large localized inter-storey drifts (Fig. 8.15). The structure is modeled as a shear-beam with regular stiffness and mass. The coherent pulse-like nature of near-source ground motion time history may cause the maximum response of the structure before a resonant mode-like response. This concept allows considering the effect of localized yielding, which produces softening in the affected stories. In this way, it is easy to see that localized yielding may lead to substantially larger interstory drifts. This method has the advantage to localize the yielding at different stories and to determine the most suitable position of the collapsed story, some times being the middle one. An example of wave-propagation in a framed structure is presented in Figure 8.16 (Kohler et al, 2007), showing that a pulse ground motion is transferred into the structure as a pulse story drift with a very high velocity (about 200m/sec).

Therefore, in far-source areas the concept of superposition of modal vibrations is the best way to understand the structural response, but in near-source areas the concept of wave propagation seems to be the most proper and simple to describe the structural response.

Concluding from the above presented aspects, the determination of the structural response involves the description of the structure movements at every time within the considered period, characterized by accelerations, velocities and displacements. The key of a proper determination of the structure response is related to the consideration of the actual characteristics of ground motions, very different in function of the source type: crustal interplate and intraplate earthquakes, and deep intraslab earthquakes. Differences come from propagation of waves, different natural periods, pulse or cyclical ground motions, earthquake duration, acceleration level, etc.

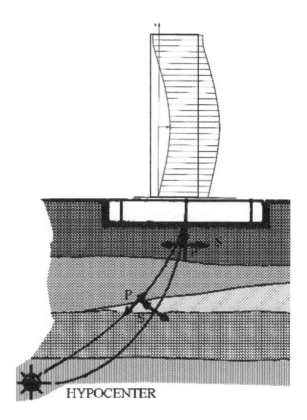

HYPOCENTER

Figure 8.15 Continuous shear-beam model (Gioncu and Mazzolani, 2002)

8.3.2 Capacity Triad: Stiffness, Strength, Ductility, plus Robustness

In order to evaluate the response of structures during earthquakes, the best method is to use a dynamic analysis for the time history representation of seismic loading, obtained from a recorded ground motions (ASCE, 2005). However, this methodology has the disadvantage to consider just the structural behavior for a given level of seismic loading, without the representation of the structure collapse. In this situation, the nonlinear static procedure is the best way to evaluate the structural response. This procedure is intended to provide a simplified approach for directly determining the nonlinear response behavior of a structure at different levels of lateral loads, ranging from the initial elastic response through the development of a failure mechanism and the initiation of collapse (Fig. 8.17). The role played in the seismic analysis by the triad of capacities (stiffness, strength and ductility) plus robustness is illustrated in the following.

Stories

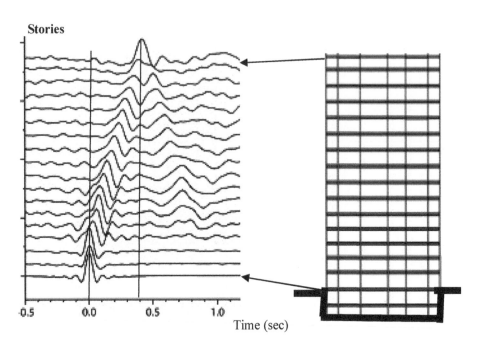

Time (sec)

Figure 8.16 Wave-propagation on a framed structure (75 m in 0.4 seconds)
(after Kohler et al, 2007)

Stiffness. Usually, the structure behaves in elastic range until the first yielding (or plastic hinge) forms and both structure and non-structural elements remain without (or with minor) damage. The structural behavior is linear and the traditional analysis methods can be used to determine the structural response. In this range, the knowledge of the actual stiffness is the main target.

Strength. When the structure is subjected to loads which overcome the elastic limit, a number of plastic hinges forms, taking advantage from the structure redundancy. The non-structural elements will be partially or totally damaged. The structural behavior is strongly nonlinear and an elastic-plastic analysis method has to be used, considering the progressive history of the plastic hinge formation.

Ductility. After the formation of a sufficient number of plastic hinges, the structure is transformed in a plastic mechanism. In this field of the structure behavior, the kinematic analysis method, which considers the element of ductility, must be used. Multiple configurations of plastic mechanisms are possible for most structures. The one caused by the smallest lateral forces is likely to appear before the others do. This mechanism is considered to be the dominant mechanism.

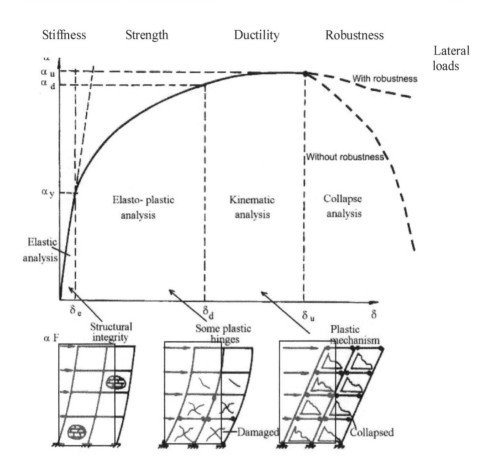

Figure 8.17 Capacity curve and characteristics of the structural response
(Gioncu and Mazzolani, 2002)

Robustness. The attainment of maximum lateral load means the beginning of the structure collapse. There are two collapse types, depending on the structural robustness: brittle or ductile collapse. The main design target is to obtain a structure with ductile collapse.

So, the seismic performance of a structure should be evaluated based on the nonlinear lateral load versus top displacement curve, which indicates its capacity to resist lateral forces. Figure 8.17 presents the four principal stages of the structural behavior: elastic, elastic-plastic, kinematic and collapse.

At the same time, this nonlinear capacity curve underlines the main attributes of the structural response: *stiffness, strength, ductility* and *robustness*.

8.3.3 Stiffness and Interstory Drifts

Stiffness is the property of a structure to resist displacement. Traditionally, strength checking is considered by designers as the primary goal of the design process. In the last time, due to the social and economical impact of the loss of functionality in buildings, there is a particular focus on the control of damage through the stiffness checking, in order to maintain the architectural integrity of non-structural elements. The stiffness checking is performed for low seismic action, when the structure works in *linear elastic range*. The top displacement of the structure can provide a good indication of the structural damage, but it cannot adequately reflect the damage of non-structural elements, which are more dependent on the relative displacement between two storys, the so-called *interstory drift* (Fig. 8.18), the location of them depending on the ground motion type. Deep earthquakes in far-field regions, the first vibration mode being dominant, produce interstory drifts with maximum values in the lower part of the structure (Fig. 8.19a). The structure deformations for ground motions in the near-source regions are qualitatively different. In this case, the second and third modes are dominant, so the maximum interstory drifts occur in the middle or at the top part of structures (Fig. 8.19b). At the same time and consequently, in the first case the plastic hinges take place in lower parts, while in the second case they are concentrated in the middle or at the top of the structure.

The condition of non-damage to non-structural elements must be verified in elastic range for given interstory drift limits, which take into account their integrity due to the interaction with structural elements. Figure 8.20 presents the effects of this interaction between a five levels framed structure with a core containing the staircase and different panel types. One can see that an important reduction of drifts can be obtained by considering this interaction, respect the behavior without panels.

8.3.4 Strength and Redundancy

The strength is the property of a structure to resist forces. Redundancy is the property to spread the damage among several elements, avoiding the concentration of failure at an individual structural element. The elastic behavior ends when the first plastic hinge occurs. It is followed by the elastic-plastic range, where a sequence of formation of plastic hinges induces a very important *non-linear mechanical behavior* (Fig. 8.17). The elastic-plastic range depends on the *structural redundancy*. Redundancy is a positive and desirable feature of a structure. A non-redundant system is the one where the failure of a component is equivalent to the failure of the entire system. Contrary, in a redundant system, all the components are participating in carrying load and the

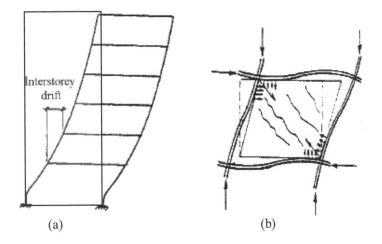

(a)   (b)

Figure 8.18 Stiffness in elastic behavior; (a) Interstory drifts; (b) Damage to non-structural elements (Gioncu and Mazzolani, 2002)

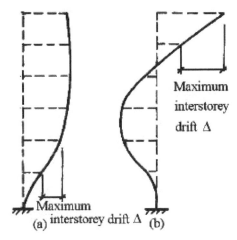

Maximum
interstorey
drift Δ

Maximum
(a) interstorey drift Δ (b)

Figure 8.19 Interstory drifts: (a) Influence of first vibration mode; (b) Influence of superior vibration modes (Gioncu and Mazzolani, 2002)

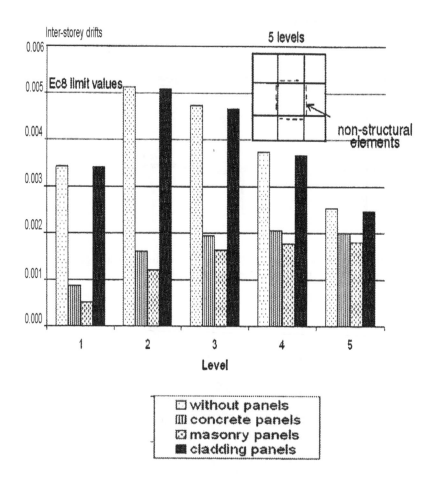

Figure 8.20 Influence of the interaction between structure and non-structural elements
(Truta et al, 2003)

system failure corresponds to the failure of all active components (Bertero and Bertero, 1999). It is important to distinguish between the structure behavior under static (or pseudo-static) lateral forces and dynamic lateral forces.

According to the static approach, the formation of the global collapse mechanism is presented in Figure 8.21, which is directly related to the degree of indeterminacy, which is defined as the number of plastic hinges formed at the member ends, up to the point of total collapse (Husain and Tsopelas, 2004). This collapse mechanism

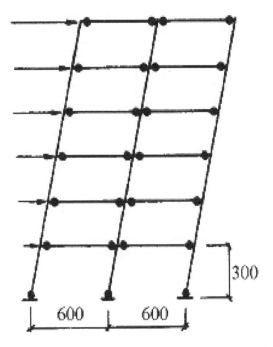

Figure 8.21 Behavior of structure under static lateral loads: global collapse
mechanism (Gioncu and Mazzolani, 2002)

considers that all the plastic hinges are ductile (the ductility capacity in the critical
range is larger than the ductility demand).

Contrary, a dynamic approach corresponding to the pulse seismic ground motion
provides the plastic mechanisms presented in Figure 8.22, which shows the succession
of plastic hinges formation (using the DRAIN-2D computer program), the collapse
mechanism being produced at the top of the structure. No global collapse mechanism
occurs if the pulse seismic action is considered. So, the determination of the collapse
mechanism, in addition to the structure dimensions, requires considering the type of
seismic action, which is essential to be taken into account.

First plastic hinges

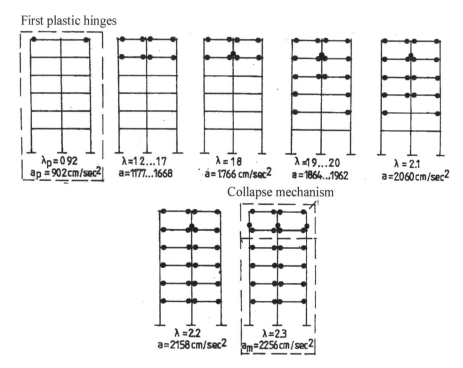

Figure 8.22 Effect of redundancy during earthquake (Gioncu and Mateescu, 1998)

8.3.5 Global and Local Ductility

Ductility is the ability of a structure to develop its bearing capacity in inelastic range and to maintain it during an earthquake. Under normal conditions, a structure experiences elastic deformations, deforming when the force is applied and returning to its original shape when the force is removed. However, extreme earthquake forces may generate inelastic deformations, from which the structure does not return to its original shape after the force is removed. The ductile structure is built by ductile elements (local ductility). Therefore, structures composed by such elements tend to withstand earthquakes in very good conditions (global ductility).

In the phase of structure pre-collapse, when the number of plastic hinges is sufficient to produce a plastic mechanism, the structural analysis can be performed by means of the *kinematic theorem of plastic collapse*, which is based on the concept of mechanism equilibrium curve (Mazzolani and Piluso, 1996). Basically, it has been observed that the collapse mechanism of frame subjected to the lateral forces can be classified in four types: global mechanism, mechanism affecting the lower or upper parts of structure or only a story of structure (Fig. 8.23). The produced mechanism type depends on the structure conformation, but at the same time on the ground motion type. Generally, the global mechanism and type 1 mechanism are accepted to be the more frequent type for intraslab earthquakes. In case of velocity pulse with short period (crustal earthquakes), the collapse is formed at the top or middle part of the structure (Types 2 and 3 mechanisms).

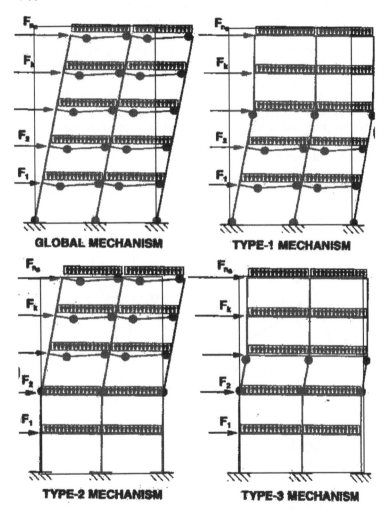

Figure 8.23 Collapse mechanism types (courtesy of Montuori and Piluso, 2000)

A measure of the overall structural ductility is given by the local material, member and joint ductility. The material ductility corresponds to the ability of a material to undergo large plastic deformation before rupture. The element or joint ductility characterizes the ability to carry out and transmit stresses to the neighbouring elements in elasto-plastic range without loss of resistance.

A framed structure cannot exhibit a ductile behavior if the plastic hinges formed at the member ends have not enough rotation capacity, being able to redistribute the bending components (Fig. 8.24). Therefore, the analysis of the behavior of plastic hinges became of primary interest (Gioncu and Mazzolani, 2002). The limitation of the rotation capacity is the result of the plastic local buckling of the compressed parts of the cross-section. There are three methodologies to evaluate the ability of plastic hinges to develop an adequate rotation capacity:

- introducing *cross-sectional classes* (Fig. 8. 25a), which limit the slenderness of the compressed parts of the cross-section, in order to eliminate local buckling effects. Four classes are proposed, corresponding to plastic, compact, semi-compact and slender sections. This methodology is used in the majority of the modern design codes, being very simple and friendly;

- using *member behavior classes,* which considers the plastic rotation capacity of members (Fig. 8.25b) (Mazzolani and Piluso, 1993). Three classes of ductility are proposed: high ductility, medium ductility and low ductility, in function of the

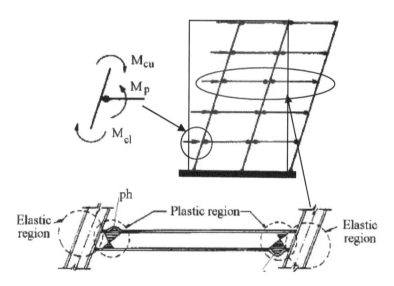

Figure 8.24 Rotation of plastic hinges and structure ductility
(Gioncu and Mazzolani, 2002)

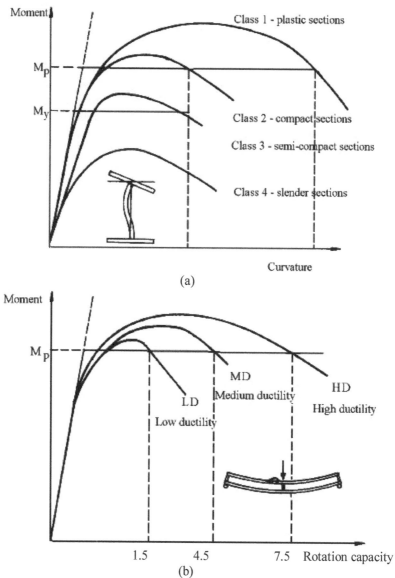

Figure 8.25 Ductility definitions: (a) Cross-sectional behavior classes;
Member behavior classes (Gioncu and Mazzolani, 2002)

member rotation capacity. This proposal has been given for the steel structures, but similar classification is proposed also for reinforced concrete elements;

- analyzing the plastic rotation capacity of steel members by using a *specialized computer program DUCTROT M* (Petcu and Gioncu, 2003, Gioncu and Mazzolani, 2002). This program is developed on the base of local plastic mechanism methodology (Fig. 8.26a) and it allows the evaluation of the plastic rotation and the rotational ductility under static or seismic actions, as a function of cross-section type, geometrical member dimensions, mechanical properties of steel. Figure 8.26b shows the rotation capacity for a wide-flange beam, using this methodology.

The rotation capacity reduces due to the influence of seismic cycles when the number of important cycles exceeds a given value. The program DUCTROT M for seismic actions considers the superposition of plastic mechanisms for positive and negative moments (Fig. 8.27a,b), resulting in the moment-rotation curve for cyclic loading. One can see that this curve is divided in two parts: (i) in the first, the cyclic loads have no influence on the rotation capacity, because the plastic rotation occurs in the moment-rotation plateau; (ii) after a number of important cycles, a significant reduction of rotation capacity arises (Fig .8.27c).

For the ground motion types characterized by short duration and reduced number of important cycles (for instance, intraplate crustal earthquakes), no reduction of rotation capacity must be considered. Contrary, for earthquakes with long duration (intraslab earthquakes) and large number of important cycles, the erosion of rotation capacity can be dramatic (Fig. 8.27c).

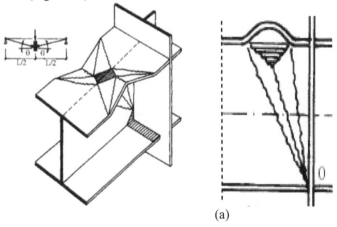

(a)

Figure 8.26 (continues)

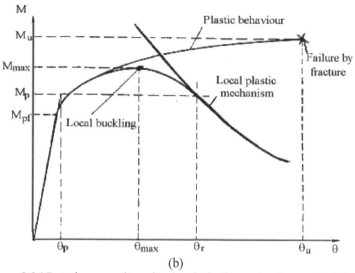

(b)

Figure 8.26 Rotation capacity using local plastic mechanism methodology:
(a) Local plastic mechanism; (b) Evaluation of rotation capacity
(Gioncu and Mazzolani, 2002)

Positive moment

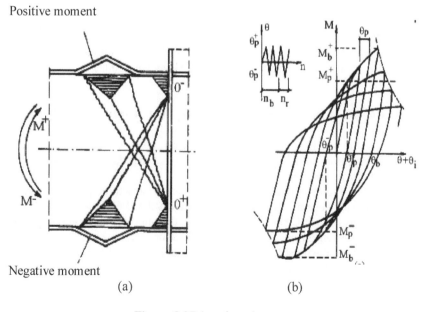

Negative moment

(a)   (b)

Figure 8.27 (continues)

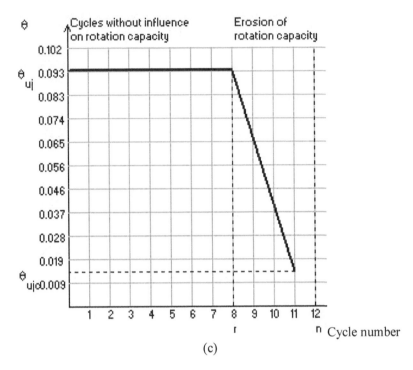

(c)

Figure 8.27 Influence of cyclic loads: (a) Plastic mechanism for cyclic loads; (b) Moment-rotation curve for cyclic loads; (c) Influence of important cycles on rotation capacity (Gioncu and Mazzolani, 2002)

8.3.6 Robustness and Ductile or Brittle Catastrophic Collapse

Robustness is the quality of a system of being able to withstand sometimes unpredictable variations of the designed forces with minimal damage, alteration or loss of functionality (Wikipedia, nd).

For structures, the importance of robustness as a structural property has been recognized following several structure failures, such as the 1968 Ronan Point Building and the 2001 World Trade Center towers. Among the unpredictable causes of failure, terrorist attacks, design errors, errors during execution, unforeseen deterioration and poor maintenance are the most frequently mentioned. But we have to consider not least the ground motions due to earthquakes, which are predicted with a large level of uncertainty or which, in many cases, are not considered in the design process, representing an exceptional loading condition. In seismic design, the concept of ductility has been introduced as the property of structures for protecting them against these exceptional situations, considering the possibility of formation of plastic hinges in elasto-plastic structures. But this property is not always sufficient to protect the structure in case of very strong earthquakes. The building collapses during the 1985 Mexico City, 1994 Northridge, 1995 Kobe, 1999 Kocaeli and 1999 Chi-Chi earthquakes clearly exemplify the cases when the predictable ground motion characteristics have been exceeded.

An important question is to know what happens to the structural system as soon as its maximum loading response is exhausted. The answer to this question belongs to a new field of research, which has not been sufficiently explored until now. According to Earthquake Engineering, the robustness can be defined as the *residual resistance* (Fig. 8.28), beyond the formation of the global plastic mechanism, when an unexpected event occurs. In the range of residual resistance, the structural systems respond in a ductile or brittle fashion, depending on their typological, geometrical and organizational configuration, as well as on the materials they are made of (Vogel, 2005).

In a robust structure, the resistance to seismic loads does not disappear suddenly after the maximum action is reached and the structure will stand up, permitting evacuation and repair with an acceptable expense. In this condition, the structure failure can be defined as *ductile collapse*. Contrary, if this condition is not satisfied by a non-robust structure, after reaching the maximum loading value, a *brittle catastrophic collapse* occurs.

The structures are designed according to the code loads, which are determined with a high level of uncertainty for the earthquake events. In many cases, these values are

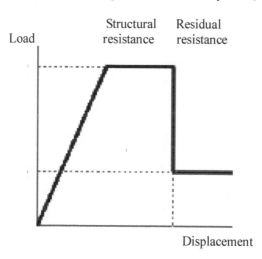

Figure 8.28 Structural and residual resistance

exceeded in practice and, therefore, this loading condition can be considered as exceptional (Mazzolani, 2002). Earthquakes can occur in non-seismic areas, where the structures are not prepared to withstand these events. The goal of robustness is to limit or mitigate the consequences of the unforeseen seismic conditions (acceleration, duration, directivity, pulse effect, etc.).

There are two main factors influencing the residual resistance of structures: constructional materials and structural configurations.

A ductile or brittle collapse is strongly dependent on the used *structural material*. Three materials, masonry, concrete and steel, those mostly used for structures, are discussed in the following.

Unreinforced masonry is typically non-elastic, non-homogeneous and anisotropic material composed of two components of quite different properties: stiff brick and relatively soft mortar. Masonry is normally designed to resist compression actions only, being very weak in tension and shear (Kaushik and Rai, 2007). Figure 8.29 shows the stress-strain curves for masonry prisms for different grades of mortar. One can see that the maximum residual compressive stress is very low: 0.2 f_m, where f_m is the compressive prism strength of masonry. Due to this fact, a structural load-displacement curve must neglect the residual resistance. In fact, the observed collapse of masonry during the strong earthquakes shows that this failure is of brittle type and there is not any residual resistance. Therefore, the masonry structures are not robust structures. The use of FRP (Fiber Reinforced Polimers) as reinforcement can improve the robustness of masonry structures (Nagy-Gyorgy, 2007)

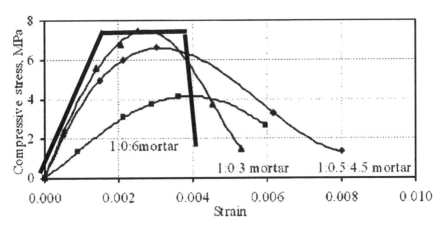

Figure 8.29 (continues)

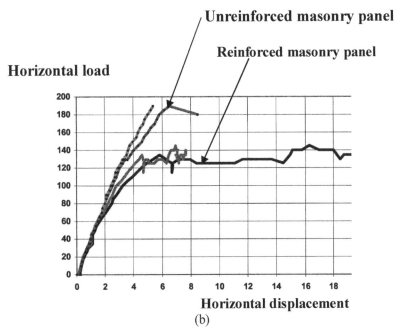

Figure 8.29 Clay brick masonry: (a) Axially compressed unreinforced masonry: (after Kaushik and Rai, 2007; (b) Horizontally loaded unreinforced and reinforced (with FRP) masonry panel (courtesy of Nagy-Gyorgy, 2007)

Reinforced concrete is the most used material for structures in seismic areas. It is composed of two materials, concrete having a good behavior in compression and bad behavior in tension. This bad behavior is compensated by the reinforcement steel bars. The concrete characteristic curves plotted as a function of strain rate are presented in Figure 8.30a. As can be observed, the concrete strength grows with the increasing of strain-rate, but, at the same time, the residual resistance decreases. Therefore, also for reinforced concrete structures it is not expected to obtain an important robustness in case of earthquake loads. This observation is confirmed by many reinforced concrete structures totally collapsing during the earthquakes. Some residual resistance can be considered only for walled reinforced concrete structures, even if it is strongly reduced by the cyclic load corresponding to an earthquake action (Fig. 8.30b). In general, the residual resistance in case of framed reinforced concrete structures must be considered with prudence in the design process.

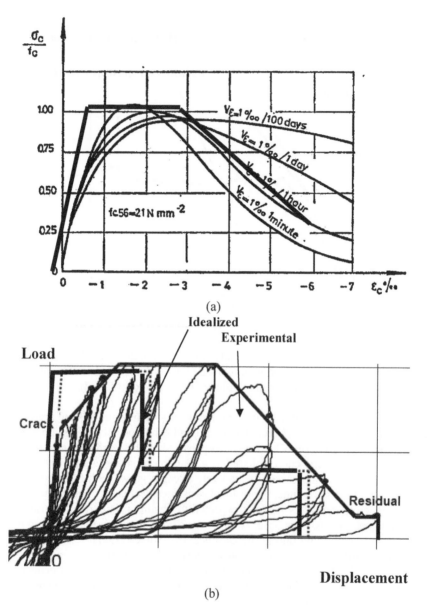

(a)

(b)

Figure 8.30 Reinforced concrete: (a) Influence of strain rate for concrete (Avram et al, 1981); (b) Idealized and experimental load-displacement curve for reinforced concrete wall (after Wallace, 2005)

Steel members. In contrast with the masonry and reinforced concrete members, where the maximum resistance is reached at the crushing of material, steel members reach the maximum capacity at the plastic buckling of the compression flange (Fig. 8.31a). This local buckling occurs before the fracture of the tension flange operates as a filter against large strains reducing the danger of section cracking (Gioncu and Petcu, 1997). After reaching the maximum value, the decreasing of resistance is moderate and the ultimate load is given by the crack of the tension flange. The difference between the loads corresponding to the plastic buckling of the compression flange and the crack of the tension flange gives the possibility to consider the residual resistance of steel members in the design process (Fig. 8.31b).

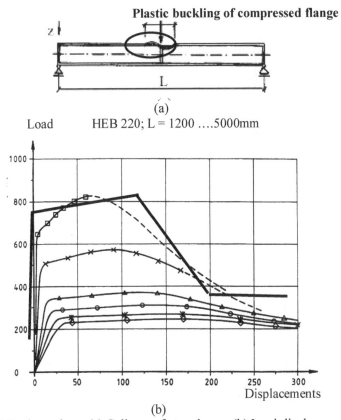

Figure 8.31 Steel members: (a) Collapse of steep beam; (b) Load-displacement curves (after Spangemacher, 1992)

This remark is valuable also for the joints of steel structures; even the much damaged connections present an important residual resistance (Fig. 8.32). This aspect is proved by the observation that during strong earthquakes many masonry and concrete structures collapse by total failure, when the steel structures just present large damage without total failure (see damaged buildings during the Northridge and Kobe earthquakes). The only case of total collapse of a steel structure was the Pino Suarez building (1985 Mexico City), due to the fact that the residual resistance could not be accounted for by the premature rupture of column welding.

The second factor influencing the residual resistance is the *structural configuration*. The structural robustness is related to the *progressive collapse* and one of the most

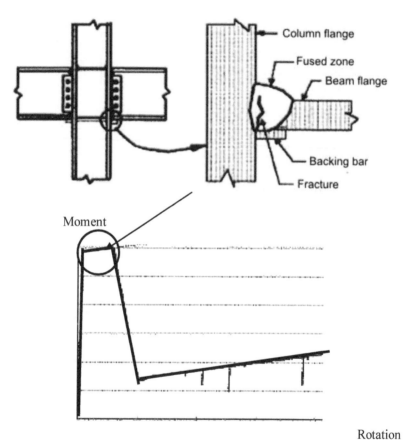

Figure 8.32 Modeling of steel fractured joint (after Kim et al, 2006)

frequent causes producing progressive collapse is the earthquake. Among the different definitions of progressive collapse, there is one which directly refers to the buildings' robustness (Mazzolani, 2002, Gioncu, 2006, 2007):

"....*design for consequences of localized failure from an undefined cause, a strategy which ensures that a building is sufficiently robust to sustain a limited extent of damage, depending on the consequence class, without collapse* ".

The most important conceptual aspect for avoiding the structural collapse during earthquake is to design the structure in such a way to allow the possibility of redistributing the internal actions when some critical members are damaged. The *Alternative Path approach* considers the residual resistance against the progressive collapse by providing a second modality to transfer the loads when the primary load-bearing members are eliminated due to local damage (Fig. 8.33a) The very recent studies on this aspect refer to the improving the catenary effects of the floor slabs in such a way to prevent progressive collapse in the event of loss of bearing capacity of columns.This improving is related to the increasing the connection's tension capacity. In order to obtain this catenary effect, the location of cables in zones susceptible to column damage is a very efficient solution for reinforced concrete buildings. In case of steel structures, it is required to increase the joint capacity for tension forces, using (for instance) the side-plate connections .

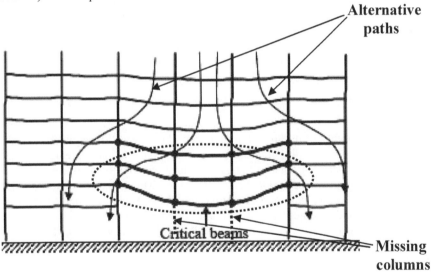

(a)

Figure 8.33 (continues)

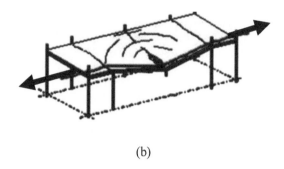

(b)

Figure 8.33 Structural residual resistance by: (a) Alternative path approach;
(b) Catenary effect (Gioncu, 2007, after Mendis and Ngo, 2003, Astaneh-Asl, 2003)

8.4 DEVELOPMENT OF MATERIALS FOR SEISMIC-RESISTANT STRUCTURES

The need to erect structures in seismic areas has led to the development of some particular typologies. Earthquake-resistant structures must absorb and dissipate induced motion through a combination of damping and inelastic deformations. Therefore, there are two main determinant factors for a good response: materials having satisfactory ductility and structural systems able to fructify this ductility in a dissipative system. The developments of structural systems will be discussed in the next section. The main materials used in seismic areas are masonry, concrete and steel.

8.4.1 Masonry

The seismic vulnerability of traditional masonry buildings is well recognized. In order to improve their behavior, the reinforced masonry construction system has been introduced in the modern masonry buildings. The most common type, used as seismic-resistant wall, is the *reinforced grouted cavity masonry*, which consists of two parallel masonry panels, which are built at a given distance (Fig. 8.34a). Inside of the cavity between the two panels, a mesh of vertical and horizontal steel rebars is located and then filled with cast concrete. The second type is the *confined masonry*, (Fig. 8.34b) which is a masonry panel confined on all four sides by reinforced concrete members, the horizontal ones being bond-beams and vertical members called tie-columns. In order to achieve an effective wall confinement, the tie-columns must be located at all corner joints and wall intersections.

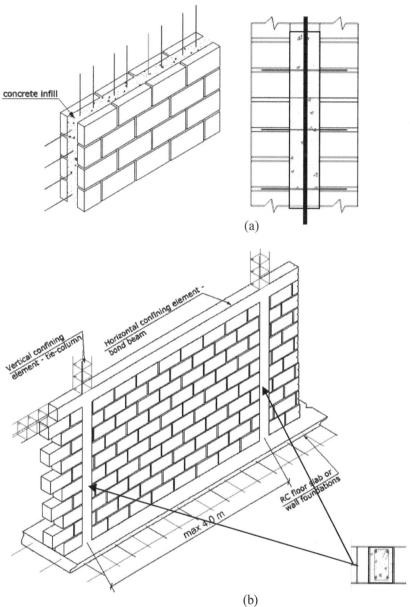

(a)

(b)

Figure 8.34 New masonry types for seismic areas: (a) Reinforced masonry;
(b) Confined masonry (after Virdi and Rashkoff, 2008)

8.4.2 Concrete

Concrete, as a material for seismic resistant structures, has two important deficiencies: low strength to unit weight ratio and very low tensile resistance. In order to improve the concrete qualities, the research is directed to obtain the so-called *high-performance concrete* with high compression and tension strengths. The performance limit of high strength concrete changed with time: 34 MPa (1950), 52 MPa (1960) and 62 MPa (1970). More recently, compressive strengths approaching 138 MPa have been obtained in cast-in-place buildings. Today, high strength concrete with 250 MPA compression strength is produced using high strength aggregates (Husem and Pul, 2007). High-performance concrete is used mainly in the columns of high-rise buildings, obtaining the most economical way to transfer the vertical loads to the foundations. Using highstrength concrete, with the strength as high as 100 MPa, the dimensions of columns are reduced, lowering, consequently, the structural weight, which is a favourable aspect in seismic design, as also the seismic forces are reduced.

There are many advances in using high strength concrete instead of the normal one. However, due to its brittleness, many engineers are still reluctant to use it in seismic-resistant structures. The theoretical and experimental research works showed that *well-confined* high strength columns in these zones, which are potentially susceptible to develop large plastic deformations, can provide a satisfactory ductile behavior with large gains both in strength and ductility, provided that adequate transverse reinforcement is used (Cusson et al, 1996, Sharma et al, 2005, Claeson, 2008, Ngo et al, 2008). The material models for unconfined and confined concrete are shown in Figure 8.35. If the ascending branch is the same for both concrete, differences consist just in strength, the descending branch for a well-confined column shows a smooth falling.

In order to assure a good behaviour against potential plastic zones in a column, the transverse column reinforced must vary along the column length, as illustrated in Figure 8.36a. Good, improved and poorly confinements are presented in Figure 8.36b.

The second important advantage in using the high-performance concrete is the increasing of the tensile strength. The *high- performance fibre reinforced concrete* is a very suitable material in the regions where a large inelastic deformation capacity is necessary to withstand the requirements induced by severe earthquakes (Canbolat et al, 2005, Parra-Montesinos, 2005). It has the important property to increase the shear resistance, ductility and energy dissipation in members subjected to reversal cyclic loadings.

There are two typical fiber reinforced concretes: a regular one, characterized by a softening response after the first cracking and the high performance one, with multiple cracking.

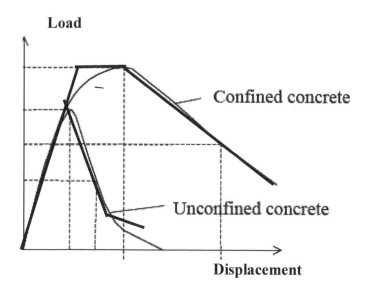

Figure 8.35 Load-displacement curves for unconfined and confined reinforced concrete (after Cusson et al, 1996)

The research activity for producing this new material is presented by Parra-Montesinos (2005), being obtained using reinforcement with fiber of steel or polyethylene plus cement, fly ash, flint sand 30-70 and water-reducing admixture. Due to the increase of construction costs associated to the addition of fibers in concrete, the high performance reinforced concretes are generally intended to be used only in critical regions, where the inelastic deformation demands may be large and fiber reinforcement is required to ensure a satisfactory behavior during an earthquake. In particular, the excellent tensile behavior of these materials makes them attractive for members with shear dominated response, as the coupling beam in structural walls (Canbolat et al, 2005).

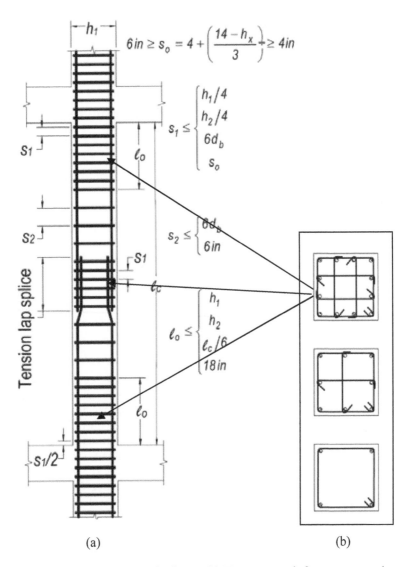

(a)   (b)

Figure 8.36 Confinement of column: (a) Transverse reinforcement spacing requirement:
(b) Hoop configuration: well, improved and poorly confinement (after NEHRP, 2008)

8.4.3 Steel

In modern design practice it is generally recognized that steel is an excellent material for seismic-resistant structures because of its performance in terms of strength and ductility, being capable of withstanding substantial inelastic deformation. In general this is true, but in the last few decades the strong earthquakes have seriously compromised this ideal image of steel as the perfect material for seismic areas. This is due to the fact that, in some cases, the good ductility of the material is lost when it is used in inadequate structural types. As a consequence, an important field of research works has been developed in the last decades, including also the improving of the material properties. This aspect can be framed in the field of high strength steels, low yield steels and aluminium alloys.

The *high strength steels* are used in structures for non-dissipative elements, which must remain in elastic range during the earthquake, due to their limited ductility, when compared to mild steels. It is generally accepted that the transition from mild steel to high strength steel occurs at the yield strength of about 350 MPa. (Bjorhovde, 2002). Figure 8.37a shows the stress-strain curves for steels exceeding yield strength of about 350 MPa and having convenient rupture strain.

The *low yield steel* is proposed to be used for dissipative elements, as the shear wall panels in moment resisting frames. This system has found a large consensus in recent years. Due to the small amount of carbon and alloying elements, the nominal yield stress is about 90 to120 MPa and the nominal ultimate stress over 50%. This enables the panels to undergo large inelastic deformations at the first stages of the loading process, enhancing the energy dissipation capability (De Matteis and Mistakidis, 2003). Recently, a new material has been proposed at the University of Naples for shear panels, the pure aluminium (Formisano et al, 2006). A comparison of the stress-strain curves for mild steel, aluminium alloy, low yield steel and pure aluminium, is presented in Figure 8.37b.

8.4.4 Shape-memory Alloys

The shape-memory alloys, also known as smart alloys, are materials which remember their initial shape. It is considered as one promising material for special devices in seismic resistant structures due to its ability of large deformation recovery, high ultimate tensile strength and dissipation of seismic energy without residual deformations. The phenomenon of pseudo-elasticity effect results from the transformation phase above austenite. When the applied stress reaches a critical value, the austenite material begins to transform into the martensite phase. Once the loading is released, the stress inducing the martensite returns to the austenite phase without residual strain (Pan and Cho, 2007). Austenite and martensite phases are the initial and the deformed molecular arrangements, respectively, which characterize the behavior of shape-memory alloy (Wikipedia, nd). As a result, one hysteretic loop can be formed by a closed stress-strain curve in the whole loading-unloading process. This hysteresis loop is representative of the energy dissipation capacity, which can be exploited in seismic resistant structures.

One must mention that this material is very expensive; so, it can be used only in small quantities as a part of some special devices.

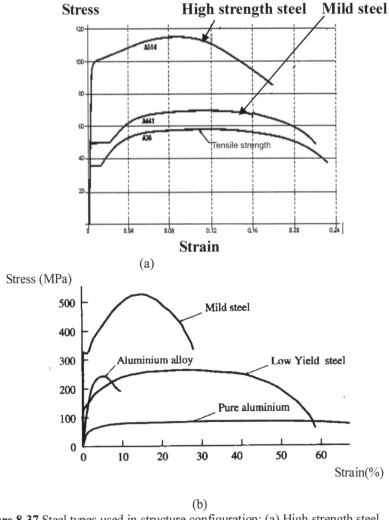

Figure 8.37 Steel types used in structure configuration; (a) High strength steel (FEMA 403, 2002); (b) Low yield steel (Formisano et al, 2006)

8.5 DEVELOPMENT OF STEEL STRUCTURAL SYSTEMS

8.5.1 Main Structural Types

In recent times, many new structural systems or devices using non-conventional civil engineering materials have been developed, either to reduce the earthquake forces acting on a structure or to absorb a part of the seismic input energy. Due to the bad behavior of some steel structures during the Northridge and Kobe earthquakes, the main important developments in structural systems are particularly observed in steel structures. Figure 8.38 shows four conventional types of techniques employed to control the structural response during earthquakes through inelastic deformations in structural members: moment resisting frame, centrically braced frames, eccentrically braced frame and infilled frames. There are many variations under each broad category and many new techniques are being developed, evaluated and implemented, as discussed below.

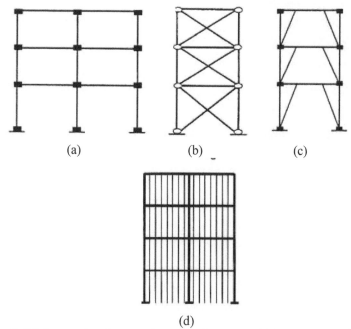

(a)   (b)   (c)

(d)

Figure 8. 38 Conventional types of structural systems: (a) Moment resisting frame; (b) Centrically braced frame; (c) Eccentrically braced frame; (d) Infilled frame (Mazzolani, 2000a,b)

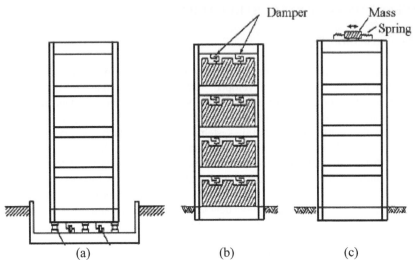

Figure 8. 39 Response control systems: (a) Base isolation; (b) Energy absorption;
(c) Mass effect mechanism (Gioncu and Mazzolani, 2002)

A significant progress has been recently made in development and application of innovative structural systems for seismic protection. The aim of these systems is the modification of the dynamic interaction between structure and ground motions, in order to minimize the structural damage and to control the structural response. So, this concept is very different from the conventional one. The response control systems can be roughly classified as shown in Figure 8.39: base isolation, energy absorption and mass effect mechanism (Gioncu and Mazzolani, 2002).

8.5.2 Moment Resisting Frames

In case of moment resisting frames, the most significant tendency is to change the traditional welded connections with bolted joints (Fig. 8.40). This tendency is the result of the bad behavior of welded connections during the 1994 Northridge and 1995 Kobe earthquakes, where the site welding type has been incriminated for the cracks occurred in joints (Gioncu and Mazzolani, 2002). These near-source earthquakes caused to steel moment resisting frames (MRF) unexpected brittle fractures in the traditional beam-to-column, welded flange-bolted web connections. In particular the full confidence in strength, stiffness and ductility of such common moment connections was seriously undermined: most connections provided small inelastic deformations and their resistance was smaller than the one of the connected elements. Major causes were recognized to be the bad quality of groove welds and the composite action with the concrete slab to the beam strength, producing a large tensile strain in the bottom flange.

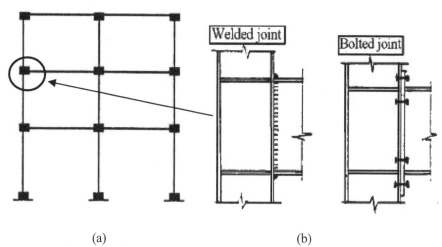

(a)   (b)

Figure 8.40 Moment resisting frames: (a) Geometrical conformation;
(b)Welded and bolted joints (Gioncu, 2000)

Among the studies of innovative solutions for moment connections, both field welded connections (i.e. welded flange-welded web connections, haunch connections, coverplate connections, reduced beam section connections) and field bolted connections (i.e. extended end plate connections, bolted flange plate connections, bolted T-stub connections) have been examined.

The *welded connection types* are presented in Figure 8.41: fully welded and welded flanges with bolted web. The last one was a standard connection type in the USA for a long time, but it was proved to be inadequate for near-source earthquakes. Some very important research works were performed both in the USA (Roeder, 2000) and in Europe (Mazzolani, 2000), in order to improve the behavior of this connection type. Among the proposed solutions, the "dog-bone" conception seems to be the most promising for protecting welded connections.

The *reduced beam section connection*, namely *"dog-bone"*, is a structural detail, which meets both the above-mentioned conception strategy, consisting in the introduction of a ductile fuse and in the improvement of the connections as well. Actually, reduced beam section allows to shift the plastic hinge from the beam ends by a controlled weakening of the beam itself in a relevant position (Fig. 8.42a), hence ensuring a better structural performance under cyclic loads at both local and global levels. In fact, the available ductility of the beam-to-column connections increases and consequently the seismic behavior of steel moment resisting frames enhances. Moreover the flange force which can be transmitted to the connection, to the panel zone, to the column continuity plates is reduced, with only a small local reduction of the beam cross section and of the overall lateral stiffness. By weakening

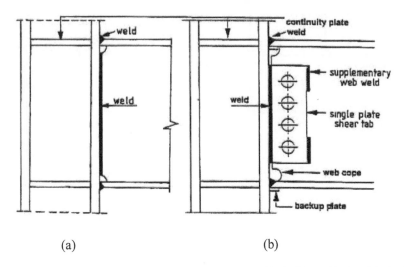

(a)   (b)

Figure 8.41 Welded connection types: (a) Fully welded; (b) Welded and bolted
connections (Gioncu et al, 2000)

specific sections of beam, the reduced beam section protects the node from plastic
deformations, producing a ductility enhancement. At local level, the inelastic
deformations in the beam do not involve the connection welds (Fig. 8.42b) and the
stress demand at the beam-to-column connection is reduced. As a consequence,
strength requirements for column flange, continuity plates and panel zone are less
stringent than for traditional beam-to-column connections. Moreover, the cross section
reduction is applied to the beam flange width, hence the b/t flange slenderness ratio
reduces and the local instability phenomena are delayed. At global level, the ductility
benefits from the increased rotational capacity; moreover "dog-bones" encourage the
attainment of more dissipative collapse mechanisms, being the capacity design for
members easier to be achieved. But, due to the large plastic deformations, the reduced
sections are much damaged and its repair may be expected to be very difficult or even
impossible. Therefore, it seems to be very rational to use a dismountable element for
the zone of reduced section (Fig. 8.42c) (Balut and Gioncu, 2003).

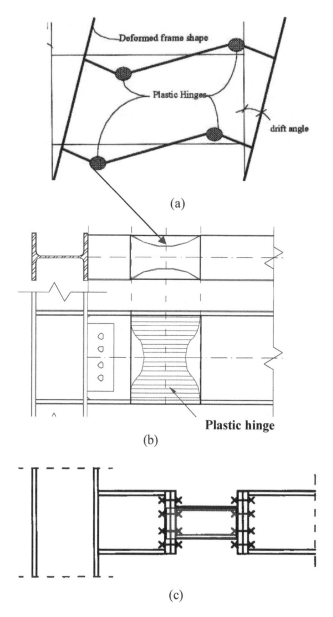

Figure 8.42 "Dog bone" concept: (a) Desired locations of plastic hinges (b) Reduced beam section; (c) Dismountable dog bone (Balut and Gioncu , 2003)

The variety of *bolted connections* is very rich in typologies: some of these types and the most used ones are presented in Figure 8. 43 as top and seat angles, flush end plates, extended end plates and welded flanges shop welded and site bolted. The behavior of these connections during earthquakes and experimental tests were good, with the exception of the case when the damage is produced for the brittle fracture of the bolts in the threaded zone. In order to improve the behavior of bolted connections, the use of high ductility bolts is proposed by Ohi et al (2000) and Balut and Gioncu (2006). The ductility of bolts could substantially be increased if the diameter in the unthreaded zone was reduced, because the bolt failure, which occurred in this zone, is ductile (Fig. 8.44).

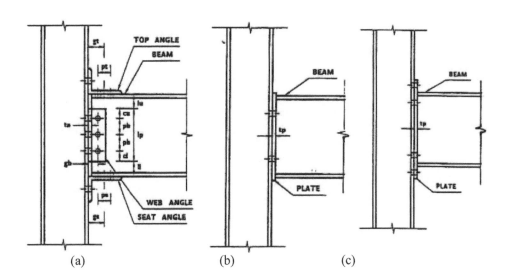

(a)   (b)   (c)

Figure 8.43 (continues)

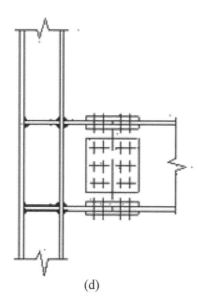

(d)

Figure 8.43 Bolted connections: (a) Top and seat angles; (b) Flush end plate
(c) Extended end plate; (d) Shop welded and site bolted

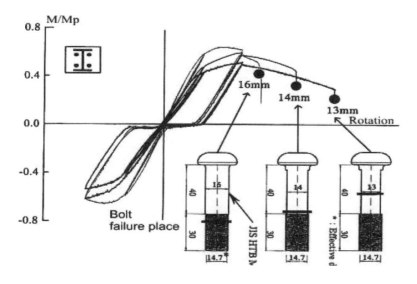

Figure 8.44 High ductility tension bolts (Balut and Gioncu, 2006, after Ohi et al, 2000)

8.5.3 Concentrically Braced Frames

The problems of concentrically braced frames (Fig. 8.45) are dominated by the buckling of compressed braces. This bad behavior was observed during the past earthquakes of Mexico City, Loma Prieta, Northridge and Kobe, where many braced frames were damaged by the buckling of braced elements. Generally, the use of steel concentric braces in seismic-resistant structures is a well-established technique. However, this system is characterized by a limited ductility capacity under cyclic loading, due to the difference between the tensile and compression behavior of braces. Due to this fact, its response to cyclic loading is not symmetric and exhibits substantial strength degradation, owing to buckling of braces under strong compression. In addition, the out-of-plane buckling of braces could cause severe damage to non-structural components. This complex response originates an actual distribution of internal forces and deformations in the overall structure which substantially deviates from the prediction based on conventional models. *Buckling restrained braces* (Fig. 8.46a) have been proposed in order to solve the above problems. The brace is conceived as a ductile brace core included in a filled tube, detailed so that the central brace can deform longitudinally independently of the mechanism which restrains lateral and local buckling. As shown in Fig. 8.46b this system is characterized by a stable hysteretic behavior and, differently from traditional braces, it permits an independent design of stiffness, strength and ductility properties. This behavior is achieved through limiting buckling of the steel core within the bracing elements. The axial strength is decoupled from the flexural buckling resistance. In fact, the axial load is confined to the steel core, while the buckling restraining mechanism resists the overall brace buckling and restrains the high-mode steel core buckling.

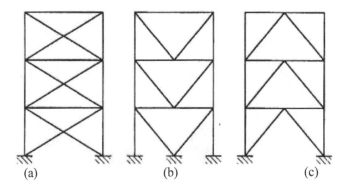

Figure 8.45 Concentrically braced systems: (a) X-braced frame; (b) V-braced frame; (c) Inverted V-braced frame (Mazzolani, 2000a,b)

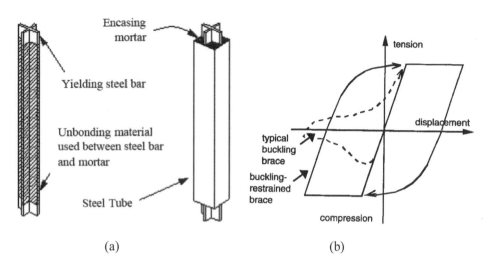

Figure 8.46 Buckling restrained braces: (a) Restrain system; (b) Hysteretic behavior
(D'Aniello et al, 2006)

8.5.4 Eccentrically Braced Frames

The eccentric braced frames (Fig. 8.47) are a hybrid lateral force-resisting system. In fact, it can be considered as the superposition of two different framing systems: the moment-resisting frame and the concentrically braced frame. They can combine the main advantages of each conventional framing system and minimize their respective disadvantages, as well. In general, eccentrically braced frames possess high elastic stiffness, stable inelastic response under cyclic lateral loading, and excellent ductility and energy dissipation capacity.

The key distinguishing feature of an eccentrically braced frame system is that at least one end of each brace is connected so as to isolate a segment of beam called "link". In each framing scheme (Fig. 8.47), the links are identified by a bold segment. The three arrangements here presented are usually named as split K-braced frame, D-braced frame, and finally inverted-Y-braced frame. Some details of these links are presented in Figure 8.48. An innovative solution for replacing the damaged links after a moderate to strong earthquake is proposed by Stratan and Dubina, (2004) (Fig. 8. 49a) for split K-braced frames, in order to reduce repair costs. The connection of the link to the beam is provided by a flush end plate and high strength bolts. Bolted connection allows the link to be fabricated by using a lower yield steel grade for assuring a good ductility. The same dismountable solution for inverted Y-braced frames is proposed by

Mazzolani et al, (2004) (Fig. 8.48b).

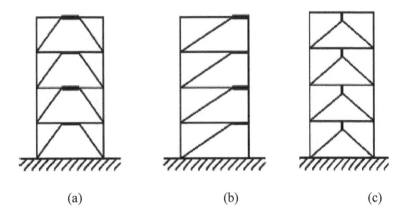

(a)   (b)   (c)

Figure 8.47 Eccentrically braced frames; (a) Split K-braced frame;
(b) D-braced frame; (c) Inverted Y-braced frame (Mazzolani et al, 2004)

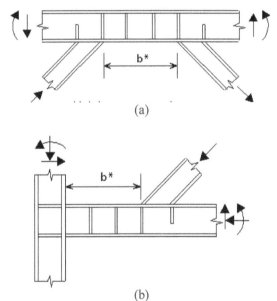

Figure 8.48 Details of active links: (a) Split K-braced frame; (b) D-braced frame
(AISC, 2006)

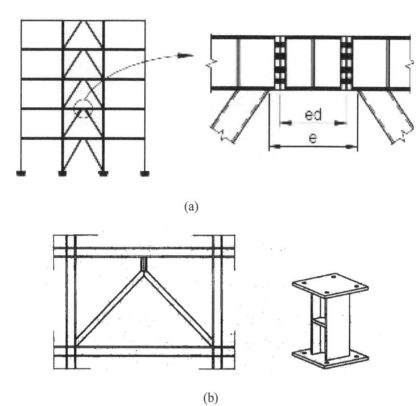

(a)

(b)

Figure 8.49 Dismountable bolted link concept: (a) For Split K-braced frames (courtesy of Stratan and Dubina, 2004); (b) For Inverted Y-braced frames (Mazzolani et al., 2004)

8.5.5 Infilled Frames

It is well known that non-structural elements may provide most of the earthquake resistance and prevent collapse of relatively flexible moment resisting frames. Infills were usually classified as non-structural elements and their influence was neglected in design. In the last period, recognizing the importance of infills, the case of infilled structures is considered as a system type, in addition to moment resisting frames and braced frames. There are two main infilled frame types: masonry infilled reinforced concrete frames and panel infilled steel frames.

Reinforced concrete frame

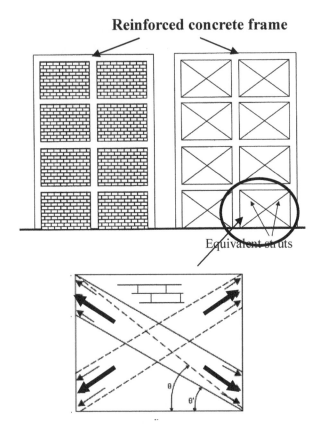

Figure 8.50 Masonry infilled reinforced concrete frames (after Decanini et al, 2004, Murty, 2004)

The *masonry infilled reinforced concrete frames* (Fig. 8.50a) is a system in which the infilled masonry gives a remarkable contribution to the initial stiffness of reinforced concrete frames, reducing the drift and the structural damage, especially for serviceability limit state. Buildings containing masonry walls have normally a very good earthquake performance. Their behavior is based on the formation of an equivalent diagonal strut. So, the analysis of masonry infilled frames is similar to the one of X-braced frames (Fig. 8.50b).

The *panel infilled steel frames* (Fig. 8.51a), where there are two main solutions for panels: to use sandwich panels (Fig. 8.51b), or shear panels (Fig. 8.51c). Long used as partition walls only, the sandwich panels are nowadays appreciated for their

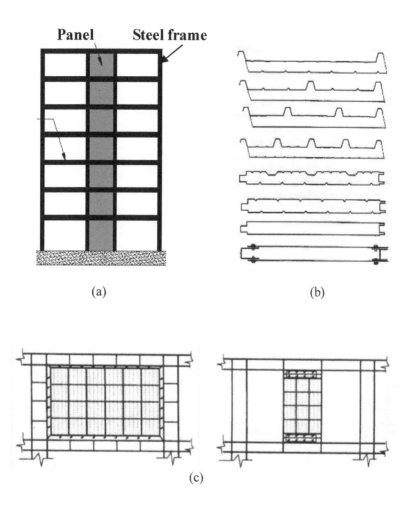

Figure 8.51 Panel infilled steel frames: (a) Typical system; (b) Typical sandwich panel typologies (De Matteis, 1998); (c) Typical shear panels (De Matteis et al, 2006)

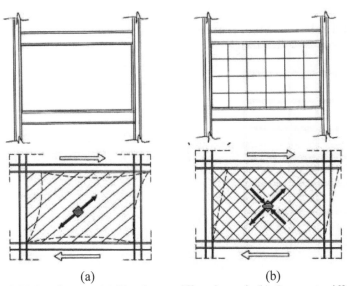

(a)   (b)
Figure 8.52 Panel types: (a) Slender unstiffened panel; (b) Compact stiffened panels
(De Matteis et al, 2006)

contribution to the structural behavior of buildings. If a suitable connecting system is adopted, such panels may, in fact, provide a remarkable increasing of both lateral stiffness and energy dissipative capacity of the structure as a whole (De Matteis, 1998). The second solution, which is alternative to concentric bracings with limited capacity to dissipate seismic energy, consists of introducing in the structure some panels made of low yield steel or aluminium (De Matteis et al, 2006). These panels are intended as a primary system in absorbing the lateral seismic actions, while the beams and columns have only the role of carrying out the gravity loads.

The classification of shear panels is based on the dissipation mechanism. In case of slender unstiffened panels, the shear panel dissipates energy by means of the tension field action, due to the fact that elastic buckling occurs in the compression field (Fig. 8.52a). For compact stiffened panels, the buckling phenomena are delayed and the dissipative mechanism corresponds to pure shear (Fig. 8.52b). Obviously, the shear buckling in elastic field occurring in slender unstiffened panels produces a poor dissipation during the cycle behavior. Therefore, the pure shear dissipative mechanism, offered by compact stiffened panels, is preferable because it allows a stable inelastic cyclic behavior and a uniform material yielding spread over the entire panel. The low yield stress ensures an amount of energy dissipation starting from small deformation levels as in case of moderate earthquakes, working as damper also during the serviceability limit state.

8.5.6 Base-isolated Systems

Base isolation is believed to be the most powerful tool in earthquake engineering, pertaining to the passive structural vibration technologies. It is meant to enable a building structure to survive a potentially devastating earthquake without important damages (Wikipedia, nd).

Conventional earthquake-resistant structural systems, presented in the previous section, are considered fixed at the ground level. Their earthquake resistance is derived from the ability to absorb seismic energy in specially designed regions of the structures, the so-called "dissipative zones", such as at the beam ends near beam-column joints. These regions should be capable of deforming into inelastic range and sustaining large reversible cycles of plastic deformations, all without losing strength and stiffness. These inelastic activities also mean large deformations in primary structural members, resulting in a significant amount of damage in structural and non-structural elements. Contrary to this behavior, the base-isolation systems provide the superstructure to be isolated from the foundation by means of certain devices, which reduce the ground motions transmitted to the structure (Fig. 8.53). These devices absorb the seismic input energy by adding significant damping. In comparison with the conventional fixed-base systems, this technique considerably reduces the structural response as well as the damage to structural and non-structural elements (Rai, 2000).

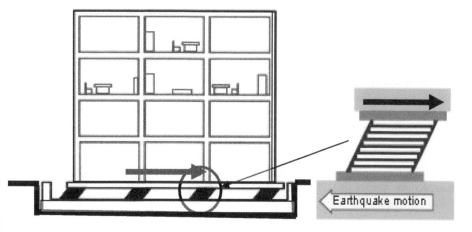

Figure 8.53 Base-isolated systems (after Rai, 2000)

(a)

In recent years the base-isolation has become the increasingly applied structural technique for buildings and bridges in high seismicity areas. A significant number of base-isolation devices have been recently developed, most of them using rubber or neoprene isolation bearings. The lead-rubber bearing is made of a layer of rubber sandwiched together with layers of steel. A solid lead plug is located in the middle of the bearing. The bearing is fitted with top and bottom steel plates which are used to attach it to the building superstructure and to the foundation, respectively. The bearing is very stiff and strong in the vertical direction, but flexible in the horizontal direction. The rubber layers allow the isolator to easily displace sideways, dissipating seismic energy and reducing the earthquake loads. They also act as a spring, ensuring that the structure returns to its original position after the end of ground motions.

8.5.7 Energy Dissipation Systems

A current strategy, widely favored for enhancing the seismic performance of fixed base systems, involves the dissipation of seismic energy through various energy dissipating devices. These devices are added to conventional structures (Fig. 8.54) and absorb seismic energy, thereby reducing the demand on primary structural members. So, a good design reduces the inelastic demand on primary structural members, leading to a significant reduction of structural and non-structural damage. Generally, these devices are introduced in the bracing member. While the conventional bracing members dissipate the input energy by means of axial plastic deformations, this energy can be dissipated on shear or flexural yielding by these devices according to some arrangement.

Many of these devices have been very recently proposed and tested. Some of them are presented in the following. Figure 8.54 shows a device based on a rectangular frame which is inserted at the intersection of the two braces in an X-braced system. This frame is made of thick steel plate shaped in order to have a uniform flexural resistance. The inverted Y-braced frames, having a vertical link, behave as a passive control system, where the link allows a large amount of input energy to dissipate, without any damage to the external framed structure. The improving of this dissipation capacity can be obtained by adding some special devices, known as ADAS (Added Damping And Stiffness Elements) systems: X-shaped, E-shaped, U-shaped, Omega-shaped, honeycomb-shaped, etc. (De Matteis et al, 2006).

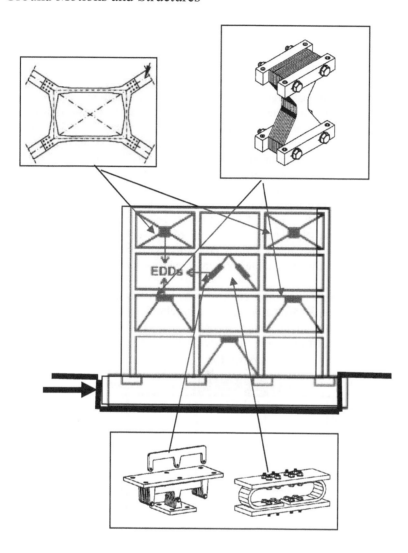

Figure 8.54 Energy dissipation systems (after Rai, 2000, De Matteis et al, 2006)

8.5.8 Active Control Systems

The philosophy of the earthquake-resistant system mentioned until now is based on a passive response to the earthquake action. In contrast, there is another expanding class of systems, which are referred to as *smart* or *active control systems* (Fig. 8.55). These systems differ from the passive systems in the sense that they control the seismic response through appropriate adjustments within the structure, as far as the seismic excitation changes. In other words, active control systems introduce elements of dynamism and adaptability into the structure, thereby augmenting the capability to resist exceptional earthquake loads. A majority of the proposed techniques involves adjusting lateral strength, stiffness and dynamic properties of the structure during the earthquake to reduce the structural response. In spite of this very promising solution to control the structural response, many serious problems exist with respect to the time delay in control actions, structural nonlinearity and very high operational costs (Rai, 2000).

The basic configuration of an active control system consists of three elements: (i) Sensors to measure external excitation and/or structural response; (ii) Computer hardware and software to compute control forces on the basis of observed excitation and/or structural response; (iii) Actuator to provide the necessary control forces.

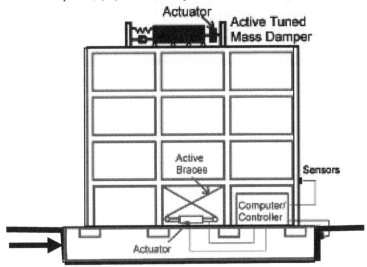

Figure 8.55 Active control systems (after Rai, 2000)

The structural control technology can be divided into *Tuned mass damper system* and *Active mass damper system*. The first is a system for absorbing the vibrations generated by high winds or moderate earthquakes, obtained by adding a pendulum or some mass on the building roof. The second uses a computer-controlled actuator, trying to suppress the oscillation of the building by activating a corresponding oscillation which reduces the ground motions effects and realizes the best performance of the structure.

One of the most promising active control systems for protecting buildings against strong earthquakes is the smart base isolation system (Yang, 1990, Wang and Dyke, 2007). It is a *hybrid system* consisting of a passive base isolation system (for instance an elastomeric bearing) connected to an active control system. This hybrid system has proved to be a very effective solution in reducing the response of buildings under strong earthquakes, easier to implement than the one of pure active control.

The conventional approach to the earthquake resistant design of buildings is based on the request of providing structures with strength, stiffness, ductility and robustness. The new technologies, such as active control systems, seem to be much more effective. But, despite the recent development, the structural control systems have some functional limitation against the earthquake actions, due to the size of the buildings.

8.6 SOIL-FOUNDATION-STRUCTURE INTERACTION

8.6.1 Main Effects of Interactions

There are significant number of controversies about the perceived effects of soil-foundation-structure interaction on the overall performance of superstructures, especially on soft soils. This beneficial role has been turned into a dogma: for instance, according EC8-5: *"For the majority of usual building structures, the effect of SSI (soil-structure interaction) tend to be beneficial, since they reduce the bending moments and shear forces acting in the various members of the superstructures..."*(Gazetas, 2006).Therefore, this interaction has been initially considered as beneficial during seismic motions, but the trend in research is changing and this causes different thinking on this phenomenon (Grondin, 2004).

The response of a structure to a shaking earthquake is affected by the interaction between three linked subsystems: soil, foundation and structure (Fig. 8.56a). A seismic soil-foundation-structure interaction analysis evaluates the collective response of this system to a specific free-field ground motion. It is considered as a natural source of energy dissipation (Crouse, 2000).

Worldwide, the seismic structural design is based on standard procedures over-simplifying the ground, where the type of foundation as well as its interaction with the soil and the structure is not taken into account. So, the structure is considered as fixed-base at the ground level (Fig. 8.56b), or at the bottom of foundation (Fig. 8.56c).

Neglecting the soil-foundation-structure interaction has often been assumed to be beneficial for the seismic response of the supported structure because its consideration is said to improve the safety margin. While this assumption is true for buildings founded on very stiff soils and rocks, it cannot be extended inconsiderately to soft or liquefiable soils (Buehler et al, 2006).

It is fundamental in seismic design the assumption that the motion experienced by the foundation of a structure during an earthquake is the same as the *free-field ground motion*, a term referring to the motion which would occur at the level of foundation if no structure were present. This assumption implies that the soil-foundation system underlying the structure is rigid and, hence, represents a fixed-base condition. Strictly speaking, this assumption never holds in practice. For structures supported on a deformable soil, the foundation motion generally is different from the free-field motion and may produce an important *rocking component*, in addition to a *translation component* (modeled by springs), due to the soil-foundation interaction (Fig. 8.57).

The rocking components and soil-foundation-structure interaction effects tend to be most significant for laterally stiff structures such as buildings with bracing systems, particularly those located on soft soil. The translation components are important for flexible structures, for which a substantial part of their seismic energy may be dissipated into the supporting soil by radiation of waves and by hysteretic action in the

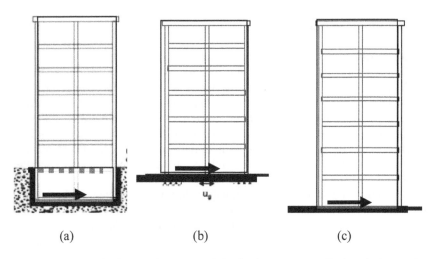

(a)   (b)   (c)

Figure 8.56 Standard procedures: (a) Linked sub-systems: soil, foundation and structure; (b) Fixed at the ground level; (c) Fixed at the bottom of foundation (after Stewart and Tileylioglu, 2007)

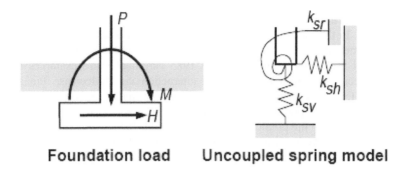

Foundation load   Uncoupled spring model

Figure 8.57 Soil-structure system: rocking and translation components
(FEMA 273, 1997)

soil.

The main soil-foundation-structure interaction effects can be important, due to the following aspects (Kramer and Stewart, 2004):

Foundation stiffness and damping. The inertia developed during the vibration of the structure gives rise to base shear, moment and torsional excitation, and these loads in turn cause displacements and rotations of the foundation. Moreover, these motions give rise to energy dissipation via radiation and hysteretic soil damping, and this energy dissipation can significantly affect the overall system damping.

The variation between free-field and foundation-level ground motions results from two processes. The first is known as *kinematic interaction* and it is caused by the presence of stiff foundation elements on or in soil, producing foundation motions, which deviate from free-field motion as a result of the foundation embedment effects. The second process is related to the *inertial interaction effects* and consists of the relative foundation displacements and rotations described above.

Foundation deformations. Flexural, axial and shear deformations of foundation elements occur as a result of loads applied by the superstructure and the supporting soil medium. These deformations can also significantly affect the overall system behavior, especially with respect to damping.

The effects on the superstructure due to the soil-foundation-structure interaction can be summarized as follows: decreasing of the natural frequency (increase the natural period) of the system, increasing of damping, increasing of lateral displacements and secondary order effects, changing of the seismic base shear, depending on the frequency content of the input motion and the dynamic features of the soil-foundation-structure system. The current seismic design philosophy regards this dynamic interaction to be beneficial to the structure behavior. The flexibility of the soil-foundation system reduces the overall stiffness of global system and therefore attenuates the peak loads caused by a given motion. This might be true in many cases.

However, there is the possibility that the soil-foundation-structure system could go into a resonance field with exciting frequency, leading to much larger inertial forces acting on the structure. In this case the soil-foundation-structure interaction is not beneficial, but rather detrimental for the seismic response of the structures. The slender frames, which collapsed in Mexico City during the 1985 earthquake, are examples of this detrimental effect. At the same time, this interaction could decrease the system ductility. Therefore, the effects of soil-foundation-structure interaction must be carefully considered in practical design, because in some cases it may lead to erroneous conclusions in the prediction of the structural behavior.

Currently, there are basically two approaches to deal with this interaction:

-*Substructure approach,* in which the system is considered as a system formed by far-field soil, near-field soil, foundation and structure, each being analyzed separately and the superposition of the results assumes that these substructures work together.

-*Direct approach,* in which soil, foundation and structure are included within the same model and analyzed in a single step. The soil is discretized with solid finite elements and the structure with beam elements.

8.6.2 Substructure Approach

In the substructure approach, the analysis is separated in three distinct parts which are combined together to formulate the complete solution. The superposition of these steps requires the assumption of linear behavior both for soil and structure (Kramer and Stewart, 2004). The three steps are presented in Figure 8.58.

Determining the foundation input motion (Fig. 8.58b), which is the motion of the ground which would occur on the structure base, considering both structure and foundation without mass. The deviation of these characteristics from the free-field motions is dependent on the type and geometry of foundation and soil and the modifications of these free-field motions are determined in the frequency (period) domain. For *shallow foundations*, the modifications result from incoherent incident waves (see Fig.7.35). In presence of these wave passage effects, translational base-foundation displacements are reduced relative to the free-field motions, but rotational motions are introduced. These effects tend to become more significant with increasing frequency, depending on the ratio between the sizes of foundation relative to the wavelengths at higher frequencies. For *embedded shallow foundations*, these effects reduce the translational movements relative to the free-fields motions and the rocking motion in comparison with the shallow foundations.

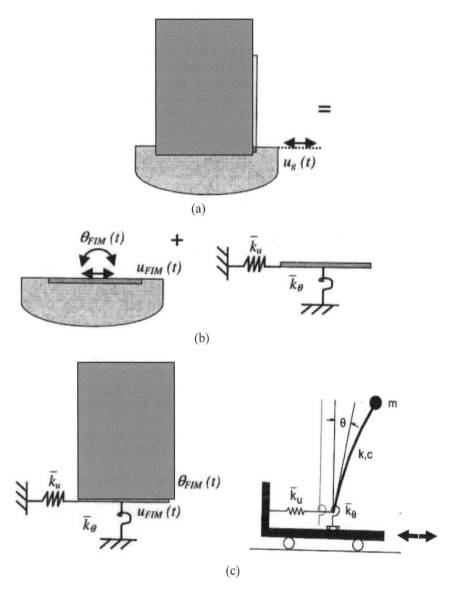

Figure 8.58 Substructure approach: (a) Soil-foundation-structure interaction; (b) Evaluation of foundation input motions and evaluation of soil characteristics; (c) Dynamic analysis of the structure (after Kramer and Stewart, 2004)

The seismic response of the *pile foundations* (used in case of soft or liquefiable soils) differs from that of shallow foundations, due to the increased stiffness and the group effect. When the dynamic analysis of piles is performed, it is important that modeling includes the following features: the variation in soil properties with depth, the nonlinear behavior of soil at the pile-soil interface, the slippage along this interface, possible liquefaction, changes in ground motions with depth and the effect of geometric damping (Romo et al, 2000).

Evaluating the soil-foundation characteristics, by taking into account soil stratigraphy, foundation stiffness and geometry. In this step the characteristics of springs for the three translational and three rotational degrees of freedom are determined, the foundation being assumed to be rigid.

Dynamic analysis of the structure supported at the base with appropriate springs and subjected to a base excitation (Fig. 8.58c).The study is performed on the simplified oscillator model. For flexible structures, the effect of interaction is to reduce the spectral acceleration. Conversely, the interaction effects tend to be significant for stiff structures, such as braced frames located on soft soil, by increasing the spectral accelerations and the base shear.

8.6.3 Direct Approach

The direct approach (also referred to as the Complete Finite Element) incorporates the far- and near-field soils and the foundation-structure system in an unique model, which is usually developed by means of numerical methods. The dynamic analysis by means of a direct approach considers the whole building, foundation and soil interaction, without splitting into substructures. This approach can take into account the influence of soft soil, liquefaction, deformation and tilting of structures.

The last few decades some sophisticated methods have been proposed, based on the numerical approaches such as Finite elements (FEM) or Boundary elements (BEM). Generally, the FEM method is mostly used for modeling the seismic soil-foundation-structure interaction under plain strain conditions for both soil and structure, what allows an easier consideration of several geometrical meshes and several strain-stress relationships. The computer program SOFIA (SOil Frame InterAction) is used by Ghersi et al (1999) for analyzing the pseudo-static behavior of the soil-foundation-structure system (Fig. 8.59a). The system consists of an asymmetric 5-floor reinforced concrete plane frame with a foundation resting on a sand deposit. The soil was subdivided by means of isoparametric quadratic elements.

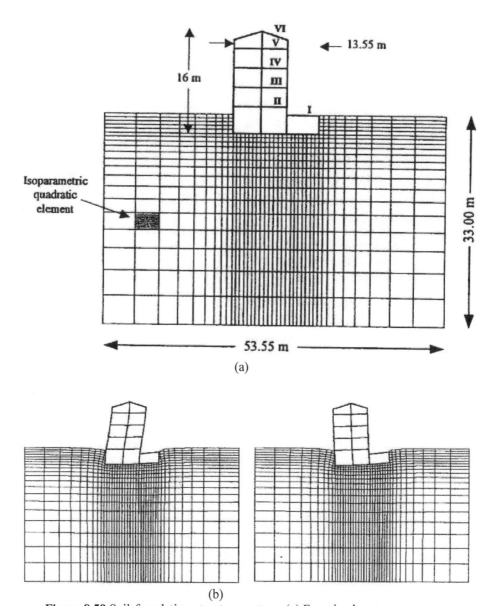

Figure 8.59 Soil-foundation-structure system; (a) Examined system;
(b) Deformations under pseudo-static condition (courtesy Ghersi et al, 1999)

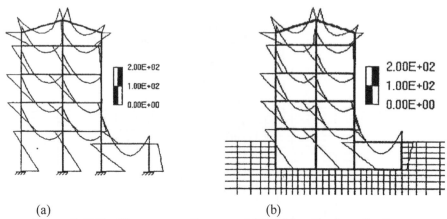

(a)   (b)

Figure 8.60 Bending moment diagrams: (a)Fixed-based schematization;
(b) Soil-foundation-structure interaction (courtesy Ghersi et al, 1999)

The soil-mass is extended in both vertical and horizontal direction, so the boundary does not influence the frame behavior. Figure 8.59b shows the deformations in right and left directions for the soil-foundation-structure system, which are different due to the asymmetry of structure. Because of the strong asymmetry of the analyzed frame, the direction of seismic action is obviously very important for the evaluation of internal actions. The soil-foundation-structure interaction analysis, made by the SOFIA computer program, shows significantly different bending moments with respect of the classical fixed-base schematization. The main effect on the structural elements is a not negligible change of bending moment values, especially in the columns of the first level (Fig. 8.60). Contrary, no very significant bending moment changes are observed in the beams. This aspect is particularly relevant, because an increase of stress in the columns, not foreseen by the usual fixed-base column model, might lead to a dangerous non-ductile story mechanism of collapse (Ghersi et al, 1999). Similar conclusions were obtained by Saez et al (2006) and Lehmann and Borsutzky (2006).

8.7 SEISMIC VULNERABILITY FACTORS

8.7.1 Consequential Structural Configuration

The configuration of buildings is related to dimensions, form, geometric proportions and location of structural components. The configuration of a building will influence its seismic performance, particularly regarding the distribution of the seismic loads. A well-designed structural configuration is expected to achieve some objectives, such as predictable behavior, simplicity of design and construction,and minimized structural cost for the required function after an earthquake (Adams, 2004).

The traditional design philosophy of seismic-resistant structures is based upon the dissipation of the earthquake input energy by means of plastic excursions, leading to structural damage which has to be controlled in order to prevent failure. The ability of a structure to withstand severe earthquakes is strictly related to the capacity of uniformly distributing the structural damage. In the classical design approach, the damage concentrations can lead to failure modes which are characterized by a reduced dissipation capacity (Mazzolani and Piluso, 1996). Therefore, one of the most critical decisions influencing the ability to withstand earthquakes is the choice of the basic structural configuration.

Building structures belong to many types and configurations and, of course, a universal ideal configuration for any particular type of building does not exist. However, there are some basic principles for seismic-resistant design which can be used as guidelines in selecting an adequate building configuration. These basic guidelines are the following (NISEE, 1997):

- Building (structure and non-structural elements) should be light and avoid unnecessary masses.
- Building and structure should be simple, symmetric and regular in plan and elevation, avoiding large height-width ratio and large plan area.
- Building and structure should have a uniform and continuous distribution of mass, stiffness, strength and ductility, avoiding formation of soft stories.
- Structure should have relatively shorter spans than non-seismic-resistant structure and avoid use of long cantilevers.
- Structure should be designed to have good details and erected respecting construction quality.
- Non-structural components should either be well separated so that they will not interact with the structure, or they should be integrated with the structure. In the last case, the structure should have sufficient lateral stiffness to avoid damage to these non-structural elements under minor or moderate earthquakes.
- In case of integrated non-structural elements, they should be distributed in the structure in such a way to avoid the introduction of any disturbance in the uniformity of structural stiffness.
- Structure should be detailed so that inelastic deformations are controlled to develop in selected regions, according to a desirable hierarchy. The structure must have a stable energy dissipating post-yield behavior.
- Structure should possess enough redundancy and robustness to have the largest possible number of defense lines, so the inelastic behavior of the whole structure finds its way to avoid the structure collapse.
- In case of neighboring building existence, the separation between the buildings must assure pounding is prevented.
- Structure should be provided with a balance of stiffness, strength, ductility and robustness among its elements, connections and supports.

-    Also soil and foundations should be included in this balance.

If these principles are respected, the structure will have a correct response during an earthquake as close as possible to the given prerequisites.

8.7.2 Typical Configuration Deficiencies

Unfortunately, in many cases, the structural engineer does not play a dominant role in the determination of the building shape. The objective of the architectural aesthetics, and/or of specialized functions, may be in direct conflict with the goals of structural design to provide a proper structural configuration, it having priority in the selection of structural planning. Therefore, it is possible that the architectural demands are not convenient from the point of view of seismic behavior. As a result of this conflict, many building structures have non-symmetric floor plans, multiple towers and setback floors or irregularities in vertical configuration. Such buildings are prone to earthquake damage due to coupled lateral and torsional movements producing non-uniform displacement demands in building elements and concentrations of stresses and forces in structural elements. In these cases, the task of the structural engineers is to limit the adverse features in a reasonable range.

Typical building configuration deficiencies include irregular geometry, a weakness in a given story, a concentration of masses or a discontinuity in the lateral force resisting system. Vertical irregularities are defined in terms of strength, stiffness, geometry and mass. Horizontal irregularities involve the horizontal distribution of lateral forces to the resisting structural system.

Geometric irregularity. As far as the *vertical structural configuration* is concerned, it is generally recognized that the compact shapes are the most favourable, so that the ideal configuration is obtained when the building has the shape of a parallelepiped (Fig. 8.61a). Irregular configurations are characterized by a variation of the boundary lines along the height, giving rise to set-backs (Fig. 8.61b) or off-sets (Fig. 8.61c). They are considered irregular, because the corresponding distribution of mass and stiffness is not uniform. Some analyses have demonstrated that not all vertical irregularities produce an increase of structural vulnerability (Mazzolani and Piluso, 1996).

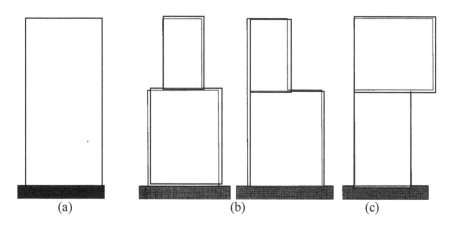

Figure 8.61 Regular and irregular vertical configurations; (a) Compact shape;
(b) Set-back configuration; (c) Off-set configuration

From the *horizontal structural configuration* point of view, it has been already pointed out the conceptual importance of the ability of a building to vibrate separately in two directions, without torsional coupling. The symmetric shapes have the ability to fulfill this demand. The ideal shape is undoubtedly the double symmetrical one (Fig. 8. 62a), because the center of gravity coincides with the center of rigidity, so the translational and torsional motions are uncoupled. Contrary, it is well known that asymmetric plan-wise buildings undergo translational as well as torsional coupled motions during seismic actions and, therefore, they are subjected to non-uniform plan distribution of damage. Figure 8.62b presents some usual unsymmetrical plans. When these shapes are proved to be unavoidable, it is desirable to use seismic separations to divide the building into independent parts, which provide a regular seismic behavior (Fig. 8.63).

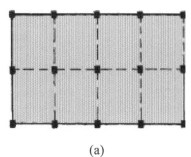

(a)
Figure 8.62 (continues)

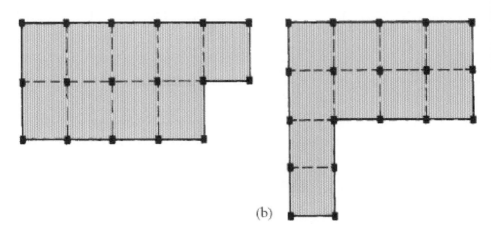

Figure 8.62 Regular and irregular horizontal conformations:
(a) Symmetric layouts; (b) Irregular layouts (Gioncu and Mazzolani, 2002)

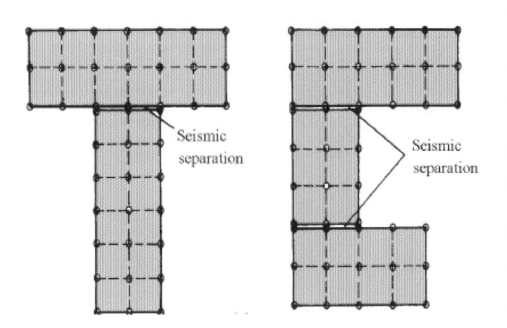

Figure 8.63 Seismic separations (Gioncu and Mazzolani, 2002)

Weak story (irregularity in strength) (Fig 8.64) The story strength is the total strength of all lateral force-resisting elements in a given story for the direction under consideration. For the shear walls, the strength is the shear capacity of walls. In the framed structures, the strength is given by the shear capacity corresponding to the flexural strength of columns. Weak stories are usually found where vertical discontinuities exist, or where member size or reinforcement has been reduced. UBC (!997) (Valmundsson and Nau, 1997, Magliulo et al, 2002) defines a weak story when its strength is less than 80% of the storey above. The weak story effect consists in the concentration of inelastic activity which can produce the total or partial collapse of the story itself (WHO, 2006). In order to prevent this situation, an over-strength is recommended, in such a way that the differences in strength do not exceed 20%.

Soft story (irregularity in stiffness) (Fig. 8.65). The increasing of story stiffness is given by the non-structural infill panels (masonry, concrete or metal). The soft story is the one where these infill panels are missing due to functional demands (Fig. 8.64). A very widespread structural configuration in existing buildings is characterized by the absence of these infill panels at the ground floor, while they are present at the elevation stories. This configuration is called *pilotis configuration,* giving rise to a soft first story, which allows a good use and distribution of the space at the ground floor, but it is very dangerous from the seismic point of view.

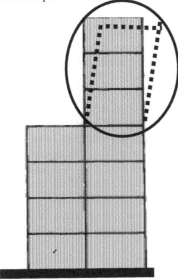

Figure 8.64 Weak story (after Gioncu and Mazzolani, 2002)

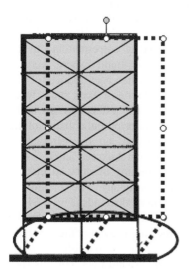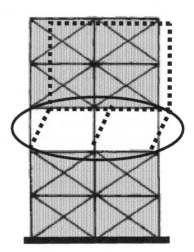

Figure 8.65 Soft stories: (a) Soft ground story (pilotis story);
(b) Soft story (Gioncu and Mazzolani, 2002)

The lateral response of these buildings is characterized by a large rotational ductility demand, which is concentrated at the end sections of the columns of the first story, while the superstructure behaves like a quasi-rigid body (Mezzi and Parducci, 2005). All the inelastic deformations are concentrated in the soft story. This structural type was the cause of many collapsed buildings during the 1999 Kocaeli earthquake (Turkey). The UBC (1997) defines the soft story when its stiffness is less than 70% of the one of the story above or less than 80% of the mean stiffness of the three storeys above (Valmundsson and Nau, 1997, Magliulo et al, 2002). A solution to avoid the failure of pilotis buildings consists in the introduction of some partial rigid ductile walls in the first story around the staircase, in order to respect the above conditions.

Mass irregularity can be detected by comparing the story weights (Fig. 8.66). The effective mass is given by the dead load of the structure at each level plus the

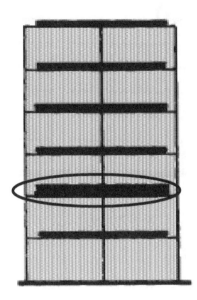

Figure 8.66 Mass irregularity (Gioncu and Mazzolani, 2002)

actual weight of partitions and permanent equipments on each floor. UBC (1997) defines as irregular a mass distribution when the effective story mass is greater than 150% of the mass of an adjacent story (Valmundsson and Nau, 1997, Magliulo et al, 2002). The study of Magliulo et al (2002) considers that this condition is too severe.

Pounding of adjacent buildings. When buildings are erected without sufficient seismic separation between them and their interaction has not been considered, the buildings may impact each other, or pound, during an earthquake. Building pounding can alter the dynamic response of both buildings and impart additional inertial loads on both structures. Buildings of the same height with matching floors will exhibit similar dynamic behavior. When the buildings pound, floors will impact with other floors, which means that the damage due to pounding usually will be limited to non-structural elements. Contrary, when the floors of adjacent buildings are at different levels, the floors will impact with the columns of the adjacent building, causing serious structural damage (Fig. 8.67a). As buildings are not in general designed for undergoing these conditions, there is a potential risk for extensive damage and possible collapse. The solution to avoid pounding is to design a suitable seismic separation between adjacent buildings (Fig. 8.67b,c).

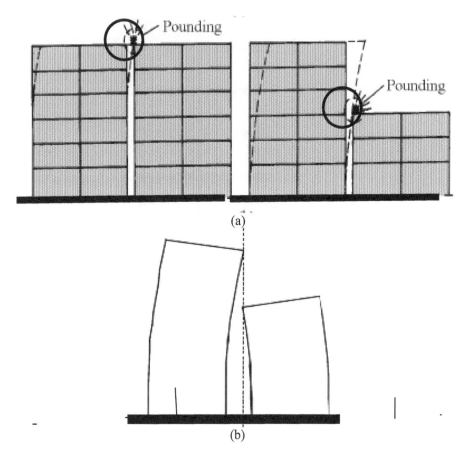

Figure 8.67 Pounding of adjacent buildings: (a) Buildings with different floor
height; (b) Sufficient seismic separation to prevent pounding
(Gioncu and Mazzolani, 2002)

8.7.3 Accidental Torsion

Damage reports on recent earthquakes have indicated that one major cause of distress in
building structures is due to the torsional motion induced by earthquakes. In many
cases, it was caused by the structural asymmetry, which produces a rotation of the
building floors around the vertical axis. However, the torsion effect can be also
observed in buildings of ideal architectural design, where a symmetrical distribution of
strength, stiffness and mass is provided in all directions. Perfectly symmetrical
buildings can undergo torsional motions, because of random unexpected irregularities.

It must be mentioned that the consequences of accidental torsion are more significant for the symmetric structures than for the asymmetrical ones. This unpredicted behavior due to accidental torsion is caused by the following factors:

Base rotational excitation, which is defined as the rotational motion around a vertical axis experienced by the building foundation as a result of spatially non-uniform motions. This spatial variability of the ground motion underneath the foundation is attributed to two effects: (i) Wave passage, because of which different points of the ground surface are excited by the same motion, but with a phase lag (Fig. 8.68); (ii) Ground motion incoherence, a term used to recognize that different points of the ground experience motions with different amplitudes and phase characteristics, because of incoming waves from different locations of an extended earthquake source, wave reflections and refractions around the building foundation, or changes produced in the waves when traveling from the source to the structure through paths of different physical properties. The increase of displacements of symmetric buildings resulting from accidental torsion due to the rotation excitation is larger for structures with small vibration periods (less than about half second) and less than 5% for long period structures (De la Llera and Chopra, 1994a,b,1995).

Uncertainties in stiffness and masses. Discrepancies between the computed and the actual values of the distribution of structural element stiffness and masses imply that a building with a nominal symmetric plan is actually asymmetric due to some unknown reason and, therefore, it will undergo torsional vibration when subjected to purely translational ground motion. The factors of uncertainty are due to:

(i) Variation of material properties and element dimensions, variability in fabrication methods and quality control, execution and erection tolerances; (ii) Uncontrolled distribution of infill panels with different stiffness; (iii) Presence of some uncontrolled hollows (mainly due to the equipments) in structural elements, such as floors and walls; (iv) Uncontrolled distribution of masses, especially for live loads.

All these aspects result in the uncertainty in the location of the centers of rigidity (CR) and mass (CM). In order to consider the effects of the accidental torsion in seismic design of buildings, some accidental eccentricities between these two centers are introduced in the calculation model (Fig. 8.69). These accidental eccentricities depend on the plan dimensions of the structure (De la Llera and Chopra, 1995, Chopra and De la Llera, 1996).

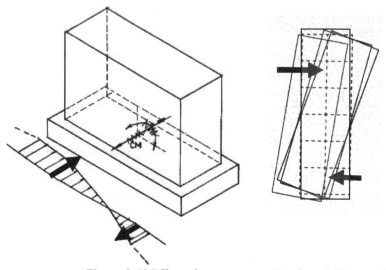

Figure 8.68 Effect of wave passage (Prada and Gioncu, 2004)

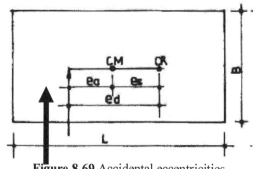

Figure 8.69 Accidental eccentricities

8.7.4 Fire Following Earthquake

The major earthquakes usually cause extreme damage to buildings and infrastructure. Fire following earthquakes can cause significant losses which, sometimes, exceed the ones produced by the earthquake shakes. Earthquakes are largely unpredictable and large fires following earthquakes are even less predictable. The risk sources of fire following earthquake are due to the damage of pipelines, electric wiring, oil systems, active and passive protection systems, interruption of transportation network and water supply system. Some historical records show that small fires are often initiated by earthquakes and they sometimes grow into large destructive fires, causing loss of life and severe damage to properties. The main concern is initially related to the damage to individual wooden buildings. A subsequent concern is the possibility of devastation resulting from a large urban conflagration. The factors affecting the probability that small fires could grow into large fires include: the amount of earthquake damage, the type and density of building, the wind conditions, the loss of water supplies and the fire fighting capabilities (Botting and Buchanan, 2000).

The fire conflagration following the 1906 San Francisco earthquake significantly exceeded the shake losses and resulted in the largest earthquake loss in American history. Similarly in Japan, the fire losses following the 1923 Great Kanto earthquake were much larger than the shake losses (Mortgat et al, 2004). Recently, during the 1994 Northridge and 1995 Kobe earthquakes, a great part of losses were attributed to fire the following earthquake.

According to the modern seismic design, ordinary structures are designed to suffer damage to some extent during strong earthquakes, exploiting their own structural ductility to avoid collapse and to safeguard human lives. The problem of fire following earthquake consists in the fact that the fire coming soon after an earthquake will find an already damaged structure, which is more vulnerable than the initial undamaged one. Therefore, the danger of global collapse due to fire for the damaged structure increases. The collapse risk depends on the extent of the earthquake damage, because the fire resistance rating of the structure could be significantly reduced. The conditions that the rating reduction is under 10% (which is considered as negligible) for steel structures are examined by Della Corte et al (2003a,b): (i) The reduction of fire resistance due to seismic damage is negligible for moment resisting frames, which are designed considering the serviceability seismic requirement at the design seismic intensity level: (ii) For earthquakes having intensity larger than the design value and/or structures not adequately designed against earthquakes, the fire resistance reduction could be significant and should be taken into account.

8.8 REFERENCES

Adams, W. (2004): Planning buildings to resist earthquakes. Joint Technical Session between Jamaica Institution of Engineers (JIE) and Carribean Division of Institution of Structural Engineers (UK), December 2004

AISC (2006): Seismic provisions for structural steel buildings. Commentary

ASCE 7 (2005): Chapter 5. Structural Design Criteria. Commentary.

Astaneh-Asl, A. (2003): Progressive collapse prevention in new and existing buildings. Emerging Technologies in Structural Engineering. Abu Dhabi Conference, 1001-1008

Avram, C. Facaoaru, I., Filimon, I., Mirsu, O., Tertea, I. (1981): Concrete Strength and Strains. Elsevier Scientific Publishing Company, Amsterdam

Bachmann, H. (2003): Seismic conceptual design of buildings - Basic principles for engineers, architects, building owners, and authorities. Swiss Federal Office for Water and Geology, BBL, Vertrieb Publikationen, CH-3003 Bern

Balut, N., Gioncu, V. (2003): Suggestion for improved "dog-bone" solution. In Behavior of Steel Structures in Seismic Areas, STESSA 2003, (ed. F.M. Mazzolani), Naples, 9-12 June 2003, Balkema, 129-134

Balut, N., Gioncu, V. (2006): Seismic design of certain axially loaded steel elements: A practical point of view. In Steel, a New and Traditional Material for Buildings (eds. D. Dubina and V. Ungureanu), Poiana Brasov, 20-22 September 2006, Taylor & Francis, London, 281-288

Bertero, R.D., Bertero, V.V. (1999): Redundancy in earthquake-resistant design. Journal of Structural Engineering, Vol. 125, 81-88

Bhattacharya, S. (2003): Pile instability during earthquake liquefaction. Ph. D. Thesis, University of Cambridge

Bhattacharya, S., Madabhushi, S.P.G., Bolton, M. (2003): Pile instability during earthquake liquefaction. ASCE Engineering Mechanics Conference (EM 2003), Seattle, 16-18 July 2003

Bjorhovde, R. (2002): Advanced steel materials for structural applications. In Stability and Ductility of Steel Structures (ed. M. Ivanyi), Budapest, 26-28 September 2002, Akademiai Kiado, Budapest, 557-564

Botting, R., Buchanan, A. (2000): Building design for fire after earthquake. 12th World Conference on Earthquake Engineering, Auckland, 26 January-4 February 2000, Paper No. 1569

Buehler, M.M., Wienbroer, H., Rebstock, D. (2006): A full seismic soil-foundation-structure interaction approach. 1st European Conference on Earthquake Engineering and Seismology, Geneva, 3-8 September 2006, Paper number 421

Canbolat, B.A., Parra-Motesinos, G.J., Wight, J.K. (2005): Experimental study on seismic behavior of high-performance fiber-reinforced cement composite coupling beams. ACI Structural Journal, January/February, 159-166

Carrubba, P., Massimino, M.R., Maugeri, M. (2000): Strain dependent impedance in shallow foundations. 12th World Conference on Earthquake Engineering, Auckland, 28 January -4 February 2000, Paper No. 1623

Chopra, A., De la Llera, L.C. (1996): Accidental and natural torsion in earthquake response and design of buildings. 11th World Conference on Earthquake Engineering, Acapulco, 23-28 June 1996, Paper No. 2005

Choudhury, D., Subba Rao, K. S. (2006): Seismic bearing capacity of shallow strip footings embedded in slope. International Journal of Geomechanics, Vol. 6, No. 3, 176-184

Claeson, C. (2008): Finite element analysis of confined concrete columns. http://www.tekna.no/KnowBase/Content/738966/doc-22-1.pdf

Crouse, C.B. (2000): Energy dissipation in soil-structure interaction. 12th World Conference on Earthquake Engineering, Auckland, 30 January-4 February 2000, Paper

No. 0366

Cusson, D, Larrard, F., Boulay, C., Paultre, P. (1996): Strain localization on confined high-strength concrete columns. Journal of Structural Engineering, Vol. 122, No. 9, 1055-1061

D'Aniello, M., Della Corte, G., Mazzolani, F.M., (2006): Steel buckling restrained braces. In Seismic Upgrading of RC Buildings by Advanced Techniques (ed. F.M. Mazzolani), Polimetrica, Milano, 179-223

Decanini, L., Mollaioli, F., Mura, A., Saragoni, R. (2004): Seismic performance of masonry infilled RC frames. 13th World Conference on Earthquake Engineering, Vancouver, 1-6 August 2006, paper No. 165

Della Corte, G., Landolfo, R., Mammana, O. (2003a): Fire resistance of MR steel frames damaged by earthquakes. In Behavior of Steel Structures in Seismic Areas, STESSA 2003 (ed. F.M. Mazzolani), Napoli, 9-12 June 2003, Balkema, Lisse, 73-79

Della Corte, G., Landolfo, R., Mazzolani, F.M. (2003b): Post-earthquake fire resistance of moment resisting steel frames. Fire Safety Journal, Vol. 38, No. 7, 593-612

De la Llera, J.C., Chopra, A.K (1994a): Accidental torsion in buildings due to stiffness uncertainty. Earthquake Engineering and Structural Dynamics, Vol. 23, 117-136

De la Llera, J.C., Chopra, A.K (1994b): Accidental torsion in buildings due to base rotational excitation. Earthquake Engineering and Structural Dynamics, Vol. 23, 1003-1021

De la Llera, J.C., Chopra, A.K. (1995): Estimation of accidental torsion effects for seismic design of buildings. Journal of Structural Engineering, Vol. 121, No. 1, 102-114

De Matteis, G. (1998): The effect of cladding panels in steel buildings under seismic actions. Ph.D. Thesis, Universita degli Studi di Napoli "Federico II"

De Matteis, G., Mistakidis, E.S. (2003): Seismic retrofitting of moment resisting frames using low yield steel panels as shear walls. In Behavior of Steel Structures in Seismic Areas, STESSA 2003, (ed. F.M. Mazzolani), Napoli, 9-12 June 2003, Balkema, Lisse, 677-662

De Matteis, G., Mazzolani, F.M., Panico, S. (2006): Dissipative metal shear panels for seismic design of steel frames. In Innovative Steel Structures for Seismic Protection of Buildings. PRIN 2001 (ed. F.M. Mazzolani), Polimetrica, Milano, 15-79

FEMA 273 (1997): NEHRP Guidelines for the Seismic Rehabilitation of Buildings. Federal Emergency Management Agency

FEMA 403 (2002): Appendix B. Structural steel and steel connection. Federal Emergency Management Agency

Formasino, A., Mazzolani, F.M., Brando, G., De Matteis, G. (2006): Numerical evaluation of the hysteretic performance of pure aluminimum shear panels. In Behavior of Steel Structures in Seismic Areas, STESSA 2006 (eds. F.M. Mazzolani and A. Wada), Yokohama,14-17 August 2006, Taylor & Francis, London, 211-217

Fukuta, T., Kitagawa, Y. (2008): Dynamic characteristics of Nickel titanium shape memory alloy as a construction material.
http://www.pwri.go.jp/eng/ujnr/33/paper/72fukuta.pdf

Garcia, L.E., Sozen, M.A. (2004): Earthquake-resistant design of reinforced concrete buildings. Earthquake Engineering. From Engineering Seismology to Performance-Based Engineering (eds. Y. Bozorgnia and V.V. Bertero), CRC Press, Boca Raton, Florida, 14.1-14.85

Gazetas, G. (2006): Seismic design of foundation and soil-structure interaction. First European Conference on Earthquake Engineering and Seismology, Geneva, 3-8 September 2006, Paper No, Keynote K7

Gazetas, G., Apostolou, M., Anastasopoulos, J. (2003): Seismic uplifting of foundations on soft soil, with examples from Adapazari (Izmit 1999 earthquake)
http://www.dundee.ac.uk/civileng/quaker/icof-gazetas.pdf

Ghersi, A., Massimino, M.R., Maugeri, M. (1999): Soil-foundation-superstructure interaction: (i) Effects of the soil; (ii) Effects of the superstructure. Earthquake Resistant Engineering Structures II, Catania, June, 1999, WIT Press, Southampton

Gioncu, V. (2000):Design criteria for seismic resistant steel structures. Seismic Resistant Steel Structures. (eds. F.M. Mazzolani and V. Gioncu), CISM Courses Udine, Springer, Wien, 19-99

Gioncu, V. (2006): Progressive collapse in space structures. Invited lecture. New Olympics. New Shell and Spatial Structures. IASS-APCS 2006 Conference, 16-19 October 2006, Beijing

Gioncu, V. (2007): Progressive collapse of steel structures. Part 1: A general view; Part 2: Methods for evaluating and prevention. XXI Congresso C.T.A, Catania, 1-3 October 2007, Dario Flaccovio Editore, 25-37, 39-48

Gioncu, V., Mateesecu, G. (1998): Comparative studies of Banat and Vrancea earthquakes. INCERC Report

Gioncu, V., Mazzolani, F.M. (2002): Ductility of Seismic Resistant Steel Structures. Spon Press, London

Gioncu, V., Mazzolani, F.M. (2006): Influence of earthquake types on the design of seismic-resistant steel structures. Part 1: Challenge for new design approaches. Part 2: Structural responses for different earthquake types. In Behavior of Steel Structures in Seismic Areas, STESSA 2006, (eds. F.M. Mazzolani and A. Wada), Yokohama, 14-17 August 2006, Taylor & Francis, London, 113-120, 121-127

Gioncu, V., Petcu, D. (1997): Available rotation capacity of wide-flange beams and beam-column. Part I: Theoretical approaches. Part II: Experimental and numerical tests. Journal of Constructional Steel Research Vol. 43, No. 1-3, 161-217, 219-244

Gioncu, V., Mateescu, G., Petcu, D., Anastasiadis, A. (2000): Prediction of available ductility by means of local plastic mechanism method: DUCTROT computer program. Moment Resisting Connections of Steel Frames in Seismic Areas (ed. F.M. Mazzolani), E&FN Spon, London, 95-146

Grondin, M. (2004): Advances simulation tools: Soil-structure interaction. REU

Final research report, Michigan State University

Gyorgyi, J., Gioncu, V,. Mosoarca, M. (2006): Behavior of steel MRFs subjected to near-fault ground motions. In Behavior of Steel structures in Seismic Areas, STESSA 2006, (eds. F.M. Mazzolani and A. Wada), Yokohama, 14-17 August 2006, Taylor & Francis, London, 129-136

Gueguen, P., Bard, P.Y., Semblat, J.F. (2002a): The site-city interaction. A new component to seismic risk. Bulletin des Laboratoires des Ponts et Chaussees, 15-32

Gueguen, P., Bard, P.Y., Semblat, J.F. (2000b): Engineering seismology: Seismic hazard and risk analysis: Seismic hazard analysis from soil-structure interaction to site-city interaction. 12th World Conference on Earthquake Engineering, Auckland, 30 January-4 February 2000, Paper 0555

Husain, M., Tsopelas, P. (2004): Measures of structural redundancy in reinforced concrete buildings. I: Redundancy indices. II: Redundancy response modification factor Rr.. Journal of Structural Engineering, Vol. 130, No. 11, 1651-1658, 1659-1666

Husem, M., Pul, S. (2007): Investigation of stress-strain models for confined high strength concrete. Sahana, Vol. 32, Part 3, June, 243-252

Ifrim, M. (1984): Dynamics of Structures and Earthquake Engineering (in Romanian). Editura Didactica si Pedagogica, Bucuresti

Ifrim, M., Macavei, .F, Demetriu, S., Vlad, I. (1986): Analysis of degradation process in structure during the earthquakes. 8th European Conference on Earthquake Engineering, Lisbon, 7-12 September 1986, 65/8-72/8

Iwan, W.D. (1996): The drift demand spectrum and its application to structural design analysis. 11th World Conference on Earthquake Engineering, Acapulco, 23-28 June 2006, Paper No. 116

Iwan, W.D. (1997): Drift spectrum: Measure of demand for earthquake ground motions. Journal of Structural Engineering, Vol. 123, No. 4, 397-404

Kaushik, H.B., Rai, D.C. (2007): Stress-strain characteristics of clay brick masonry under uniaxial compression. Journal of Materials in Civil Engineering, Vol. 19. No. 9, 728-739

Kim, S.M., Oh, M.H., Kim, M.H., Kim, S.D. (2006): Analytical modelling and nonlinear analysis of beam-column connection in steel moment resisting frame. Journal of Asian Architecture and Building Engineering, Vol. 5, No. 2, 309-316

Kohler, M., Heaton, T.H., Bradford, S.C. (2007): Propagating waves in the steel, moment-frame building recorded during earthquakes. Bulletin of Seismological Society of America, Vol. 97, No. 4, 1334-1345

Kramer, S.L., Stewart, J.P. (2004): Geotechnical aspects of seismic hazard. Earthquake Engineering: From Engineering Seismology to Performance-Based Engineering (eds. Y. Bozorgnia and V.V. Bertero) CRC Press, Boca Raton, Florida, 4.1-4.85

Lehmann, L., Borsutzky, R. (2006): Seismic analysis of structures: Influence of the soil. 1st European Conference on Earthquake Engineering and Seismology, Geneva, 3-8 September 2006, Paper Number 933

Magliulo, G., Ramasco, R., Realfonzo, R. (2002): Seismic behavior of irregular in elevation plane frames. 12th European Conference on Earthquake Engineering, London, 9-13 September, 2002, Paper No. 219

Mazzolani. F.M. (2000a): Steel structures in seismic zones. Seismic Resistant Steel Structures (eds. F.M. Mazzolani and V. Gioncu), CISM courses Udine, Springer, Wien, 1-18

Mazzolani, F.M. (ed.) (2000b): Moment Resistant Connections of Steel Frames in Seismic Areas. E&FN Spon, London

Mazzolani, F.M. (2002): Structural integrity under exceptional actions: basic definitions and field of activity. Improvement of Building's Structural Quality by New Technologies. COST 12, Workshop, Lisbon, 19-20 April 2002, 67-80

Mazzolani, F.M. (coord. & ed.) (2006): Seismic Upgrading of RC Buildings by Advanced Technics. Polimetrice, Milano

Mazzolani, F.M., Della Corte, G., Faggiano, G. (2004): Seismic upgrading of RC buildings by means of Advanced Techniques: The ILVA-IDEM project. In Proceedings of the 13th World Conference of Earthquake Engineering, B.C. Canada

Mazzolani, F.M., Piluso, V. (1993): Member behavior classes of steel beams and beam-columns. Giornate Italiane delle Costruzioni in Acciaio, Viareggio, 24-27 October 1993, 405-416

Mazzolani, F.M., Piluso, V. (1996): Theory and Design of Seismic Resistant Steel Frames. E & FN Spon, London

Mazzolani, F.M., Mandara, A., Faggiano, B. (2007): Structural integrity and robustness assessment of historical buildings under exceptional situations. In Performance, Protection and Strengthening of Structures under Extreme Loading (PROTECT 2007), Whistler, Canada, 20-22 August 2007.

MCEER (nd): How earthquakes affect buildings http://mceer.buffalo.edu/infoservice/reference_services/EQaffectBuildings.asp

Mendis, P., Ngo, T. (2003): Vulnerability assessment of concrete tall buildings subjected to extreme loading conditions. CIB-CTBUH International Conference on Tall Buildings, Malaysia, 8-10 May 2003

Mezzi, M., Parducci, A. (2005): Preservation of existing soft-first-storey configurations by improving the seismic performance. 3rd Conference on Conceptual Approach to Structural Design, Singapore

Montuori, R., Piluso, V. (2000): Plastic design of steel frames with dog-bone beam-to-column joints. In Behavior of Steel Structures in Seismic Areas (eds. F.M. Mazzolani and R. Tremblay), Montreal, 21-24 August 2000, Balkema, Rotterdam, 627-634

Mortgat, C.P., Zaghw, A., Singhal, A. (2004): Fire following earthquake loss estimation. 13th World Conference on Earthquake Engineering, Vancouver, 1-6 August 2004, Paper No. 2191

Murty, C.V.R. (2004): Brick masonry infills in seismic design of RC framed buildings: Part 2-Behaviour, The Indian Concrete Journal, August, 31-38

Mylonakis, G. (2002): Simplified model for seismic pile bending at soil layer interfaces. 7th U.S. National Conference on Earthquake Engineering, 7NCEE, Boston, 21-25 July 2002

Nagy-Gyorgy, T. (2007): Polimeric Composite Materials (in Romanian). Editura Politehnica

NEHRP (2008): Seismic design of reinforced concrete special moment frames. A guide for practicing engineers. Seismic Design Technical Brief No. 1

Ngo, T.D., Mendis, P.A., Teo, D., Kusuma, G. (2008): Behavior of high-strength concrete columns subjected to blast loading
http://www.cinenv.unimelb.edu.au/aptes/publications/hsc_columns_blast.pdf

NISEE (1997): Proper selection for superstructure. Berkeley University.
http://niseee. Berkeley.edu/bertero.html

Ohi, K., Lee, S.L., Shimawaki, Y., Ohtsuka, H., Guzman, R. (2000): Inelastic behavior of end-plate connections during earthquakes and improvement on their rotation capacity. Bulletin of Earthquake Resistant Structure Research Center, University of Tokyo Vol. 33, 81-86

Pan, Q., Cho, C. (2007): The investigation of a shape memory alloy micro-damper for MEMS applications Sensors, Vol. 7, 1887-1900

Pappin, J.W. (2002): Earthquake engineering of foundations and lifelines. 12th European Conference on Earthquake Engineering, London, 9-13 September 2002, paper 846

Parra-Montesinos, G..J. (2005): High-performance fiber-reinforced cement composites: An alternative for seismic design of structures. ACI Structural Journal, September/October, 668-675

Pecker, A. (1996): Seismic bearing capacity of shallow foundations. 11th World Conference on Earthquake Engineering, Acapulco, 23-28 June 1996, Paper No. 207

Petcu, D., Gioncu, V. (2003): Computer program for available ductility analysis of steel structures. Computer and Structures, Vol. 81, 2149-2164

Pitilakis, K.D. (1995): Seismic microzonation practice in Greece: A critical review of some important factors. 10th European Conference on Earthquake Engineering (ed. G. Duma), Vienna, 28 August-2 September 1994, Balkema, Rotterdam, 2537-2545

Prada, M. Gioncu, V. (2004): Influence of Torsion Rigidity on Spatial Analysis of RC Structures (in Romanian), Oradea University Publishing House

Rai, D.C. (2000): Future trends in earthquake-resistant design of structures. Current Science, Vol. 79, No. 9, 1291-1300

Roeder, C. (2000): State of the art report in connection performance SAC Joint Venture

Romo, M.P., Mendoza, M.J., Garcia, S.R. (2000): Geotechnical factors in seismic design of foundations: State-of-the-art report. 12th World Conference of Earthquake Engineering, Auckland, 30 January-4 February 2000, paper 2832

Saez, E., Lopez-Caballero, F., Farahmand-Razavi, A.M .(2006): Effects of SSI on the capacity spectrum method. 1st European Conference on Earthquake Engineering and Seismology, Geneva, 3-8 September 2006, Paper Number 1073

Safak, E (1999): Wave-propagation formulation of seismic response of multistory buildings. Journal of Structural Engineering, Vol. 125, No. 4, 426-437

Savidis, S.A., Bode, C., Hirschauer, R. (2000): Three-dimensional structure-soil-structure interaction under seismic excitation with partial uplift. 12th World Conference on Earthquake Engineering, Aucland, 30 January – 4 February 2000, Paper No. 0290

Semblat, J.F., Duval, A.M., Dangla, P. (2000): Analysis of seismic site effects: BEM and modal approach vs experiments. 12th World Conference on Earthquake Engineering, Auckland, 30 January- 4 February 2000, paper 1257

Semblat, J.F., Gueguen, P., Kham, M., Bard, P.Y., Duval, A.M. (2002a): Site-city interaction at local global scales. 12th European Conference on Earthquake Engineering, London, 9-13 September 2002, Paper 807

Semblat, J.F., Kham, M., Gueguen, P., Bard, P.Y., Duval, A.M. (2002b): Site-city interaction through modifications of site effects. 7th National Conference on Earthquake Engineering, Boston, 21-25 July 2002

Semblat, J.F., Kham, M., Bard, P.Y., Gueguen, P. (2004): Could "Site-city interaction" modify site effects in urban areas? 13th World Conference on Earthquake Engineering, Vancouver, 1-6 August 2004, Paper No. 1978

Sharma, U.K., Bhargava, P., Kaushik, S.K. (2005): Behavior of confined high strength concrete columns under axial compression. Journal of Advanced Concrete Technology, Vol. 3, No. 2, 267-281

Spangemacher, R. (1992): Zum Rotationachweis von Stahlkonstruktionen, die nach Traglastverfahren berechnet warden. Ph.D. Thesis, Aachen University

Stewart, J.P., Tileylioglu, S. (2007): Input ground motions for tall buildings with subterranean levels. PEER TBI –Task 8 Final Report
http://peer.berkeley.edu/tbi-work/documents/TRI-Task8_final-eport_May2007.1.pdf

Stratan, A., Dubina, D. (2004): Bolted links for eccentrically braced steel frames. Connections in Steel Structures V, Amsterdam, 3-4 June 2004, 223-232

Teguh, M., Duffied, C.F., Mendis, P.A., Hutchinson, G.L. (2006): Seismic performance of pile-to-pile cap connections: An investigation of design issues. Electronic Journal of Structural Engineering, Vol. 6, 8-18

Truta, M., Mosoarca, M., Gioncu, V., Anastasiadis, A. (2003): Optimal design of steel structures for multi-level criteria. In Behavior of Steel Structures in Seismic Areas, STESSA 2003 (ed. F.M. Mazzolani), Napoli, 9-12 June 2003, Balkema, Lisse, 63-69

UBC (1997): Uniform Building Code. Structural Design Requirements. Vol. 2, Chapter 16

Valmundsson, E.V., Nau, J.M. (1997): Seismic response of building frames with vertical structural irregularities. Journal of Structural Engineering, Vol. 123, No. 1, 30-41

Virdi, K.S., Rashkoff, R.D. (2008): Low-rise residential constructions detailing to resist earthquakes.

http://www.staff.city.ac.uk/earthquakes/MasonryBrick/ReinforcedBrick
Masonry/ConfinedBrickMasonry.htm

Vogel, K. (2005): Designing robust structures. A classic as well as very modern approach to structural design.

http://www.jcss.ethz.ch/events/WS_2005_11/PTT/Knoll_Vogel.-ppt.pdf

Wallace, J. (2005): Lightly-reinforced wall segments

http://peer.berkeley.edu/research/pdf/Wallace-all_Segments_Final_V3_Present.pdf

Wang Y., Dyke, S. (2007): Smart system design for a #D base-isolation benchmark building. Structural Control and Health Monitoring, 18 September 2007

WHO (2006): Seismic vulnerability factors. In Guidelines for Seismic Vulnerability Assessment of Hospitals. World Health Organization

Wikipedia (nd): Base isolation.

http://en.wikipedia.org/wiki/Base_isolation

Wikipedia (nd): Robustness.

http://en.wikipedia.org/wiki/Robust

Wikipedia (nd): Shape memory alloy.

http://en. Wikipedia.org/wiki/Shape_memory_alloy

Yang, J.N. (1990): Aseismic hybrid control system for building structures under strong earthquake. Journal of Intelligent material Systems and Structures, Vol. 1, No. 4, 432-445

Chapter 9

Advances in Seismic Design Methodologies

9.1 CHALLENGES IN SEISMIC DESIGN

9.1.1 Design Concepts before Northridge and Kobe

The beginning of modern seismic design may be fixed during the 1950s, when the dissipation of seismic energy through plastic deformations was considered by Housner. His method develops a limit design type analysis for ensuring a sufficient energy-absorbing capacity to guarantee an adequate safety factor against collapse in case of extremely strong ground motions. The first study on the inelastic response spectrum was made by Velestos and Newmark in 1960. They obtained the maximum response deformation for elastic-perfectly plastic structures. Since this first application in seismic design, the response spectrum has become a standard measure of the demand of ground motion. The utility of the response spectrum lies in the fact that it gives a simple and direct indication of the overall displacement and acceleration demands of the earthquake ground motion, for structures having different period and damping characteristics, without needing to perform detailed numerical analysis. A new concept was proposed in 1969 by Newmark and Hall, by constructing spectra based on accelerations, velocities, and displacements, in the short, medium and long period ranges, respectively (Gioncu and Mazzolani, 2002).

Based on these developments, the first practical seismic design philosophy stated the requirements for a minimum level of safety of building. Specifically, the designed structures are expected:
- To resist minor level of earthquake without significant damage.
- To resist moderate level with some non-structural damage and repairable structural damage.
- To resist major level of earthquake without collapse.

However, the majority of the seismic design methodologies today explicitly consider only one performance objective, in case of rare major earthquakes, defined as protection of occupants against injury or loss of life. Criteria for checking buildings against minor or moderate earthquakes, which may occur relatively frequently in the life of the building, are not explicitly specified.

The recent decades have seen a dramatic rise in economic losses related to earthquakes. In the period 1988-1997 (including Loma Prieta, Northridge and Kobe earthquakes, which cannot be considered catastrophic from the point of view of magnitude) the economic losses were estimated as twenty times larger than in the previous 30 years (see Section 3.1). These losses are due to several factors, which include:

- Dense population of buildings located in seismically active regions.
- Stock of aged and non-engineering buildings, which were designed or constructed without respecting any anti-seismic rules or based on inadequate concepts.
- Deficiencies in knowledge of regional seismic hazard, behavior of structural materials and structural systems under dynamic loads.
- Increasing cost of business interruption, due to the loss of building functions.
- Large amount of damage in non-structural elements and contents.

Contrary, the structures, which were designed in conformance with the modern seismic design, performed as expected, the loss of lives being minimal, in agreement with the intent of the adopted methodology. However, there is a misperception from many owners, insurers and government agencies about the expected performance of seismic designed buildings (FEMA 349, 2000). The economic losses of the past earthquakes raised many questions regarding the adequacy of the current seismic design rules to prevent these losses.

So, in the recent years, a new philosophy for the seismic design of constructions has been discussed within the engineering community.

9.1.2 After Northridge and Kobe: Appropriate Performance Levels

The main characteristic of the new seismic design is the participation of the owners and users for establishing target and appropriate performance levels. They are determined not only in terms of structural design, but from the demands of owners, users and society. As a consequence, the new methodology must be very explicit to be easily understandable also for people ignoring structural design (Aoki et al, 2000).

The bases of the new seismic design started in the United States, with the Vision 2000 Document (SEAOC, 1995), which provides the main concepts of this approach (Bertero, 1996a,b). The goal of this design philosophy is to produce structures, that have predictable seismic performance under multiple levels of earthquake intensity.

Performance - based seismic design is defined as:

"...consisting of selection of design criteria and structural systems such that at the specified levels of ground motion and with defined levels of reliability, the structure will not be damaged beyond certain limiting states or other useful limits " (Bertero and Bertero, 2000).

It means that a comprehensive performance-based seismic design involves several steps (Gioncu and Mazzolani, 2002):

- Selection of performance objectives and definition of the acceptable damage level.

- Definition of multi-level appropriate design criteria and specification of ground motion levels, corresponding to the different design criteria.
- Consideration of a conceptual overall seismic design in function of these levels and option for a suitable structural analysis method for each level.

The first step in the performance-based design philosophy is to define an acceptable damage level due to a given earthquake. There is no general agreement on this damage level, but there are some generally accepted criteria for determining these performances:

- *Collapse prevention* directly related to the life safety in order to prevent loss of life, injuries and damage of the contents of buildings. The structure can undergo serious damage, but it must stand after the earthquake.
- *Reparable damage.* A distinction is made between the structural damage which cannot be repaired and that which can be repaired. Irreparable damage is a specific subject for individual engineering judgment of experts. The damage refers both to structural and non-structural elements.
- *Acceptable business interruption,* which can be borne by the owner. In some cases, the value of the business is more important than the value of the buildings themselves and the interruption of this activity is intolerable. If the building owner wishes to avoid losses due to these interruptions, a stronger and stiffer design than the minimum required by design codes is necessary. But the owner must be fully aware about the supplementary cost for this requirement.

The second step in a comprehensive seismic design is the definition of multi-levels appropriate design criteria to satisfy the performance objectives. A building can be subjected to low, moderate or severe earthquakes. It may survive these events undamaged; it can undergo slight, moderate or heavy damage; it may be partially destroyed or it can collapse. These levels of damage depend on the earthquake intensity. The low intensity earthquakes occur frequently; the moderate earthquakes are rarer, while the strong earthquakes occur once or maximum two times during the structure life period. It is also possible that a devastating earthquake will not affect the structure during its life.

Under these conditions, the relation between multi-level design criteria and the performance objectives can be summarized in the following:

- *Frequent low earthquakes* must be associated with the acceptable business interruption.
- *Rare moderate earthquakes* must be considered for reparable damage.
- *Very rare strong earthquakes* are related to collapse prevention.

The next step is to adopt an appropriate conception for each level and a suitable structural analysis method. The strategies for each performance level are:

- *Elastic analysis* of structure for frequent low earthquakes, so the non-structural elements suffer only minor damage. For assuring their integrity, interaction between structure and non-structural elements must be considered.

- *Elasto-plastic analysis* for rare moderate earthquakes. The non-structural elements are partially damaged, so the analysis must consider only the structure behavior, without any interaction with the non-structural elements.
- *Kinematic analysis* for the very rare earthquakes, considering the behavior due to the possible formation of plastic mechanisms, must be performed.

9.1.3 Required Performances and Seismic Hazards

A key feature of seismic design is the presentation of the earthquake actions to be considered in design. The required structural performance can be established in function to the given earthquake actions. Generally, it is accepted the following framing for relating seismic hazard to magnitude M:
- Low seismic hazard, M < 4.5.
- Moderate seismic hazard, M = 4.5 – 6.5.
- High seismic hazard, M > 6.5.

Looking to the world maps with the seismic areas (see Section 6.5), one can see that the high seismic hazard areas are concentrated mainly near to known faults, especially around the tectonics plate boundaries, where interplate earthquakes occur. The remaining areas, diffuse seismic zones, characterized by intraplate earthquakes, can by framed into low to moderate seismic hazard areas. The required structural performances must consider these differences in seismic hazards. Therefore, the seismic design must be divided in to two different approaches:
- *Seismic design for high seismic risk areas*, where the required performance must be very high. Especially after the seismic decade 1989-1999, a very important stock of valuable research works has been developed, giving rise to consistent improvements in seismic design methodologies for these areas. High performance levels are required for structures situated in these areas.
- *Seismic design for low to moderate seismic risk areas*, for which the research results are not at the same level as for the high seismic risk areas. A wrong tendency to use the results obtained for the high risk areas was noticed. But in the last decade it was remarked that the design concept in low and moderate seismic areas cannot be the same as in the high seismicity areas and, as a consequence, a very important change has been observed in this aspect. Limited performance levels are required for the structures situated in these areas.

This division in seismic design approaches is crucial from the economical point of view, allowing the designer to optimize the costs of seismic protection measures. Design considerations for the seismic resistance of buildings in areas of low and moderate seismic risk are in many ways different than the ones in areas of high seismic risk. As the earthquake design procedures are generally based on considerations coming from areas of high seismicity, the current seismic design rules are not readily adaptable to areas of low to moderate seismic risk.

9.2 PERFORMANCE BASED SEISMIC DESIGN:
IMPLICATION OF OWNERS, USERS AND SOCIETY

9.2.1 Generic Performance-based Engineering

Generic performance-based engineering creates a process which can be applied in any aspect of building design. It implies design, evaluation and construction of engineering facilities, whose performance under common and extreme loads responds to the diverse needs and objectives of owners, users and society. Following the performance-based design, the designer has to perform some given activities during the following stages (Yamawaki et al, 2000):

-   To clarify the actual performance demand of the owner through discussion with him or her (*preliminary design stage*).
-   To determine the target performance based on the agreement among them and to confirm that the results of design satisfy the target (*developed design stage*).
-   To confirm the construction as-built performance (*construction supervision stage*).
-   To provide support to maintain the as-built performance *(maintenance support stage)*.

The premise is that the performance levels can be quantified, that performances can be predicted analytically, and that the cost of the improved performance, requested by the owner, can be evaluated. The quantification refers to:

-   *Prediction of hazard* for extreme actions produced by earthquakes, tsunamis, winds, fires, explosions, landslides, etc.
-   *Prediction of demands*, by structural modeling and non-linear analysis.
-   *Prediction of damage*, by studying the component and the structure fragilities, as well as by estimating the losses.

The performance-based seismic design should follow the procedures developed for generic performance-based engineering, but it has some distinct features discussed in the next sections.

9.2.2 Performance-based Seismic Design in Earthquake Engineering

By now, it is widely acknowledged that seismic design is not a one-step process with a single set of criteria for universal level of protection. There is a minimum level of protection demanded by society in order to safeguard against the structure collapse which endangers human lives. But, in addition to the life safety, the society has other responsibilities, including the continuous operation of critical facilities and protection against excessive damage which may have far-reaching consequences for the society at local, regional or international level. Moreover, the owners want options for maximizing the return on their investments and/or for providing life safety protection to the inhabitants of their facilities beyond the minimum required by the society (Krawinkler, 1996, 1997).

"Performance-based seismic design implies design, evaluation and construction of engineering facilities whose performance under different earthquake levels responds to the diverse needs and objectives of owners, users and society" (Krawinkler, 1999*)*.

The new performance-based seismic design procedures intend to design structures which perform at appropriate levels for all earthquakes. By defining various performance levels, performance-based seismic design allows business interruption losses after a seismic event to be controlled and, when appropriate, minimized. With the development of this new methodology, building owners now have choices which were never available before.

"Building owners can now tailor the performance of their structure to their needs" (Poland and Hom, 1997).

For instance, hospitals, where the functionality of the facilities after a major seismic event is more important than the actual structures, can use a performance basic level to attain a higher performance level than the one for life safety. Production companies, such hi-tech manufactures, as well as important banks, where the values of the business is more important than the value of the buildings themselves, are able to design and upgrade their facilities to maintain continued operations, even after a major earthquake.

In order to implement the performance based seismic design, it is necessary to select the performance objectives. A performance objective is a simple statement of the desired building behavior, giving that it experiences earthquake demands of specified severity. The important parameters for owners and users include the potential loss of life, the cost of repairing any sustained damage, as well as the amount of time during which the building is out of service while it is repaired or, in extreme cases, replaced. A most practical approach would seem to be the adoption of a limited series of standard behavioral states, from which the design performance objectives could be developed. The Vision 2000 projects (SEAOC, 1995, Hamburger, 1996) have identified the following series of standard behavioral state definitions with the appreciation of the service disruption time (Giuliano et al, 2004):

- *Fully operational*. Only very minor structural or non-structural damage occurred. The building retains its original stiffness and strength. Non-structural components operate and the building is available for normal use without any service interruption. Repairs, if required, may be done at the convenience of the building users. The risk of life threatening injury during the earthquake is negligible.
- *Operational*. Only minor structural damage occurred. The building structure retains nearly its original stiffness and strength. Non-structural components are secured and, if utilities are available, most of them would function. Life safety systems are operational. Repairs may be done at the convenience of the building users. The risk of life threatening injury during the earthquake is very low. The service interruption is less than 3 days.

-   *Life safe*. Significant structural and non-structural damage occurred. The lateral strength has still a margin against collapse. Non-structural components are secure, but cannot operate. The building may not be safe for occupancy until repaired. The risk of life threatening injury during the earthquake is low. The service interruption is less than 3 months.
-   *Near collapse*. Substantial damage occurred. The building has lost most of its original stiffness and strength, having a very little margin against collapse. Non-structural components may become dislodged and present a falling hazard. In case the experts decide that the building can be repaired, the service interruption is longer than 3 months. But in many cases the repair is not practical.

The proposed levels of damage for structure and non-structural elements corresponding to these standard behavior states are shown in Figure 9.1. These states represent (Gioncu and Mazzolani, 2002):

-   No damage for both structural and non-structural elements for *fully operational*.
-   No damage in structural elements, minor damage in non-structural elements for *functional*.
-   Repairable damage of structural elements, important damage of non-structural elements for *life safe*.
-   Not-repairable damage of structural elements, complete failure of non-structural elements for *near collapse*.

The relation between the four limit states and the four probabilities of earthquake occurrence is presented in Figure 9.2, as a performance objective matrix. The basic objectives of the earthquake design are illustrated in this matrix as the diagonal line which represents the minimum objectives required by the codes. The unacceptable performances are located under these minimum objectives. At the same time, there are enhanced objectives, if the owner consents to a supplementary payment for providing better performance or lower risk than the one corresponding to minimum objectives.

9.3 DEVELOPMENT OF MULTI-LEVEL BASE SEISMIC DESIGN

9.3.1 Different Multi-Level Approaches

Performance-based seismic design is a desirable concept, whose implementation has a long way to go. It appears to promise engineered structures, whose performance can be quantified and conforms to the owner's desires. But this methodology will be a lost cause, if rigorously held to this promise, as it is well known that it is not possible to predict all important seismic demands and capacities with perfect confidence (Krawinkler, 1999). Unfortunately, while the above parameters are quite meaningful to the owners and users and, therefore, can serve as a basis for selecting among the given building performance alternatives, they are not particularly useful as a basis for design.

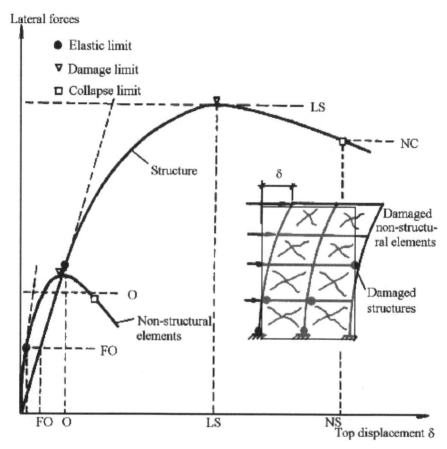

Figure 9.1 Performance and damage levels (Gioncu and Mazzolani, 2002)

There is no direct way that engineers could design for such performance specifications, such as for instance "a business interruption of 3 weeks". Therefore, there is a need to establish the corresponding relationships among the behavioral parameters, which are meaningful for building owners and users respectively, and the reliability analysis for the design professionals (Hamburger, 1996).

Under these conditions, another methodology for seismic design is developed. As Figure 9.2 shows, there is a minimum level of protection demanded by society and introduced in the codes in order to safeguard buildings against collapse endangering human lives and to reduce economic losses at the minimum level. In the frame of

Vision 2000, these minimum levels are four, but in the design practice this methodology can be used only for very important buildings, for very strong earthquakes and for special demands of building owners.

For current cases, the checks required to guarantee the good behavior of a structure during an earthquake must be examined in the light of multi-level base seismic design, which is limited, from practical reasons, at maximum three levels. Due to the fact that a specific analysis must be performed for each level, no one designer will accept more than two or maximum three levels for the structure design.

In the seismic load-top displacement curve (Fig. 9.1), there are three very important points:
- Limit of elastic behavior without any damage.
- Limit of damage with major damage.
- Limit of collapse, for which the structure is at the threshold of breakdown.

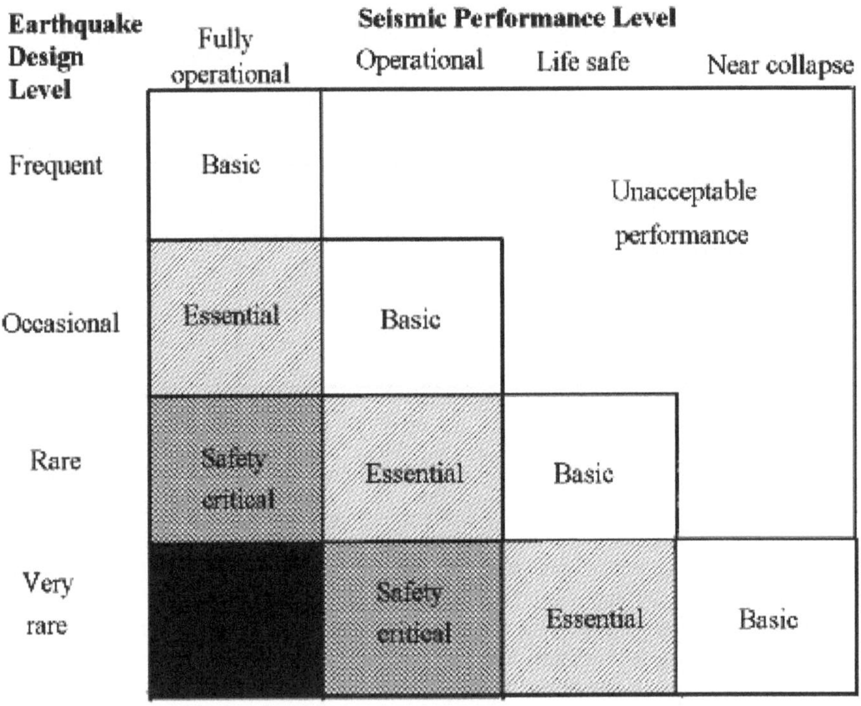

Earthquake Design Level	Seismic Performance Level			
	Fully operational	Operational	Life safe	Near collapse
Frequent	Basic	Unacceptable performance		
Occasional	Essential	Basic		
Rare	Safety critical	Essential	Basic	
Very rare		Safety critical	Essential	Basic

Figure 9.2 Performance objectives (Gioncu and Mazzolani, 2002)

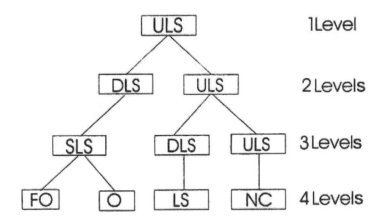

Figure 9.3 Evolution of multi-level design approaches (Truta et al, 2003)

In function of taking different limit states for structure and non-structural elements, some multi-level base approaches have been developed (Fig. 9.3):

One level design approach, in which only the *ultimate limit state* (ULS) is considered, for the prevention of structure collapse. This approach was the official design approach for a long time in many codes.

Two levels design approach, which is today a current code methodology and the seismic design philosophy of Eurocode 8. These two levels are:

-*Damage limit state* (DLS), for which structures are designed to remain elastic, or with minor plastic deformations and non-structural elements remain undamaged or have minor damage.

-*Ultimate limit state* (ULS), for which structures exploit their capability to deform beyond the elastic range, non-structural elements being partially or totally damaged.

Three levels design approach, which can be considered as the objective for the next generation of codes. Verification at the three levels has been proposed by Mazzolani and Piluso (1993, 1996), Gioncu and Mazzolani, (2002), being defined as the combination of the followings limit states:

-*Serviceability limit state* (SLS), for frequent earthquakes. This limit state imposes that the structure, together with non-structural elements, could suffer minimum damage and the discomfort for inhabitants should be reduced to the minimum. So, for this level, the structure must remain within the elastic range or it may undergo just unimportant plastic deformations.

-*Damageability limit state* (DLS), for occasional earthquakes. This limit state considers an earthquake intensity which produces damage in non-structural elements and moderate damage in the structure, which can be repaired without great technical difficulties.

-*Ultimate limit state* (ULS), for earthquakes which may rarely occur, representing the strongest possible ground shaking. For these earthquakes, both structural and non-structural damage is expected, but the safety of inhabitants has to be guaranteed. In many cases damage is so substantial that the structures are not susceptible to be repaired and their demolition is recommended.

One can see that the four design levels, developed by Vision 2000 and presented in Figure 9.2, practically correspond to the three design levels, with the exception of the serviceability limit state, which is divided in two (fully operational and operational limit states).

9.3.2 Definition of Earthquake Design Levels

There are two approaches to define the main characteristics of earthquake design for different limit states.
-    *Probability approach*. Design earthquakes are determined for a given probability of exceeding in 50 years (Fig. 9.4a).
-    *Recurrence period approach*. The level of acceleration or magnitude is determined in function of the recurrence period (Fig. 9.4b).

The annual probability of exceeding the peak ground accelerations is considered for low, moderate and high seismicity regions.

The relationship between the two approaches is presented in the following:

Earthquake design level	Probability of exceedance	Recurrence interval
Frequent	50% in 50 years	43 years
Occasional	20% in 50 years	72 years
Rare	10% in 50 years	475 years
Very rare	2% in 50 years	970 years

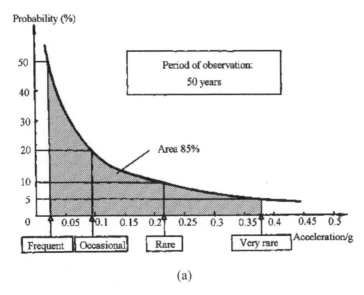

(a)

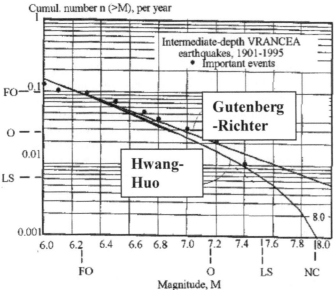

(b)

Figure 9.4 Definition of earthquake design levels: (a) Probability approach; (b) Recurrence period approach (Gioncu and Mazzolani, 2002)

9.3.3 Structural Design for Multi-level Approaches

One limit level procedure was, for a long time, the basis of code provisions. Recently the modern codes (EC8 and UBC 97) introduced the checks for two limit levels. The future codes are expected to consider three limit levels.

Two levels design. This seismic design philosophy is connected to the traditional design methodology for other loading types, for which two levels are considered. But for seismic loadings there are some features in design criteria, which must be underlined. The structural demands are presented in Figure 9.5. For two levels design, checking the two demands, rigidity and strength, must be considered as a basic verification. The other checks, such as strength for limitation of damage, or ductility for ultimate limit state, are optional.

Three levels design. This methodology differs from the previous approach not only for the number of levels, but also for a more clear definition of the performance levels. The design criteria are presented in Figure 9.6: (i) for *serviceability limit level* under frequent low earthquakes, the strategy calls for the complete elastic response of the structure. The lateral deformations are limited by the interstory drift limits, given for the non-structural elements. In order to guarantee their integrity, the interaction between structure and non-structural elements must be considered. The basic verification of structural rigidity and strength verification (condition for the elastic behavior) is only optional; (ii) for *damageability level* under occasional moderate earthquakes, an elasto-plastic analysis must be performed. The produced damage can be repaired without great economical and technical difficulties. The basic verification refers to the structural member strengths, the checking of rigidity and ductility being optional. The non-structural elements are just partially damaged, so the analysis must consider only the structure, without any interaction with the non-structural elements; (iii) for *ultimate level* under rare severe earthquakes, a kinematic analysis, which considers the behavior of a possible formation of plastic collapse mechanisms, must be performed. The basic verification refers to the ductility, the strength verification being only optional. The design strategy refers to the control of the formation of pre-selected plastic collapse mechanism and the rotation capacities of plastic hinges. The non-structural elements are completely damaged.

9.3.4 Optimal Solutions for Seismic Design

In order to identify the optimal solutions, three steel moment resisting frames with 3, 6 and 9 stories were examined by Gioncu and Mazzolani (2002) and Tirca et al (1997) (Fig. 9.7). For the 6 and 9 story buildings, eccentrically braced frames with long link beams are also considered. The effect of the variation of the following parameters is studied: ground acceleration (0.08g for low seismicity, 0.20g for moderate seismicity, 0.32g for high seismicity), normalized spectral value for corner period 0.7 sec and

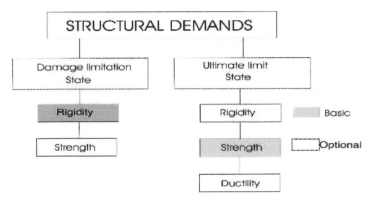

Figure 9.5 Design criteria for two levels (Truta et al, 2003)

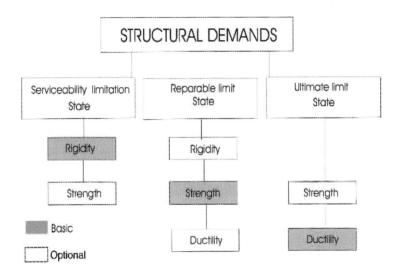

Figure 9.6 Design criteria for three levels (Truta et al, 2003)

behavior factor (q = 4 for medium ductility, q = 6 for high ductility and q = 8 for very high ductility).

Figure 9.8 presents the structure weight of 3 and 6 story buildings, in function of the behavior factor for different ground accelerations (for ultimate limit state) and for different values of strength and corresponding ductility. The different ground accelerations have no great influence on the structure weight. Contrary, using a high value of the behavior factor, an increase of requested ductility produces the increasing of the structure weight. So, the benefit of designing buildings with large behavior factor values may be lost. From Figure 9.9, it is very clear that it is rational to use low values of the behavior factor (q = 3 to 4.5) for reduced number of stories and for low seismicity. High values of the behavior factor are recommended only for a large number of storys and high seismicity.

Figure 9.10 shows the structure weight for damage limitation (by assuming an interstory drift limit of 0.005h) and ultimate limit state. It is clear that these two limit levels give the same structure only for very low seismicity. The ratio between the shear forces corresponding to serviceability and ultimate limit levels is plotted in Figure 9.11, in function of the behavior factor and the structural natural periods. For behavior factor greater than 4.5, the control of design is given by the serviceability limit level. So, for moderate and high seismicity, the controlling criterion is the damage limitation. For these seismicity levels, the benefit of designing buildings with large values of behavior factor may be useless, as the building could be too flexible for limiting damage in non-structural elements.

Considering the above results, there are some important remarks, which must be taken into account in structural design:

- For low seismicity it is possible to design the structure by using one level only (ultimate or damage limitation levels).

- For moderate and high seismicity the damage limitation is the dominant criterion. In order to attenuate this condition and to reduce the interstory drifts, the interaction between structure and non-structural elements must be considered.

- Different criteria must be used when designing for low, moderate and high seismicity levels.

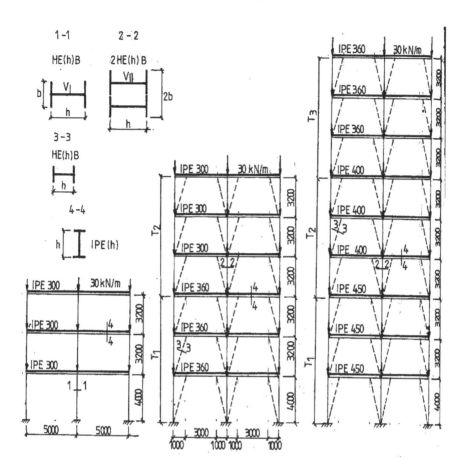

Figure 9.7 Examined frames (Gioncu and Mazzolani, 2002)

9.3.5 Simplified Strategies for Seismic Design

In order to simplify the design procedures, two main characteristics must be considered in design: the seismicity of the site and the importance of the building (Kennedy and Medhekar, 1999).

A building can be located in a zone with *very low, low, moderate* and *high* or *strong earthquakes* according to the following values of acceleration:

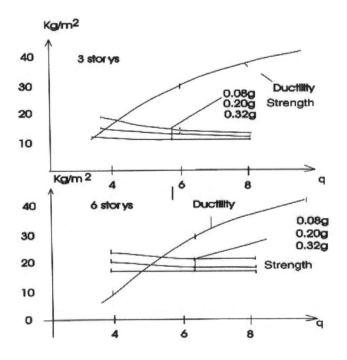

Figure 9.8 Strength-ductility relations for ultimate limit levels (Truta et al, 2003)

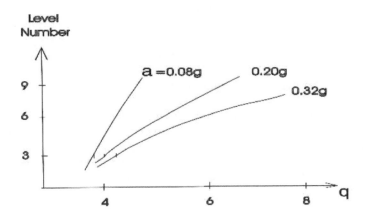

Figure 9.9 Recommended reduction factor q (Truta et, 2003)

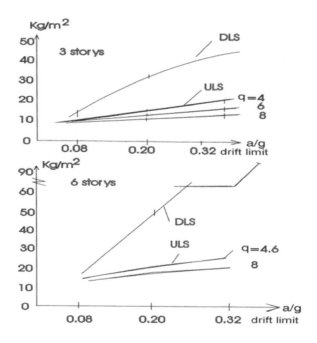

Figure 9.10 Damage limitation-ultimate limit states relation (Truta et al, 2003)

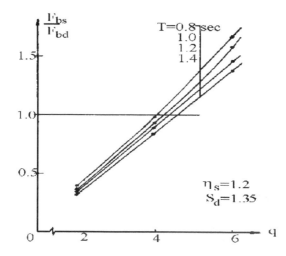

Figure 9.11 Influence of serviceability limit level (Truta et al, 2003)

-      Very low seismicity, accelerations less than 0.08g.
-      Low seismicity, acceleration less than 0.20g.
-      Moderate seismicity, accelerations less than 0.32g.
-      High seismicity, accelerations less than 0.50g.
-      Strong seismicity, accelerations larger than 0.50g.

     The building can be framed in two main typologies:
-      *Ordinary*, when the damage occurring during an earthquake does not produce important disturbance in the community;
-      *Important*, when damage or collapse can produce important social perturbations.

In spite of these differences, the current seismic design methodologies propose the same procedures to be followed in all seismic zones. It should be logical to spend more design effort on obtaining structures, that adequately perform in zones of high seismicity and for important buildings. Contrary, less design effort must be required for zones of low seismicity and for ordinary buildings.

Considering these aspects, a strategy is proposed by Truta et al (2003) in order to reduce the design effort:

Building	**Seismicity**				
	Very low	**Low**	**Moderate**	**High**	**Strong**
Ordinary	1L	2L	2L	3L	3L
Important	2L	2L	3L	3L	4L

For very low seismicity and ordinary buildings, the design can consider only one level. For very low seismicity, but important buildings and low seismicity or moderate seismicity, but ordinary buildings, two levels must be considered. Three levels are required for moderate seismicity and important buildings or always for high seismicity or for strong seismicity and ordinary buildings. Four levels must be used only for strong seismicity and important buildings.

9.4 RESPONSE SPECTRA AS REPRESENTATION OF GROUND MOTIONS

9.4.1 Response Spectrum Types

Since the concept of response spectrum was introduced into Earthquake Engineering, this technique has been widely used to estimate force and deformation demands, which are imposed on structures by the earthquake ground motions. Today, response spectra form the basis of seismic design forces in the majority of seismic codes. Response spectrum represents the maximum response of a single degree of freedom (SDOF) system to a given input motion, as a function of natural frequency and damping.

Determining the structural response requires that the ground motions be translated into forces acting on the building. The forces have their origin in the fast change of actions during a short time. These actions present dynamic characteristics, because the variations are sufficiently rapid to involve inertial forces. Therefore, the dynamic analysis must evaluate the structural response by taking into account these forces. The general method for determining the seismic forces to apply to a structure is based on a simple equation $F = ma$, where m is the mass of the building and a is the acceleration of the structure. An elastic response spectrum is a simple plot of the peak response in terms of acceleration, velocity or displacement of a series of frequency which are forced into motion by the same base vibration. The resulting plot can then be used to pick off the response of any linear system, its natural frequency of oscillation being given. If the input used in calculating a response spectrum is steady-state periodic, then the steady-state result is recorded.

After digitizing an accelerogram of a particular earthquake (Fig. 9.12a) and assuming a numerical value for damping, the response of a SDOF elastic system can be calculated. The dynamic motion is applied at the base of a cantilever having different natural periods, which models the case of a structure restrained at the ground level (Fig. 9.12b). The complete history of response for this elastic system can be computed. The maximum values of accelerations, velocities and displacements are determined from the obtained system response. By repeating the above process for a great number of cantilever systems, for a given value of damping, a plot of the response spectrum can be obtained (Fig. 9.12c) in function of system periods. One can see that, due to the resonance effects, the spectra have the tendency to amplify the response in the range of low periods.

Examining the resulting plot of response, one can see some amplification due to the *resonance effects*, when the frequency of input excitation corresponds to the frequency of the system. The main scope in constructing the spectra is to emphasize these amplifications.

Separate plots for accelerations, velocities and displacements can be obtained by using the same procedure but the most common formats for spectra are (Fig. 9.13):
- Spectral acceleration vs. period (frequency).
- Spectral displacement vs. period (frequency).

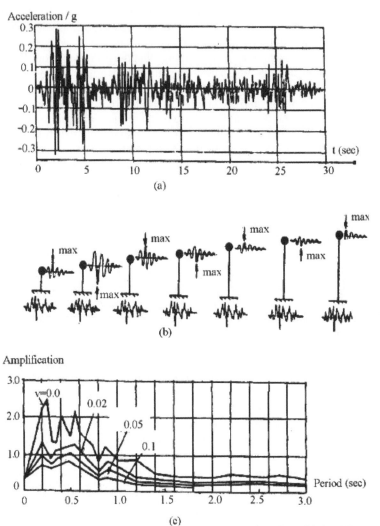

Figure 9.12 Elastic response spectra: (a) Recorded acceleration; (b) SDOF system; (c) Resulted spectra (Gioncu and Mazzolani, 2002)

Because the actual elastic spectra present many peaks which cannot be used in practice, the constructed elastic spectra must be *smoothed* (Fig. 9.13a). So, the actual variation is substituted by a combination of linear or hyperbolic variations. A very important characteristic of smoothed spectra is the *corner period* (Fig. 9.13b), which is equivalent to the predominant period of the ground motion. It can be determined also as the period corresponding to peak acceleration associated to the maximum relative velocity spectrum. It is customary to employ the spectra normalized to the value of the peak ground acceleration.

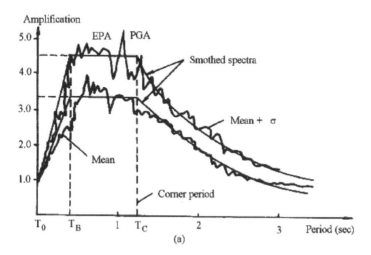

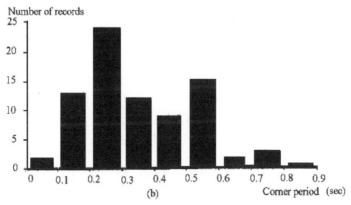

Figure 9.13 Smoothes spectra: (a) PGA and EPA values;
(b) Corner period distribution (Gioncu and Mazzolani, 2002)

Ground motions are time process affected by a high degree of uncertainty, deriving from many source types and site conditions. Therefore, the modeling of seismic excitation turns out to be a random process. So, the average value and the standard deviation can be determined from the processing of the recorded data. The proposed values, generally admitted, are the average value plus one standard deviation (Fig. 9.13a). The parameters defining the spectra are the peak ground acceleration, velocity or displacement (PGA, PGV and PGD), as well as the effective peak acceleration and velocity (EPA and EPV). EPA is the average of the maximum ordinates in the period range from 0.1 to 0.5 sec, divided by a mean value of 2.5. EPV is the same average of the maximum ordinates of velocity in the period range of 0.8 to 1.2, divided by the same mean value of 2.5.

Separate plots for accelerations, velocities and displacements can be obtained by using the same procedure but the most common formats for spectra are (Fig. 9.14):
- Spectral acceleration vs. period (frequency).
- Spectral displacement vs. period (frequency).

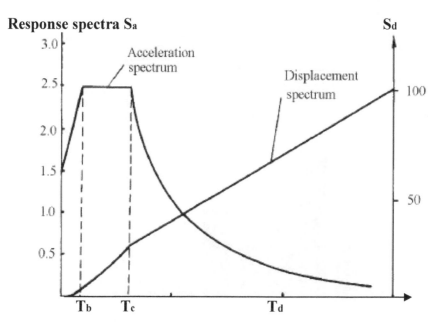

Figure 9.14 Elastic and displacement spectra (Gioncu and Mazzolani, 2002)

9.4.2 Main Characteristics of Acceleration Spectra

The form of the elastic response spectrum, $S_e(T)$, which is introduced in the building codes, is shown in Figure 9.14. It presents the amplification of ground accelerations in function of T(s), which is the vibration period of a linear single–degree-of-freedom system. The main parameters of this spectrum are:

- a_g, ground acceleration ;
- a_{max}, amplification of ground acceleration;
- T_b, corner period at the beginning of the plateau (generally $T_b = 0.2\ T_c$). For periods shorter than about 0.03 sec, the response spectrum follows the constant PGA line;
- T_c, corner period at the end of the plateau, equivalent to the predominant period of ground motion (ranging approximately from 0.3 to 0.7 sec);
- T_d, value defining the beginning of constant displacement response range of the spectrum (ranging approximately from 3 to 4 sec).

All the values of these parameters depend on the ground motion values. The main factors influencing the shape of the response spectra will be presented in the followings sections.

9.4.3 Influence of Different Factors on Spectra

There are many factors which influence the form of the response spectral acceleration. They are:

Structural damping. The reduction of elastic response due to damping has been studied since long ago. The important fact worth pointing out here is that the effect of damping depends on the vibration period of the system, being more pronounced for short periods (0.2 to 0.7 sec) (Fig. 9.15). Most of building codes have design spectra implicitly based on 5% of the critical damping.

Earthquake types. It is very important to notice the great difference between intraplate and interplate earthquakes and the importance of the distance from source (especially for intraplate earthquakes) (Fig. 8.16), which must be reflected into the elastic spectra (Lam et al, 1996). The differences are given by the natural period of ground motions. For intraplate earthquakes, characterized by low to moderate earthquakes and short ground periods, an important increasing of amplification is noticed for short structure periods. At the same time, the corner periods are much reduced for intraplate earthquakes (0.3 to 0.5 sec) in comparison with the interplate earthquakes, for which the natural periods are longer (0.5 to 1.5 sec) (especially in function of site soil conditions).

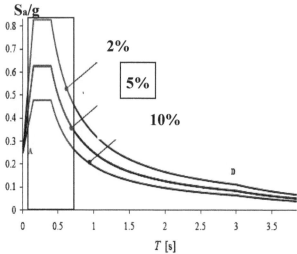

Figure 9.15 Influence of structural damping (after Oliveto and Marletta, 2005)

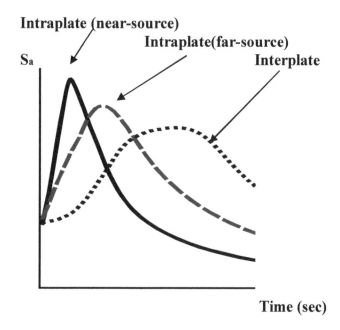

Figure 9.16 Spectra for intraplate (near and far-source) and interplate earthquakes

Effects of soil conditions. Earthquakes of the last decades, especially those of Mexico City, Armenia and San Francisco Bay regions, have reemphasized the importance of local geologic deposits on the amount of damage and the resultant loss of life. In general, damage and loss of life in each of these earthquakes were concentrated in areas with deposits of soft soils. These concentrations of damage have emphasized the need to modify the response spectra patterns to better account for the amplification effects of local geologic deposits (Borcherdt, 1994). Therefore, the influence of local geological conditions on the features of response spectra is widely recognized and most codes provide specifications for these relationships, usually by categorizing soil profiles into soil classes. The ground motion experienced on the surface results from the transmission of energy waves released from the bedrock source, which is transmitted first through the bedrock itself and then undergoes significant modifications in the soil layers, as soon as the energy waves come close to the Earth's surface. Typically rock sites experience short period response, but with a rapid decay. Thus, short duration high intensity motion may be expected in such locations. Conversely, soft soils, particularly when they extend to moderate depth, are likely to filter out some of the short period motion and usually amplify longer period responses, particularly when the soil mass has a natural period similar to the high energy component of the earthquake. Soft soil response spectra have a flatter and broader plateau.

Considering these aspects, Eurocode 8 introduced five soil classes, defined by the average shear wave velocity (V_s) in the upper meters of soil. Soil class A corresponds to the rock or other rock-like geological formations; class B to deposits of very dense sand, gravel or very stiff clay; class C to deep deposits of dense or medium sand, gravel or stiff clay; class D to deposits of loose-to-medium cohesionless soil; and class E to soil consisting of surface alluvium The effects of the soil conditions depend on the earthquake type, being different for interplate (Type 1) and intraplate (Type 2) earthquakes (Fig. 9.17). One must mention that Eurocode 8 (2002) does not explicitly specify the framing of the two magnitude earthquake types (Type 1 and Type 2) in the two typical earthquakes (interplate and intraplate). The spectral amplification factors, for interplate (Type 1, M > 5.5) and intraplate (Type 2, M < 5.5) earthquakes are presented in Figure 9.18, where the influence of soil classes is identified. Examining the spectrum for interplate earthquakes (M >5.5) and the corresponding characteristics given in Figures 9.17a and 9.18a, one can see that the weak soils produce an increasing of acceleration amplification and corner periods.

EC8-00 TYPE 1	S	Tb	Tc	Td
soil A Vs > 800 m/s	1,00	0,15	0,4	2,0
soil B 360<Vs<800 m/s	1,10	0,15	0,5	2,0
soil C 180<Vs<360 m/s	1,35	0,20	0,6	2,0
soil D Vs < 180 m/s	1,35	0,20	0,8	2,0
soil E (h <= 20 m)	1,40	0,15	0,4	2,0

(a)

EC8-00 TYPE 2	S	Tb	Tc	Td
soil A Vs > 800 m/s	1,0	0,05	0,25	1,2
soil B 360<Vs<800 m/s	1,2	0,05	0,25	1,2
soil C 180<Vs<360 m/s	1,5	0,10	0,25	1,2
soil D Vs < 180 m/s	1,8	0,10	0,30	1,2
soil E (h <= 20 m)	1,6	0,05	0,25	1,2

(b)

Figure 9.17 Characteristics of soil classes: (a) Type 1, interplate earthquakes (M >5.5); (b) Type 2, intraplate earthquakes (M <5. 5)

Spectral amplification

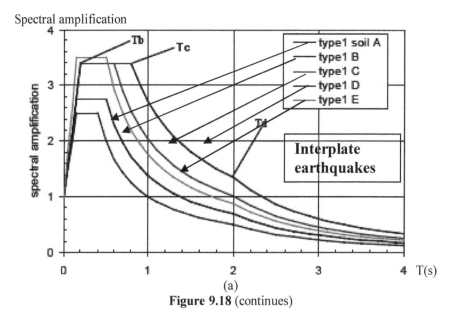

(a)
Figure 9.18 (continues)

Spectral amplification

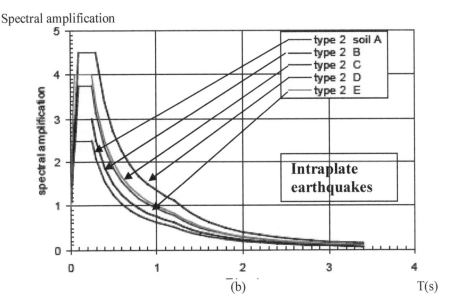

Figure 9.18 Elastic response spectra for soil classes: (a) Interplate (Type 1)
earthquakes; (b) Intraplate (Type 2) earthquakes (Gioncu, 2006, after Sabetta and
Bommer, 2002)

In Figures 9.17b and 9.18b, corresponding to the intraplate earthquakes (M <5.5), a
very important increasing in acceleration amplification is observed, larger than for the
interplate one. But the influence of soil conditions on corner periods is very reduced,
contrary to the observed increasing in case of interplate earthquakes.

Inelastic response spectra. Following the design philosophy for structures in
seismic area, the sizing of individual members, connections and supports is typically
based on the distribution of internal forces, which are computed by means of the
design response spectrum, substantially reduced from the elastic one, determined using
the recorded ground motions. As a result of this reduction, the severe ground motions
demand large deformations in the structural systems and, therefore, inelastic structural
response and energy dissipation are inevitable requirements. As a consequence, the
elastic response spectrum has a limited capability to predict the structure behavior
during severe earthquakes and an inelastic response spectrum must be considered. The
inelastic response spectrum is obtained by reducing the elastic response spectrum,
taking into account the structure capacity to dissipate seismic energy (Fig. 9.19). So,

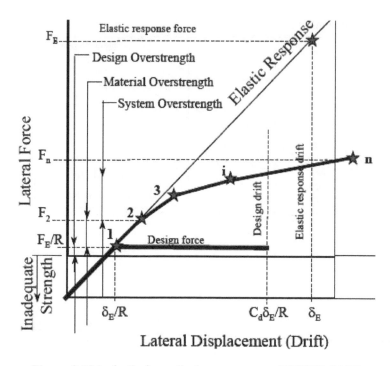

Figure 9.19 Inelastic force-displacement curve (NEHRP, 2003)

the design seismic lateral forces correspond to the level causing the formation of a first plastic hinge in the most heavily loaded element in the structure. The loading increasing causes the formation of additional plastic hinges and the structure capacity increases until the maximum is reached. The *overstrength* capacity obtained by this inelastic action provides the reserve strength necessary for the structure to resist the extreme motions of the seismic forces which may be generated by the actual ground motions. It may be noted that the structural overstrength described above results from the development of a sequential plastic hinging in a redundant structure. Figure 9.19 explores some of the factors which contribute to produce the structural overstrength: material overstrength (actual material strength being higher than the reduced nominal strength) and additional overstrength (by selecting sections exceeding those strictly required by the computations). The result is that structures typically have a much higher lateral resistance than the one specified by the codes (NEHRP, 2003).

The main problem is how to optimize these aspects in structural design. The transformation of the elastic response spectrum into an inelastic response one is the best way to follow in a simplified methodology, which uses the response spectra. The inelastic response spectra can be obtained in seismic codes by modifying the elastic response spectra by means of a factor, namely the *q-factor*, which takes into account

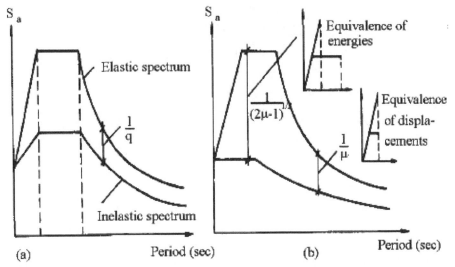

Figure 9. 20 Inelastic response spectra: (a) Using a constant q factor;
(b) Using a variable q factor (Gioncu and Mazzolani, 2002)

the dissipative capacity of the structure up to failure. The two ways to transform the elastic response spectrum into the inelastic one are presented in Figure 9.20: by reducing the elastic response spectra by means of a reduction factor q, which can be constant or variable, in function of the structure natural period and considering some equivalence conditions (based on energy or displacement).

9.4.4 Displacement Response Spectra

The conventional seismic design is entirely a force-based approach, which assumes that an acceleration response spectrum reliably provides, for a given degree of damping, the elastic force level acting on the structure and, therefore, it represents an adequate indicator of the seismic design. Recently, however, displacement-based design has attracted growing interest among engineers, because it is recognized that displacements describe in a more explicit way the structural response under seismic actions, compared to forces (Priestley, 1997, Tolis and Faccioli, 1999). The correlation between acceleration and displacement spectra is presented in Figure 9.14.

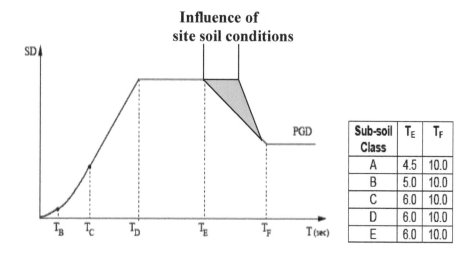

Figure 9.21 Elastic displacement response spectrum (after Sabetta and Bommer, 2002)

In this approach, the elastic acceleration spectrum is replaced by the elastic displacement spectrum. Since damage of structures subjected to earthquakes is certainly expressed in terms deformations (strains at fibers, curvatures at sections, rotations at members and drift at story levels), the displacement-based approach is conceptually more appealing (Bommer and Elnashai, 1999).

For design purposes, Sabetta and Bommer (2002) proposed an elastic displacement spectrum format, presented in Figure 9.21. The proposed values of T_b, T_c and T_d are presented in Figure 9.17, while for T_e and T_f are in Figure 9.21, in function of the site soil classes. The influence of these conditions is to enlarge the field of maximum amplification of spectrum.

The displacement response design method is considered to offer the following advantages with respect to the spectral acceleration response design method (Medhekar and Kennedy, 2000):

- Displacements play a major role at the preliminary design stage itself, resulting in a good control on displacements over the entire design process. Target displacement criteria are selected for the serviceability and ultimate limit states and thus the damage control is achieved directly.
- Strength and stiffness of the lateral load resisting system are chosen to satisfy the desired deformation criteria.
- The estimation of the fundamental period of the structure is not required for the preliminary design.
- The displaced shape may be explicitly linked to the member ductility demands.

- The drift of non-structural elements at various levels of damage can be directly used in design.
- The structural factor q, used in spectral acceleration response spectrum, it not needed.

9.4.5 Capacity Response Spectra

The conventional response spectrum is represented by the acceleration versus period relationship for different damping levels. In the capacity response spectrum format, the period axis is converted in the displacement axis (Fig. 9.22). The structure period T is represented by radial lines, instead of being the horizontal axis.

The advantage of this representation is that both strength and displacement demands are evident in a single graph. The elastic displacement demand and the structure period may be determined for elastic forces. Considering the inelastic behavior, the inelastic displacement at the reduced force is obtained in the horizontal branch of the capacity curve corresponding to the intersection with inelastic reduced spectra. At the same time, the reduced period due to inelastic deformation may be determined. Another advantage of this representation is that both earthquake demands and structure capacity may be compared directly from the same figure.

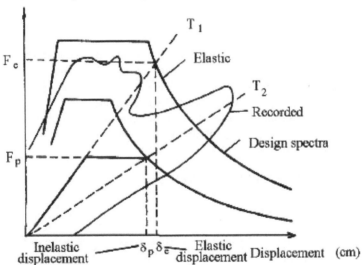

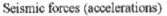

Figure 9.22 Capacity spectrum (Gioncu and Mazzolani, 2002)

9.4.6 Criticism of Using Response Spectra

Due to its simplicity, the response spectra representation provides a very reliable tool to estimate the ground motion effects on the ordinary structures. The criticism of the use of this representation is based on the following points (Gioncu and Mazzolani, 2006):

-The assumption that the use of only one or a reduced number of ground motions is sufficient to describe the earthquake effects is wrong, because other records can provide a response spectra shape, which is considerably different from the one of the considered records. This is, especially, the case of intraplate earthquakes, for which the number of records on the same location is very small.

- The shapes of spectra for distinct ground motion levels are very different on the same site. Therefore, the use of the same spectra for structural checking at different levels is an unsuitable procedure.

-The records near to source are very different from the ones recorded far from the epicenter, both for interplate and intraplate earthquakes. Unfortunately, there is a reduced number of near source records.

-The damping effects are very different in function of the level of the considered earthquake. For reduced level, the effects are also much reduced, so use of the standardized 5% critical damping ratio is an inaccurate procedure.

-One of the most important factors in determining the structural response, the ground motion duration, is completely ignored in the set-up of response spectra. Generally, the duration is very different for interplate, intraplate or intraslab earthquakes, so the number of cycles able to dissipate seismic energy is also very different and the reducing factor q must consider this fact.

-The proper consideration of superior vibration modes is very difficult, because the response spectra are constructed on the base of the first vibration mode. Some particular response spectra must be considered, especially in case of intraplate earthquakes, where the superior vibration modes are dominant.

Despite these criticisms, the use of response spectra remains the most practical method. Therefore, the improving of this methodology is a task for research (see Chapter 10).

9.5 SEISMIC ANALYSIS PROCEDURES

9.5.1 Main Characteristics

Seismic analysis is a subset of structural analysis, whose scope is the evaluation of the response of buildings under earthquakes. It is part of the process of structural design in regions where earthquakes are prevalent (Wikipedia, nd). The primary objective of seismic analysis is to determine the dynamic response of structures under seismic actions.

When the ground motions move the base of the structure, its masses experience forces of inertia as the structure attempts to follow the ground motions. The resultant movements of the structure depend in a complex way on the characteristics of ground motions, the dynamic properties of the structure, the characteristics of the structural materials and the type of foundations. Earthquake forces are generated by the inertia of buildings when they dynamically respond to ground motions.

The dynamic nature of the response makes earthquake loading different from other building static loads. As a consequence, the need to consider the dynamic behavior of buildings sometimes generates discomfort within the designers. The designer temptation to consider earthquakes as static loads is a trap which must be avoided since the dynamic characteristics of the building are fundamental for evaluating the structural response. Nevertheless, the effective earthquake design methodologies can be, and usually are, simplified without detracting from the effectiveness of the design. Indeed, the high level of uncertainty about the ground motions generated by earthquakes seldom justifies the use of complex and sophisticated methodologies. Instead of an intricate methodology, it is preferable to use a simplified one, but considering some specific features of seismic design. In this way, a good seismic design is the one where the designer takes control of the building, by dictating how the building has to respond (King, 2000).

9.5.2 Methods of Analysis

The analysis of a structural system to evaluate both deformations and internal forces induced by applied loads or ground excitation is an essential step in the design of a structure to resist earthquakes. A structural analysis procedure requires:
- Representation of earthquake ground motions.
- Model of structure.
- Method of analysis.

For the representation of earthquake ground motions, there are the following procedures:
- Response spectrum is a simple plot of the peak of steady-state response (displacement, velocity or acceleration) of a series of oscillations with varying natural frequency or period, which are forced into motion by the same base vibration or shock (see Section 9.4).
- Lateral load distribution in function of ground and structure natural periods, based on the assumptions that the structure response is controlled by the vibration modes.
- Representation of time-history acceleration in function of site-recorded ground motions or artificial accelerogram compatible with the design response spectrum.

For modeling the structures, there are the following approaches:
- 2-D modeling of structure, when only the plain structural response is considered, for regular structures.

- 3-D modeling of structure, for which the spatial structural response is considered, for important and irregular structures.

Clearly, the inelastic time history analysis, which predicts both force and deformation (damage) demands with sufficient reliability in many elements of the structural system, is the desired solution. The implementation of this solution requires the availability of a set of ground motion records (each with three components), which account for the uncertainties and differences in severity, frequency characteristics and duration due to rupture characteristics and distances of the various faults, potentially causing ground motions in the considered site. It requires further the capability to adequately model the cyclic load deformation characteristics of all important elements of the three-dimensional soil-foundation-structure system. Moreover, it requires the adequate knowledge of deformation capacities including the deterioration characteristics, which define the limit state of acceptable performance (Krawinkler and Seneviratna, 1998).

It is fair to recognize that at this time none of the afore mentioned requirements for the desired solution have been adequately developed and that efficient tools for implementation do not yet exist. Recognizing these limitations, the main task is to perform an evaluation process, which must be relatively simple, but able to capture the essential features significantly affecting the performance goal. In this context, the accuracy of demand prediction is desirable, but it may be not essential, since neither seismic input nor capacities are accurately known.

In this perspective, a set of structural analysis methods was developed, from the simplest to the most complex ones. It must be kept in mind that they are not necessarily physically realistic methods, but represent a design analysis which is thought to ensure an adequate safety for an engineered structure. These simplified analyses call for a series of engineering judgments, from the seismological to the structural aspects (Studer and Koller, 1994). The differences between these methods lie in the way they incorporate the seismic input and on the idealization of the structure response. All methods of analysis have to comply with the current design philosophy, which requires that the a structure must not collapse and must retain its structural integrity under rare strong earthquakes and must also not be damaged in use under frequently moderate earthquakes. Concerning the methods of structural analysis, they can be divided into five distinct analytical procedures:

- Linear equivalent static analysis.
- Linear response spectrum analysis.
- Linear dynamic analysis.
- Non-linear static (push-over) analysis
- Non-linear dynamic (time-history) analysis.

The choice of the analytical method is subjected to limitations based on building characteristics. The linear procedures maintain the traditional use of a linear stress-strain relationship, but incorporate adjustments to overall building deformations and material acceptance criteria to permit better consideration of the probable non-linear characteristics of seismic response. The non-linear static procedure, often called push-over analysis, uses simplified non-linear techniques to estimate seismic structural deformations. The non-linear dynamic procedure, commonly known as time-history analysis, due to the dynamic nature of seismic excitation, is the most available method to determine the structural behavior during an earthquake. But the cost of such analysis is generally high, while the results correspond to a particular excitation only, which does not offer a reliable basis for design.

9.5.3 Equivalent Static Analysis

The more simple procedure is the so-called equivalent static force analysis, sometimes also called the simplified dynamic analysis. This method may be available when one particular mode of vibration is predominant as compared to others and the system is accurately modeled by a single degree of freedom system. In this case the design spectrum method is reduced to one mode of vibration to express the dynamic behavior of the system. Usually, the first mode shape is considered as a primary mode of vibration, which can be simplified further into a simple line. The equivalent static forces are computed as shown in Figure 9.23 for the fundamental period of vibration. Then, a classical static analysis can be performed under the action of these equivalent static forces. This method is an approximate method, which is adequate for certain types of structures (regular buildings) and ground motions (having natural periods

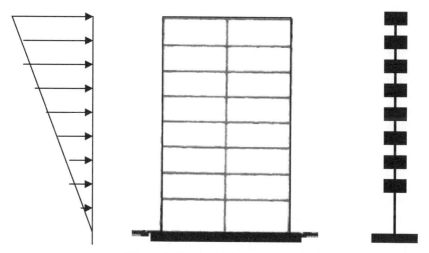

Figure 9. 23 Equivalent static analysis

close to the first vibration of the structure). Contrary, the results of this procedure can by very inaccurate when applied to a building with a highly irregular structure system, unless the building is capable of response to seismic loads in a nearly elastic manner. When the contribution of higher modes of vibration is significant, this method is not conservative. In these cases, a complete dynamic response spectrum analysis is advisable (ESDEP, 2008).

9.5.4 Response Spectrum Analysis: Modal Response Method

The response spectrum analysis is the standard procedure of the modern seismic design codes. It aims to directly give the maximum effects of earthquake in various elements of the structure. The general method, called also the multi-modal method, consists on computing the various modes of vibration of the structure (Fig. 9.24a) and the maximum response of each mode with reference to a response spectrum (Fig .9.24b), by determining the lateral distribution of seismic forces for each mode and the corresponding internal forces. The structure response can be defined as a combination of many modes. A rule is then used to combine the responses of these different modes. For this reason the method is also known as the superposition of modal response method. For the combination of these modal responses, three methods are available (Wikipedia, nd):

-   The sum of absolute values of the modal response. This is the most conservative method, because it assumes that the maximum modal values, for all modes, occur at the same time.
-   The square root of the sum of the squares (SRSS) of forces and displacements. This method assumes that all of the maximum modal values are statistically independent. For three-dimensional structures, in which a large number of frequencies are almost identical, this assumption is not justified.
-   The complete quadratic combination (CQC) of modal responses, which is based on random vibration theories in order to minimize the introduction of avoidable errors. This method has been incorporated as an option in most modern computer programs for seismic analysis.

There are computational advantages in using the response spectrum method of seismic analysis for the prediction of member forces and displacements in structural systems. But the use of the response spectrum method has some limitations, being only an approximate method. The first approximation refers to the use of spectra given for a single degree of freedom system, valuable only for the first vibration mode, to determine the structural response for the superior vibration modes. The second one is that it is restricted to linear elastic analysis, in which the damping properties can only be estimated with low degree of confidence. The third one refers to the procedure of superposition of different response modes, in which, due to the sum of square values, the sign of the values disappears.

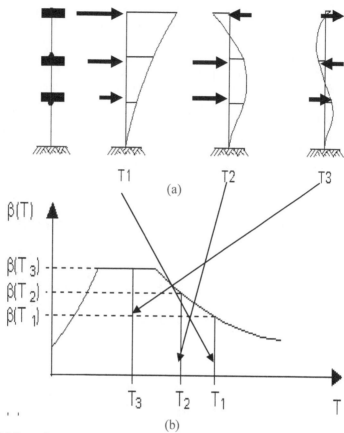

Figure 9.24 Steps in response spectrum analysis: (a) Computation of mode shapes and periods; (b) Reading the response spectrum (after ESDEP, 2008)

9.5.5 Non-linear Static Analysis: Push-over Method

Due to the development in computing programs, the current engineering practice for seismic analysis of buildings is moving away from simplified linear-elastic methods of analysis towards a more complex non-linear-inelastic technique. The non-linear static (push-over) analysis is a simple way to evaluate the inelastic ultimate capacity of structures. In this method, the designed structure is subjected to incremental lateral loads, using one or more predominant load patterns. Under incremental loads, various structural members may yield sequentially. Consequently, at each load level, the structure experiences a loss in stiffness. Using the push-over analysis, a characteristic non-linear force-displacement relationship can be determined. There are two different approaches in push-over analysis: traditional and adaptive (modal) procedures.

Traditional push-over procedure. The analysis can be described as applying lateral loads in invariant pattern (pattern remain constant during the increasing loads) (Fig. 9.25a), which approximately represents the relative inertial forces generated at each floor level and pushing the structure under monotonically increasing lateral loads. Basically, a push-over analysis is a series of incremental static analysis carried out to develop the capacity curve of the building from the elastic until the collapse state (Fig. 9.25b).This procedure provides a shear versus displacement relationship and indicates the inelastic limit as well as lateral capacity of the structure until collapse under this constant pattern loads. So, the load distribution for the push-over analysis, in function of the presumed main vibration modes and ground motions, plays a crucial role in this procedure. Generally, constant, triangular or parabolic distributions are the most used patterns.

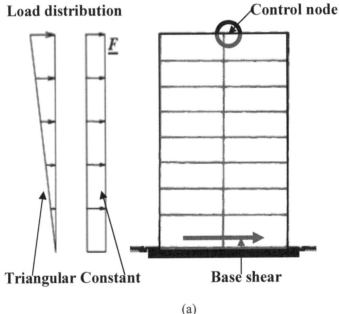

(a)

Figure 9.25 (continues)

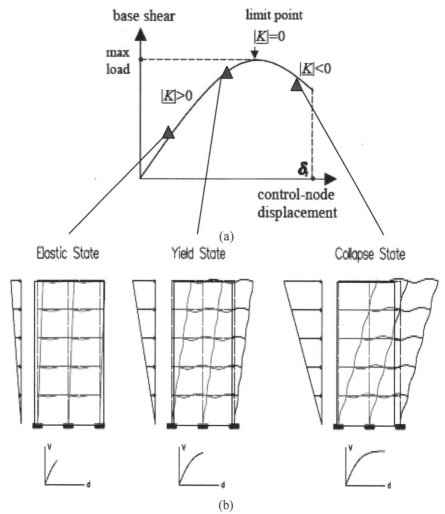

Figure 9.25 Non-linear static (push-over) analysis: (a) Base-shear and controlled displacement; (b) Results of push-over analysis (after Archer, 2001)

The push-over method is expected to provide information on many response characteristics, which cannot be obtained from an elastic static or dynamic analysis (Krawinkler and Seneviratna, 1998); they are:
-   Identification of the critical regions where the deformation demands are expected to be higher and which have to become the focus of thorough detailing.

- The method neglects the duration effects and cumulative energy dissipation, by assuming that damage is function only of lateral deflection of the structure.
- The push-over analysis is a static analysis which neglects the dynamic effects as kinetic and viscous damping energies.
- The method is developed only as 2-D procedure, the incorporating of 3-D (including torsional effects) being very difficult.
- The procedure considers only the lateral earthquake loading, the vertical earthquake components being ignored. Identification of the strength discontinuities in plan or elevation will lead to changes in the dynamic characteristics in the inelastic range.
- Estimate of inter-story drifts which account for strength or stiffness discontinuities.
- Evaluation of the consequences of strength deterioration of individual elements on the behavior of the structural system.
- Evaluation of the realistic force demands on potentially brittle elements, as columns, beams, braces, connections, non-structural elements.
- Estimate of the ductility demands for elements which have to deform inelastically in order to dissipate the induced seismic energy.
- Consideration of the completeness and adequateness of load path, considering all the elements of the structural system and non-structural elements, which significantly contribute to the lateral load distribution.

    In spite of these performances, the push-over method suffers from several fundamental deficiencies (Kim and D'Amore, 1999):

- The method implies that there is a separation between the structural capacity and earthquake demand. Non-linear structural behavior is load path dependent and it is not possible to separate the loading input from the structural response. So, at each loading step, the lateral load distribution must be changed.
- The push-over procedure is overly simplifying, considering that the structure behavior can be characterized only by two parameters, base shear and top level displacement. It is difficult to capture all possible structural responses in the vertical plan direction using only two parameters.
- The traditional push-over procedure does not account for the progressive changes in modal properties due to the influence of superior vibration modes and, consequently, the shear distribution on the structure height, which take place in a structure as it experiences cyclic non-linear yielding during an earthquake. Figure 9.26 show the modification in lateral load distribution due to the changing of structure properties as result of plastic hinges formation.
- Even if the top level displacement can be determined with some accepted approximation using a constant load pattern (triangular or uniform), the inter-story drifts estimation can be highly inaccurate, if higher mode effects are of importance (Fig. 9.27). In the case presented in this figure, for the triangular distribution, the error is about 51% and for uniform distribution, 92% (Pinho et al, 2005).

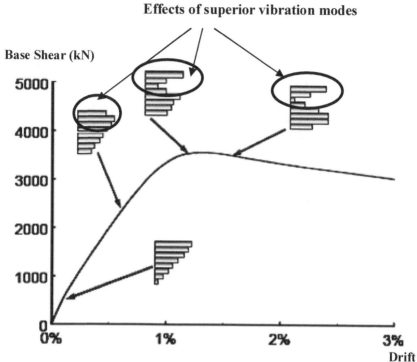

Figure 9.26 Influence of superior vibration modes due to the changing of structural
properties (after Pinho, 2007)

Therefore, the results obtained using the push-over method demonstrate that it is a very useful one, but not an infallible tool for assessing inelastic strength and deformation demands and for exposing the design weaknesses. On the positive side, the method provides good estimations of global as well as local inelastic deformation demands for structures which primarily vibrate in the fundamental mode. On the negative side, deformation estimates obtained from a push-over analysis may be grossly inaccurate for structures where higher mode effects are significant (Figure 9.27). This problem can be mitigated by applying more than one load pattern, including load patterns, which account for higher mode effects. Even with the application of multiple load patterns, this analysis could provide misleading results (Lawson et al, 1994).

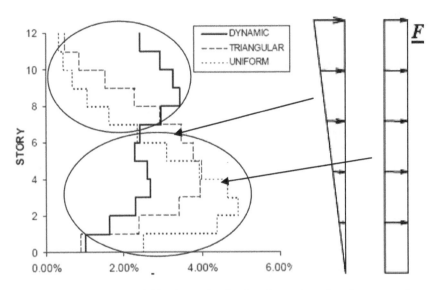

Figure 9.27 Interstory drifts determined using dynamic and push-over analyses
(after Pinho et al, 2005)

The traditional push-over procedure is confronted with many criticisms, mainly concerning:

(i)   The load pattern with which the push-over analysis is carried out, based only on the fundamental vibration mode of the elastic system;

(ii)   The ignorance of structural rigidity changing due to higher vibration modes, and consequently, the changing of load pattern features.

Due to these weaknesses of the traditional push-over procedure, some new methodologies were developed in the last time.

Multimode load pattern of push-over procedure. A methodology to determine the structure response by combination of the vibration modes using different participation factors is presented in Section 9.5.4 for elastic analysis. The same idea is extended for static inelastic push-over method, by a combination of push-over values determined for each vibration mode using different participation factors. The dynamic behavior of the majority of the structures is only affected by a small number of vibration modes. This procedure can be reduced to the determination of a load pattern by the combination of the ones corresponding to each vibration mode, obtained from a linear elastic analysis of the structure, using the different participation factors (Almeida and Caneiro-Barros, 2003). This procedure considers the lateral loads in predetermined pattern of the most important vibration mode, but ignores the redistribution of inertia forces due to structural yielding and the associated changes in the vibration properties of the structure.

Modal push-over procedure is developed for cases when the influence of high vibration modes is significant. The principal objective of this procedure is to develop a push-over analysis procedure based on the structural dynamic theory, which retains the conceptual simplicity and computational attractiveness of current procedures with invariant force distribution, but provides superior accuracy in estimating seismic demands on buildings (Chopra and Goel, 2001, 2002). This procedure determines the first three vibration modes and the traditional push-over procedure are applied for each of them. At the same time, the non-linear response history analysis (RHA), which is considered as etalon for the "exact" analysis, is performed. The peak responses are determined for each three modes and a combination of these values using the square-root-of-sum-of-squares (SRSS) procedure for inelastic analysis is used (Fig. 9.28a). So, the load pattern and the structural rigidity changing are determined for each vibration mode separately. The main problem is the methodology for the combination of these values. The modal push-over procedure results (MPA 3 modes) have been compared with the "exact" results obtained using non-linear response history analysis (NL-RHA) for many structures. Figure 9.28b presents this comparison for the story drifts of the 9–story building, showing a very good correspondence. In addition, a comparison with the proposed procedures recommended by FEMA-273 (1997) for lateral load distributions: uniform, equivalent lateral force (ELF) and square root of sum of squares (SRSS). As clearly demonstrated in this figure, the height-wise variation of story drifts determined by using the FEMA force distribution considerably differs from the non-linear RHA, clearly showing that the FEMA force distributions are not adequate, because they underestimate the story drift demands. For uniform distribution, the largest errors are in the upper stories, reaching 64%. For ELF distribution, the largest errors are noted both in the upper and the lower stories, reaching 35%. For SRSS distribution, the largest errors are in the lower stories, reaching 31%. Contrary, the MPA-3 modes procedure is more accurate than all the FEMA force distributions, with story drifts, which are underestimated by, at most, 7% and overestimated by no more than 32%. At the same time, the plastic hinge rotations estimated by all three FEMA force distributions contain unacceptably large errors (Chopra and Goel, 2001).

The modal push-over analysis was implemented for many frames with 3, 6, 9, 12, 15 and 18 stories, including two or three vibration modes, different ductility factors (1, 1.5, 2, 4 and 6) and 20 ground motions corresponding to Californian earthquakes (crustal interplate earthquakes) with magnitudes ranging from 6.6. to 6.9 (Chintanapakdee and Chopra, 2003). The main conclusion of this study is that the first

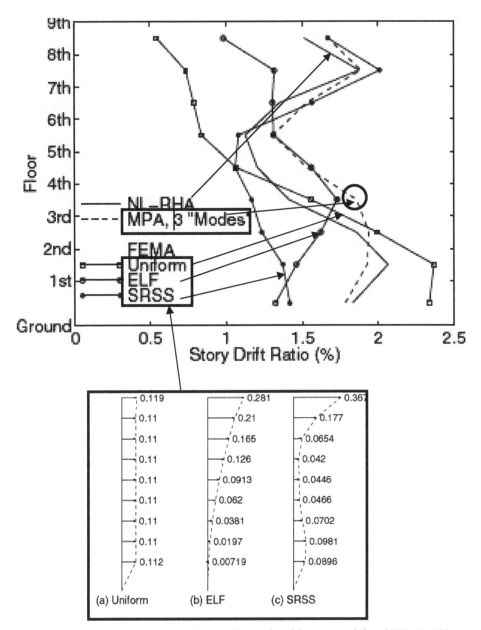

Figure 9.28 Comparison of dynamic non-linear time history, modal and FEMA-273 (1997) push-over analyses (after Chopra and Goel, 2001)

mode alone, which is the basis for push-over procedures currently used in engineering practice, does not adequately estimate the seismic demands. If two or three vibration modes are included in analysis, the height-wise variation of story drifts determined using modal push-over procedure is generally similar to the "exact" results from non-linear response history analysis. For low-rise to mid-rise buildings, it is sufficient to include, beside the first mode, only the second vibration mode. The third vibration mode contribution should be included only for taller frames.

The criticisms of this procedure refer to the used quadratic modal combination rules (SRSS or CQC) used in computing the results. These rules inevitably lead to monotonically increase the obtained values, since the possibility of sign changing is precluded, whilst it may be needed to represent the uneven redistribution of forces after an inelastic mechanism is triggered at some location (Pinho et al, 2005).

Adaptive push-over procedure considers the changing of both characteristics which determine the structural response, the load pattern and the structural rigidity. When plastic hinges form in the structure, constant distribution of lateral forces cannot be any longer used. In this procedure, the dynamic analysis is transformed into a static analysis, which is repeated many times as soon as the stiffness changes due to the formation of new plastic hinges (Fig. 9.29a). For each step, a new distribution of lateral loads is changed according to the mode shapes of the structure. So, these loadings are updated at each analysis step, reflecting progressive damage accumulation and resulting modification of modal parameters (Fig. 9.29b), which characterize the structural response at increasing loading levels.

According to the procedure developed by Papanikolaou (Papanikolaou et al, 2005), the lateral load pattern is not kept constant during the analysis, but it is continuously updated, based on a combination of the instantaneous mode shapes and spectral amplifications corresponding to the inelastic periods of the structure (Fig. 9.30). One can see that for each step, the structural periods are changed (especially for the first mode) and, correspondingly, the configuration of lateral loads is modified, this change being obtained from the response spectra. After defining the lateral profiles for all different modes, a modal combination (SRSS or CQC) is carried out. The base shear remains independent of the load pattern, being controlled by the current loading level only. Thus, the magnitude of the push-over load is still applied incrementally, as formed in the conventional push-over approach. The main criticisms regarding this procedure refer to the same quadratic modal combination rules (SRSS, CQC), as has been discussed for the modal push-over procedure.

Within the adaptive push-over procedure, the application of a displacement method, as opposed to the force method used in the previous procedure, seems to be more attractive, being in line with the present tendency for the development and code implementation of deformation-based design and assessment methods. Further, such a displacement-based design procedure seems to lead to superior response predictions,

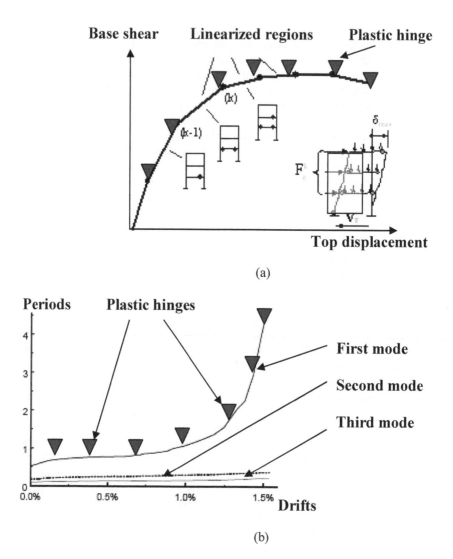

(a)

(b)

Figure 9.29 Adaptive push-over procedure: (a) Steps in order to applied a new response spectrum (after Turker and Irtem, 2006); (b) Modification of periods (after Pinho et al, 2006)

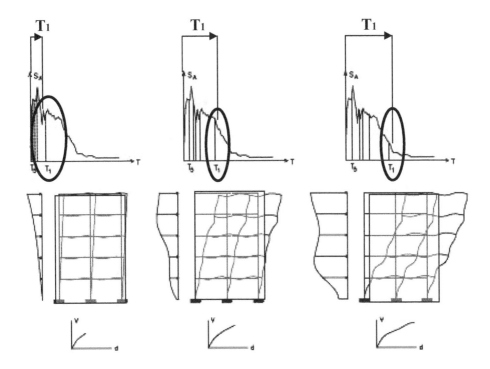

Figure 9.30 Adaptive push-over procedure (after Papanikolaou et al, 2005)

with no additional modeling and computational effort, with respect to conventional push-over procedures (Pinho et al, 2006). In this method, the SRSS combination has reduced detrimental effects. In Figure 9.31, the interstory drifts profiles for a 12-story building are presented for an "exact" dynamic analysis and for displacement-based adaptive push-over procedure. This procedure seems to have the potential to provide accurate predictions, throughout the entire deformation range, both for the capacity curve (Fig. 9.31a) and for the interstory drifts profile (Fig. 9.31b).

9.5.6 Non-linear Dynamic Earthquake Analysis

Non-linear dynamic earthquake analysis is a powerful tool for the study of the structural seismic response. A set of ground motion records or artificially generated time-history representations can give an accurate evaluation of the anticipated seismic performance of structures (Mwafy and Elnashai, 2001). It is the most adequate and comprehensive analysis procedure to evaluate the non-linear seismic response of

Base shear (kN)

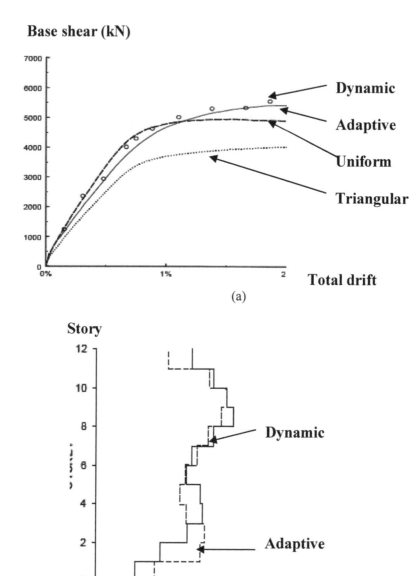

(a)

(b)

Figure 9.31 Displacement-based adaptive push-over: (a) Capacity curve;
(b) Interstory drifts profile (after Pinho et al, 2006)

structures, but the currently available computer hardware and design software effectively limit both size and complexity of structures, which may be analyzed using this procedure. At present, there is no general purpose non-linear analysis software which permits practical evaluation of large structures (Lee, 2008). There are three barriers preventing the engineer from the application of non-linear dynamic earthquake analysis:

-   It is too complex to be solved by means of personal computers, currently used in structural practice.
-   The results are too sensitive to the properties of structural material non-linearities.
-   The analysis results significantly vary in function of the earthquake records used.

Therefore, until these barriers will be eliminated, this procedure remains a very useful tool for research works only. The main problems of using this methodology are related to times history representation of ground motion modeling, hysteretic models for structural material and element behavior, time-history analysis and incremental dynamic analysis.

Time-history representation of ground motions. The ground motions are very difficult to predict in terms of time of occurrence, intensity and duration. The definition of spatial characteristics, like acceleration velocity and displacement in three directions, together with the temporal ones, like period and duration, is also considerably difficult. With respect to Earthquake Engineering, there are two ways in which ground motions can be represented (Gioncu and Mazzolani, 2002):

(i)   Using recorded digitized time-history representation from a data bank, classified in function of source type, distance of the source-recording station and station soil characteristics. So, the designer can select the most representative time-history records, corresponding to the involved site. This way presents many difficulties, due to the fact that the codes ask for the use of a minimum of 5 records for the structure analysis. In spite of the large number of available records, it is very difficult to have a sufficient number of records with characteristics similar to those expected at the site of the structure. On the other hand, the large variability in the characteristics of ground motions, like peaks of acceleration or velocity, period, duration, etc., which are able to influence in very large measure the structure response, produces an important scattering, without the certainty that the structural response to selected ground motions represents the actual one.

(ii)   Using artificially generated time-history representations is a very convenient way to solve the above-mentioned problems. These ones can be created by the combination of actual recording portions. The artificially generated representations are not necessarily the representation of a physically realistic earthquake; they should be considered as encompassing all possible seismic loadings, corresponding to the source type, dominant period, soil conditions and duration. So, it is possible to generate an earthquake scenario in the form of a time-history representation, different for interplate earthquakes (with differences in pulse velocity types resulted from subduction, collision or strike-slip), intraplate earthquakes (produced by ancient or rift faults) or intraslab earthquakes.

Between these two ways, in the perspective of introducing the influence of earthquake types in design practice, the use of the artificial time-history representation seems to be very promising for structural design.

Hysteretic models for components behavior, which must include the strength and stiffness deterioration. Results of seismic evaluation of various frames demonstrate that the strength deterioration becomes a dominant factor, when the response of the structure approaches the limit state of collapse. Therefore, the implementation of the non-linear dynamic earthquake analysis is not complete without models capable of tracking the history of beams, columns and connections damage until or at least close to collapse, and incorporate, in an explicit manner, the effects of deterioration in the seismic response. The main hysteretic models for steel, reinforced concrete and plywood are presented by Ibarra et al (2005).

Time-history analysis. The method is based on the direct numerical integration of the motion differential equations. In this aim, different algorithms can be adopted, where the elasto-plastic deformation of the structure must be considered. The analysis of the structure (Fig. 9.32a) may be performed by using, as base excitation, an actual or artificially generated ground motion, in the form of acceleration time-history (Fig. 9.32b). The variation of displacements (or forces) at different levels (or points) in the structure are presented in Figure 9.32c, also in the time-history form.

.One can see that an amplification due to resonance effects, an increasing of motion duration and the tendency of regularization of movements result as far as the level increases from the bottom to the top.

Incremental dynamic analysis. The time-history analysis determines the dynamic behavior of a structure for a given level of lateral loads. By analogy, by passing from a single static analysis to the incremental static push-over, one obtains the extension of a single time-history analysis into an incremental one, where the seismic loading is scaled. So, the structure is submitted to a succession of dynamique analyses under an increased load intensity (Fig. 9.33) with the objective to obtain more realistic information about the structural behavior under seismic actions (Pinho, 2007). This involves successive scaling and application of each accelerogram followed by assessment of the maximum response, up to the achievement of the structural collapse.

Incremental dynamic analysis (IDA) (Vamvatsikos and Cornell, 2002) is a parametric analysis method which allows obtaining the capacity curve from a dynamic analysis. This procedure offers to investigate, by scaling the induced base acceleration, a continuous picture of the structural behavior from elastic to yield range and eventually to collapse. The method allows tracing also the sequence of yielding and

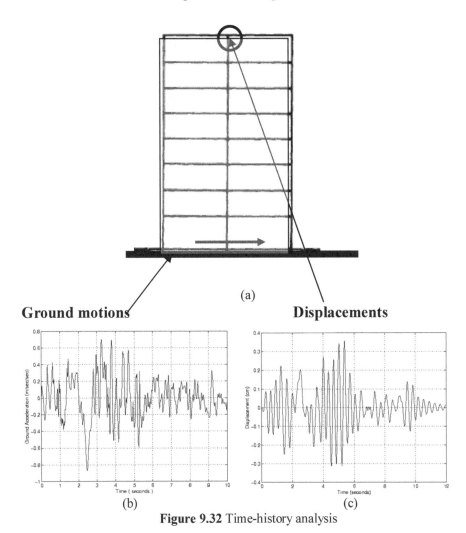

(a)

Ground motions   **Displacements**

(b)   (c)

Figure 9.32 Time-history analysis

failure in the structural members. Figure 9.34 displays, for a 20-story building (Vamvatsikos and Cornell, 2002), the interstory drift profile for several acceleration levels (0.01g, 0.1g and 0.2g). One can see that, for reduced intensities, the maximum deformations are concentrated at the top of the frame. Contrary, as the intensity increases, producing the structure collapse at 0.23g, the 3rd floor seems to accumulate the largest amount of deformation.

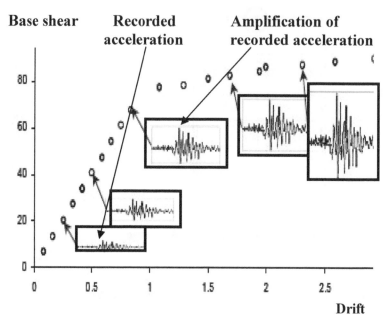

Figure 9.33 Capacity curve obtained using incremental dynamic analysis
(after Pinho, 2007)

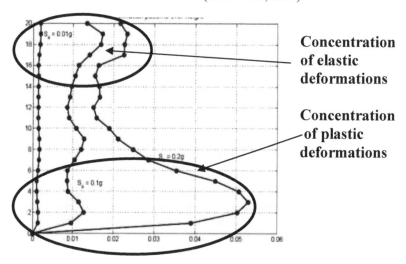

Figure 9.34 Incremental dynamic analysis for a steel building
loaded by 1940 El Centro earthquake (after Vamvatsikos and Cornell, 2002)

The incremental dynamic analysis is used for evaluating the behaviour of the q-factor (Mazzolani and Piluso, 1996). The use of this factor allows performing a simple elastic analysis which is very easy for design purposes, instead of very complex non-linear inelastic analyses. For a given structure under a specific acceleration (Fig. 9.35), according to the Ballio-Setti method, a series of computations of non-linear dynamic response is performed by scaling step by step the initial ground acceleration shape (Ballio, 1985). The q-factor is defined as the ratio between the ground acceleration leading to collapse and the one corresponding to the first plastic hinge formed in the structure. The maximum value which can be assigned to the q-factor is given by the intersection of the behavior curve coming from the incremental dynamic analysis and the bisectrix of the orthogonal axes, which represents the elastic behavior of the structure. This point of intersection between the two forms of behavior, elastic and inelastic, allows a direct link between the linear and non-linear computations.

The main criticisms referring to the incremental dynamic analysis are based on the observation that a correlation exists between the earthquake magnitude and natural vibration periods. The reduced magnitudes are associated with the reduction of natural periods and these periods significantly increase with the amplification of acceleration. In the incremental dynamic analysis these periods remain constant, and, therefore, the results are affected by this shortcoming of the IPA procedure. The challenge for improving this methodology is to increase the natural periods simultaneously with the scaling of the initial acceleration.

9.5.7 Estimation of Progress in Seismic Analysis

Despite the fact that the accuracy and efficiency of the computational tools have been substantially increased, there are still many reservations, mainly about their complexity and suitability for practical design applications. Moreover, the calculated inelastic dynamic response is quite sensitive to the characteristics of the input motion, thus the selection of a suitable set of representative acceleration time histories is mandatory. In addition, there are many difficulties in adequately representing the cyclic load-deformation characteristics of all the important structural elements (foundations, columns, beams, connections, infilled elements, etc.). The computational effort significantly increases to consider all these aspects in a correct way, raising doubts on its suitability in the everyday practice (Mwafy and Elnashai, 2001).

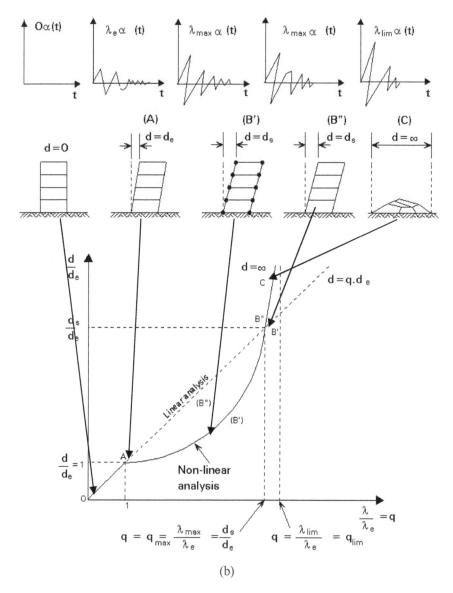

(b)

Figure 9. 35 Correlation between elastic and inelastic analysis, using IDA analysis
(after Ballio, 1985)

9.6 BEHAVIOR OF NON-STRUCTURAL COMPONENTS DURING EARTHQUAKES

9.6.1 Importance of Non-structural Components

Non-structural components are, by definition, not intended to resist any seismic forces than those resulting from their mass. They are also, in the main, elements which are not designed by structural engineers, but architects and mechanical or electrical engineers take primary responsibility for them.

After many years of efforts by researchers and practitioners to develop performance-based earthquake engineering methodologies, it is now obvious that the problem of non-structural components is one of the most critical issues of the seismic design. Unlike structural components, most of the non-structural components are vulnerable to a relatively low level of earthquake. According to reconnaissance reports on the earthquakes of 1989 Loma Prieta and 1994 Northridge, the economic loss due to non-structural components generally exceeds the one due to structural elements (Lee et al, 2006). In the case of the 1995 Kobe earthquake, the cost of non-structural components damage equalizes the cost of structural damage. Poor performance of non-structural components is the main contributor to damage, losses and business interruption for the majority of facilities after an earthquake. This subject received special attention after the 1971 San Fernando earthquake, when it became clear that the damage to non-structural components can not only result in major economic loss (79% from damage costs was due to non-structural components), but also can pose a real threat to life safety. For example, an evaluation of damage to hospitals following this earthquake revealed that many facilities still structurally intact were no longer functional, because of loss of essential equipment and supplies.

The value of non-structural components is, generally, expressed as a percentage of the total cost of a building, depending on the type of the building considered. In a storage building with few mechanical services and non-structural elements, non-structural elements can be 20 to 30% of the total investment. For industrial buildings, with important production equipments, this percentage increases until 40 to 50%. In complex civil buildings, the cost of non-structural components increases drastically. The average incidence on the total investment of non-structural components in case of offices, hotels and hospitals is given in the following table as a percentage of the total investment (Whittaker and Soong, 2003):

	Office	Hotel	Hospital
Structural	18%	13%	8%
Non-structural	62%	70%	48%
Contents	20%	17%	44%

One can see that, in case of these types of civil buildings, the cost of structural systems is the lowest percentage of investments (8 to 18%), followed by contents (17 to 44%); the non-structural component costs are the majority (48 to 70%), in function of the building functions. Clearly, the investment in non-structural components and building contents (82 to 92%) is far greater than the one for structural components and framing. In some cases, such as art museums, high-tech laboratories or computer centers, the values of non-structural components, especially the building contents, exceed by far the other investment costs. Thus, a large portion of the potential losses for owners, occupant insurance companies or financial institutions is associated with non-structural components. Therefore, the increasing interest on the development of protection measures to reduce this kind of losses is not surprising.

9.6.2 Non-structural Component Types

The different types of non-structural components can be subdivided into three groups (Fig. 9.36):
-   *Architectural elements*, like exterior curtain wall, prefabricated panel, glazing, claddings, interior non-load bearing partition, suspended ceilings, parapets, penthouse, chimney, appendages and ornamentations, etc.
-   *Mechanical and electrical utility components*, like elevators or escalators, air conditioning equipments, boilers, storage tanks, heating systems, ventilations, piping systems, plumbing, fire protection systems, electric motors, etc.
-   *Contents,* like record storages, movable partitions, production equipment and systems, computer and communication equipments, furniture, etc.

9.6.3 Primary Causes of Non-structural Damage

The seismic risk for a particular non-structural component at a given facility is governed by the regional seismicity, proximity to an active fault, local soil condition, dynamic characteristics of the building structure and non-structural components, their bracing and anchorage to the structure, location of the non-structural component within the structure, function of the facility, importance of the particular component to the operation of the facility (FEMA 74, 2005).

The earthquake shakes the ground in all directions (horizontal and vertical). Therefore, the non-structural components must be specially designed to resist the earthquake forces in a variety of directions. Earthquake ground shaking has several primary effects on non-structural components in buildings. These are inertial or shaking effects on the non-structural elements themselves, distortions imposed on non-structural components when building structure sways back and forth, the effect of the structure on non-structural components, and the pounding effects at the interface between adjacent structures.

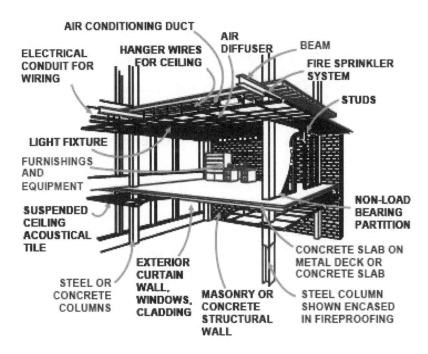

Figure 9.36 Structural and non-structural components (FEMA 74, 2005)

Direct effects. When a building is shaken during an earthquake, the base of the building moves in harmony with the ground, but the entire building and the building contents above the base will experience inertial forces. These forces depend on the base acceleration of the site, the amplification of this acceleration with the height of the point of attachment of the non-structural component above the foundation and its weight. All non-structural components can be damaged due to the inertial forces, which may cause overturning of slender objects or sliding of stocky objects (Fig. 9.37), if these are not anchored or the attachment is inadequate. During a moderate earthquake, damage to critical equipments and contents may be more important than damage to the building itself. In addition, damage to such equipments can lead to an extended business interruption, which may pose the greatest earthquake financial risks. In these conditions, the proper design of connections between non-structural elements and structure is of first importance. Frictional resistance of components and attachments cannot be relied upon in seismic events.

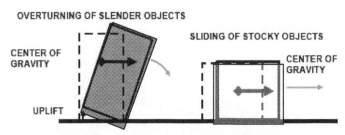

Figure 9.37 Direct effect on non-structural components (FEMA 74, 2005)

Building distortion. During an earthquake, building structures distort or bend from side to side in response to the earthquake. This produces an interaction between structural elements and non-structural components. The effects of this interaction can be grouped into two categories: first, the effect of the response of the structural system on the non-structural components and second, the effect of non-structural components on the response of the structural system (NISEE, 1997).

Effect of structure to non-structural components. The displacement over the height of each story, known as story drift, depends on the size of the earthquake and the characteristics of the particular building structure. Windows, partitions, claddings and other items, which are tightly located into the structure, are forced to distort with the same amount of the story drift (Fig. 9.38). Elements made of brittle materials, like glass, plaster, drywall partitions, and masonry infill, cannot tolerate significant distortions and they will crack when the building structure pushes directly on them. Most architectural components, such as façade masonry infilled panels and internal partitions, are damaged because of this type of action, not because they themselves are shaken or damaged by inertial forces.

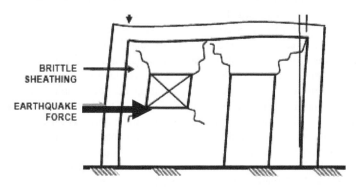

Figure 9.38 Effect of building deformation on architectural component
(FEMA 74, 2005)

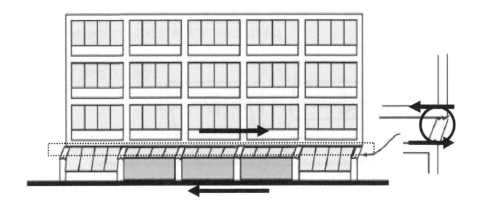

Figure 9.39 Short columns produced by partial infillsd masonry (FEMA 389, 2004)

Effects of non-structural components to structure. There are notable cases of structural and non-structural interaction, that has been the cause of structural damage leading to collapse. These cases involves rigid, strong architectural components, like masonry infill or concrete walls, which inhibit the movement or distortion of the structural framing and cause premature failure of columns or beams. The short column resulting from a partial infilled masonry (Fig. 9.39) are especially very dangerous, producing the column collapses due to shear forces.

Pounding or movement effects. Another source of non-structural damage involves pounding or movement between adjacent buildings across the separation joints. A separation joint is the distance between two different buildings which allows the structures to move independently of each other. In order to provide functional continuity between separate buildings, the utilities must often extend across these building separations (Fig. 9.40). Flashing, piping, fire sprinkler lines and all equipment services have to be detailed to accommodate the seismic movements expected at these locations, when the two structures move closer together or further apart. In case such details are not adequately solved, the danger of fire after an earthquake significantly increases, due to the fracture of gas pipes. Damage to items crossing seismic gaps is a common type of earthquake damage. If the size of the gap is insufficient, pounding between adjacent structures may produce damage to structural components, such as parapets or cornices.

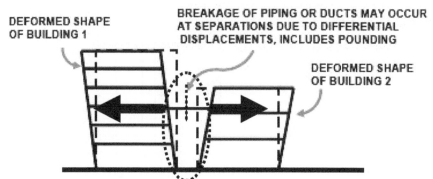

Figure 9.40 Pounding effect on buildings (FEMA 74, 2005)

9.6 REFERENCES

Almeida, R., Caneiro-Barros, R. (2003). A "new" multimode load pattern for pushover analysis: the effect of higher modes of vibration. Earthquake Resistant Engineering Structures IV, Ancona , 3-13

Aoki,Y., Ohashi, Y., Fujitani, H., Saito, T., Kanda, J., Emoto, T., Kohno, M. (2000): Target seismic performance levels in structural design for buildings. In 12th World Conference on Earthquake Engineering, Auckland, 30 January-4 February 2000, Paper No. 0652

Archer, G.C. (2001): A constant displacement iteration algorithm for non-linear static push-over analysis. Electronic Journal of Structural Engineering, Vol. 2

Ballio, G. (1985): ECCS approach for the design of steel structures to resist earthquakes. Symposium on Steel Buildings, Luxembourg, IASE-AIPC-IVBH Report, Vol. 48, 313-380

Bertero, V.V. (1996a): State-of-the-art in design criteria. In 11th World Conference on Earthquake Engineering, Acapulco, 23-28 June 1996, Paper No. 2005

Bertero, V.V. (1996b): The need for multi-level seismic criteria. In 11th World Conference on Earthquake Engineering, Acapulco, 23-28 June 1996, Paper No. 2120

Bertero R.D., Bertero, V.V. (2000): Application of a comprehensive approach for the performance-based earthquake–resistant design buildings. In 12th World Conference on Earthquake Engineering, Auckland, 30 January-4 February 2000, Paper No. 0847

Bommer J.J., Elnashai, A.S. (1999): Displacement spectra for seismic design. Journal of Earthquake Engineering, Vol. 3, No. 1, 1-32

Borcherdt R.D. (1994): Estimate of site-dependent response spectra for design (methodology and justification). Earthquake Spectra, Vol. 10, No. 4, 617-653

Chintanapakdee, C., Chopra, A.K. (2003): Evaluation of modal pushover analysis using generic frames. Earthquake Engineering and Structural Dynamics, Vol. 32, 417-442

Chopra, A.K., Goel, R.K. (2001): A modal pushover analysis procedure to estimate seismic demands for buildings: Theory and preliminary evaluation. Pacific Earthquake Engineering Research Center, Report PEER 2001/03

Chopra, A.K., Goel, R.K. (2002): Modal pushover seismic analysis of SAC buildings excluding gravity loads.   12th European Conference on Earthquake Engineering, London, 9-13 September 2002, Paper No. 522

ESDEP (2008): Lecture 17.4: Structure analysis for seismic action. ESDEP LECTURE NOTE, WG 17, Seismic Design
http://www.esdep.org/4ccr/members/master/wg17/10400.htm

EUROCODE 8 (2002): Design of structures for earthquake resistance

FEMA-273 (1997): NEHRP guidelines for the seismic rehabilitation of building

FEMA-349 (2000): Action plan for performance based seismic design. Earthquake Engineering Research Institute

FEMA 389 (2004): Primer for design professionals. Chap. 5. Improving performances to reduce seismic risk. FEMA Risk Management Series

FEMA-74-FM (2005): Earthquake hazard mitigation for non-structural elements. Field Manual

Gioncu, V. (2006): Advances in seismic codification for steel structures. Costruzioni Metalliche, Vol. 43, No. 6, 69-87

Gioncu, V., Mazzolani, F.M. (2002): Ductility of Seismic Resistant Steel Structures. Spon Press, London

Gioncu, V., Mazzolani, F.M. (2006): Inluence of earthquake types on the esign of seismic-resistant steel structures. Part 1: Challenges for new design approaches; Part 2: Structural responses for different earthquake types. Behavior of Steel Structures in Seismic Areas, STESSA 2006 (F.M Mazzolani and A. Waida) Yokohama 17-20 August 2006, Taylor and Francis, London, 113-120, 121-127

Giuliano, A., Martinez, M., Rubinstein, M., Mollerm, O. (2004): Performance-based preliminary seismic analysis approach of new reinforced concrete frame structures. In 13th World Conference on Earthquake Engineering, Vancouver, 1-6 August 2004, Paper No. 2901

Hamburger, R.O. (1996): Implementing performance based seismic design in structural engineering practice. In 11th World Conference on Earthquake Engineering, Acapulco, 23-28 June 1996, Paper No. 2121

Ibarra, L.F., Medina, R.A., Krawinkler, H. (2005): Hysteretic models that incorporate strength and stiffness deterioration. Earthquake Engineering and Structural Dynamics, Vol. 34, No. 12, 1489-1511

Kennedy, D.J.L., Medhekar, M.S. .(1999): A proposed strategy for seismic design of steel buildings. Canadian Journal of Civil Engineering, Vol. 26, 564-571

King , A. (2000): Earthquake Loads & Earthquake Resistant Design of Buildings. Building Research Association of New Zeeland (BRANZ) Report

Kim, S., D'Amore, E. (1999): Push-over analysis procedure in earthquake engineering. Earthquake Spectra, Vol. 15, No. 3, 417-434

Krawinkler, H. (1996): A few basic concepts for performance based seismic design. In 11th World Conference on Earthquake Engineering, Acapulco, 23-28 June 1996, Paper No. 1133

Krawinkler, H. (1997): Research issue in performance based seismic engineering. In Seismic Design Methodologies for the Next Generation of Codes (eds. P. Fajfar and H. Krawinkler), Bled, 24-27 June 1997, Balkema, Rotterdam, 47-58

Krawinkler, H. (1999): Challenges and progresses in performance-based earthquake engineering. In International Seminar on Seismic Engineering for Tomorrow. Tokyo, 26 November 1999

Krawinkler, H., Seneviratna, G.D.P.K. (1998): Pros and cons of pushover analysis of seismic performance evaluation. Engineering Structures, Vol. 20, Nos. 4-6, 452-464

Lam, N., Wilson, J., Hutchinson, G. (1996): Building ductility demand: Interplate versus intraplate earthquakes. Earthquake Engineering and Structural Dynamics, Vol. 27, 749-746

Lawson, R.S., Vance, V., Krawinkler, H. (1994): Non-linear static push-over analysis – why, when and how? 5th US National Conference on Earthquake Engineering, Vol. 1, 283-292

Lee, S (2008): Non-linear dynamic earthquake analysis of skyscrapers. 8th World Congress of Council on Tall Buildings and Urban Habitat, Dubai, 3-5 March 2008

Lee, T.H., Kato, T., Matsumiya, T., Suita, K., Nakashima, M. (2006): Seismic performance evaluation of nonstructural components: drywall partitions. Annals of Disasater Prevention, Research Institute , Kyoto University, No. 49, 177-188

Mazzolani, F.M., Piluso, V. (1993): Design of steel structures in seismic zones. ECCS Manual, TC13, Seismic Design Report

Mazzolani, F.M., Piluso, V. (1996): Theory and Design of Seismic Resistant Steel Frames. E&FN Spon, London

Medhekar, M.S., Kennedy, D.J.L. (2000): Displacement-based seismic design of buildings. Part 1: Theory; Part 2: Application Engineering Structures, Vol. 22, 201-209, 210-221

Mwafy, A.M., Elnashai, A.S. (2001): Static pushover versus dynamic collapse analysis of RC buildings. Engineering Structures, Vol. 23, No. 5, 407-424

NEHRP (2003): Structural design criteria. Chapter 4, Commentary http://www.bssconline.org/NEHRP2003/C4.pdf

NISEE (1997): Non-structural components. In Guidelines for Achieving Efficient Seismic-Resistant Design. National Information Service for Earthquake Engineering, University of California, Berkely

Oliveto, G., Marletta, M. (2005): Seismic retrofitting of reinforced concrete buildings using traditional and innovative techniques.   Journal of Earthquake Technology, Vol. 42, No. 2-3, 21-46

Papanikolaou, V.K., Elnashai, A.S., Pareja, J.F. (2005): Limits of applicability of conventional and adaptive pushover analysis for seismic response assessment. Report of University of Illinois at Urbana-Champaign
http://mae.cee.edu/documents/cd_rom_series/05-02/Report 05-02.pdf

Pinho, R. (2007): Analisi pushover adattiva in spostamento. ANIDIS 2007, XII Convegno L'Ingegneria Sismica in Italia, Pisa, 10-14 June 2007

Pinho, R., Antoniou, S., Casarotti, C., Lopez, M. (2005): A deplacement-based adaptive pushover for assessment of buildings and bridges. Seismic Assesment and rehabilitation of existing buildings, Istanbul, 30 May-1 June 2005

Pinho, R., Antoniou, S., Pietra, D. (2006): A displacement-based adaptive pushover for seismic assessment of steel and reinforced concrete buildings. 8th US National Conference in Earthquake Engineering, San Francisco, 17-21 April 2006, Paper No. 1701

Poland, C.D., Hom, D.B. (1997): Opportunities and pitfalls of performance based seismic engineering. In Seismic Design Methodologies in the Next Generation of Codes, Bled, 24-27 June 1997, Balkema, Rotterdam, 69-78

Priestley, M.J.N. (1997): Displacement-based seismic assessment of reinforced concrete buildings. Journal of Earthquake Engineering, Vol. 1, No. 1, 157-192

Sabetta, F., Bommer, J. (2002): Modification of the spectral shapes and subsoil conditions in Eurocode 8.   12th European Conference on Earthquake Engineering, London, 9-13 September 2002, Paper No. 518

SEAOC, Structural Engineering Association of California (1995): Vision 2000 - A framework for performance based engineering. Vision 2000 Committee, Sacramento, California

Studer, J., Koller, M.G. (1994): Design earthquake: The importance of engineering judgements. 10th European Conference on Earthquake Engineering, Vienna, 28 August - 2 September 1994, Balkema, Rotterdam, 167-176

Tirca, L, Gioncu, V., Mazzolani, F.M. (1997): Influence of design criteria for multi-story steel MR frames. Behavior of Steel Structures in Seismic Areas, STESSA 97 (eds. F.M. Mazzolani and H. Akiyama), Kyoto, 3-8 August 1997, 10/17 Salerno, 266-275

Tolis, S.V., Faccioli, E. (1999): Displacement design spectra. Journal of Earthquake Engineering, Vol. 3, No. 1, 107-125

Truta, M., Mosoarca, M., Gioncu, V., Anastasiadis, A. (2003): Optimal design of steel structures for multi-level criteria. Behavior of Steel Structures in Seismic Areas, STESSA 2003 (ed. F.M. Mazzolani). Naples, 9–12 June 2003, Balkema, Lisse, 63-69

Turker, K., Irtem, E. (2006): An effective multi-modal and adaptive pushover procedure for buildings. Bulletin of the Istanbul Technical University, Vol. 54, No. 5, 34-45

UBC (1997): Uniform building code

Vamvatsikos, D., Cornell, C.A. (2002): Incremental dynamic analysis. Earthquake Engineering and Structural Dynamics, Vol. 31, No. 3, 491-514

Wikipedia (nd): Seismic analysis.
http://en.wikipedia.org/wiki/Seismic_analysis

Wikipedia (nd): Response spectrum
http://en.wikipedia.org/wiki/Response_spectrum

Whittaker, A.S., Soong, T.T. (2003): An overview of nonstructural components research at three U.S. earthquake engineering research centers. Proceedings of Seminar on Seismic Design, Performance, and Retrofit of Nonstructural Components in Critical Facilities. Session V: Risk and Performance Evaluation, 271- 280

Yamawaki, K., Kitamura, H., Tsuneki, Y., Mori, N., Fukai, S. (2000): Introduction of a performance-based design. In 12th World Conference on Earthquake Engineering, Auckland, 30 January–4 February 2000, Paper 1511

Chapter 10

Challenges for the Next Generation of Seismic Codes

10.1 DEVELOPMENTS OF SEISMIC DESIGN CODES

10.1.1 From the Hammurabi's Code to the Modern International Codes

A building code is a document containing standardized requirements for the design and the erection of most building types. These codes regulate the building life from the conception to the maintenance, in order to protect health, safety and welfare of the occupants. Codes address all aspects of construction, including structural integrity, fire resistance, lighting, electrical and heating plants, air conditioning, energy conservation, plumbing, sanitary facilities, ventilation, as well as the correct use of construction materials.

The design of seismic resistant structures is a relatively new branch, in comparison with the other branches of structural engineering, and the first attempts to develop it as an engineering science were observed at the beginning of the last century, the most important concepts being achieved during just the last 40 years (Gioncu and Mazzolani, 2002). Therefore, the codification in seismic design is a quite recent activity, whose maturation period developed during the last years only. The history of development in construction codification can be subdivided in the following principal periods.

Period of intuitive design, during which the building erection was based on intuition, empirical rules, accumulated experience and knowledge transmission from generation to generation. No official rules existed in construction activity during this period, all the responsibility of construction erection belonging to the builder. It is well known that the oldest code containing specifications about the relationship between owners and builders, including the safety responsibilities of the latter, is due to Hammurabi, almost four thousand years ago. A collection of a number of Babylonian legal decisions is today known as the Code of Hammurabi (1795-1750 BC). There are nearly three hundred ancient cuneiform inscriptions dealing with the regularization of various governing matters, among them being the responsibility of builders in terms of safety (Petroski, 1985, Aktan et al, 2007):

- *If a builder build a house for some one, and does not construct it properly and house which he built fall in and kill its owner, then the builder shall be put to death.*
- *If it kills the son of the owner, the son of that builder shall be put to death.*
- *If it kills a slave of the owner, than he shall give a slave of equal value.*

-   *If it ruins good, the builder shall make compensation for all that has been ruined and, because he did not construct properly this house which collapsed, he shall rebuilt the house from his own means.*
-   *If the builder built a house and does not make its constructions meeting the requirement and a wall fall in, that builder shall strengthen the wall at his own expense.*

Period of observations. This period introduced for the first time the observations of damage that occurred during an earthquake. The effects of some disastrous events were examined in order to learn lessons about the damage and to elaborate some rules to prevent undesirable behaviors. The tragic 1755 Lisbon earthquake constrained the authorities to elaborate the first building code provisions for seismic resistance, focused on prohibiting certain types of construction, which were observed to poorly behave in past earthquakes, and requiring the use of given constructional details and techniques, which were observed to provide a better performance (Hamburger, 2003). After the 1908 Messina earthquake, a special commission of practicing engineers and engineering professors was commissioned to recommend improved constructional requirements. The resulting report included, for the first time, recommendations for the lateral distribution of seismic loads. Similar activities were developed in Japan, after the 1923 Kanto (Tokyo) earthquake, when a 33m building height limit was imposed. In the USA, it was observed that certain types of constructions (especially unreinforced masonry buildings) consistently performed in a poor way, so rules were developed to regulate the features of these construction types in order to improve their performance. This period is characterized by the fact that the regulations were strictly based on experiences learned during the earthquakes.

Early prescriptive codes. The Californian earthquakes that occurred in the early 20th Century (especially the 1906 San Francisco earthquake) were the reason for the start of the early seismic code developments. The basic requirement was introduced that structures intended to resist earthquakes must be provided with sufficient strength to resist specific lateral forces. Therefore, the use of lateral seismic forces in design became widely used and the code development refers to the evaluation of these lateral forces. During the first period, these forces depended on the building weight only, but this approach was improved in the further development of provisions. For the first time in the 1943 provisions for the Los Angeles City, the lateral forces were determined in function of the first vibration mode and the response spectra of ground motion. These preliminary concepts were based on grossly simplified physical models, engineering judgments and a number of empirical coefficients. Influenced by the conventional design concepts, the earthquake actions were considered as static loads and the structures as elastic systems. This simple concept was the standard design methodology for several decades. There are good reasons to understand the success of this design approach. This methodology has been well understood by structural engineers, because it is relatively easy to be implemented (Gioncu and Mazzolani, 2002).

Modern period. The next step in code development was the study of the dynamic response of structures, leading to the shear base, being distributed through the height of the building according to the shape of the fundamental mode. This was an evolution from an arbitrary set of forces based on the earthquake damage to a set of forces applied as static loads. It would approximately reproduce the peak dynamic response of the structure to the design earthquake. Linear elastic response spectra provided to estimate the level of forces and deformations present in structures. The following step in the code development was the dissipation of seismic energy through plastic deformations and the use of the inelastic spectra. The control factor of seismic energy dissipation was the ductility, defined as the ability of a structure to undergo deformations after its initial yield, without any significant reduction of the ultimate strength. These two items, i.e. the calculation of lateral design forces and the means of providing sufficient ductility, constituted the two most effective provisions in seismic design. After the Northridge and Kobe earthquakes, new design aspects were recognized for structures situated in near-source areas: the importance of velocity and displacement spectra, instead of acceleration spectra, as well as the importance of superior modes of vibrations.

Period of personal computers. Since the early 1970s, a drastic change in seismic codes has taken place, thanks to the availability of personal computers and the implementation of a great number of programs for structural engineering, which very easily perform static and dynamic analyses in elastic and elasto-plastic ranges. These technological advances allow obtaining more refined results, giving to the researchers the perspective to improve the design methodology. The code recommended structural analysis, instead of providing practical procedures, turns to the examination of phenomenological aspects. As a consequence, the modern codes have to contain recommendations for the proper use of different methodologies of analysis, in function of structure and earthquake types. In this situation, the principal task of the seismic codes is to provide the seismic loads and general rules for designing an optimal structural configuration.

International codes. The new initiative to create an international system of codes started in the last period. This international code system includes the principles for the determination of seismic actions, the seismic design methodology and the main factors to be considered. But the seismic codes today, in comparison with the codes for other structural actions, have some very important peculiarities. Different from other codes, the development of seismic provisions is strongly based on the observations of some damaging earthquakes. Consequently, the current set of design factors found in national standards are based on a measured combination of historical seismic events, state-of-the-art research works and engineering judgments, very different in each country, being a function of its local constructional experience in seismic areas, coming from the nature of ground motions, as well as from traditions and jurisdictions. Recognizing the specific aspects of seismic design, connected to the regional particularities, all documents elaborated in the frame of an international action could serve as a guideline for all issues belonging to regional or national standards.

10.1.2 The Long Way from Theory to Practice

Today, thanks to the development of computers, the actual performance of a structure during a strong ground motion can be satisfactorily evaluated by numerical modeling, provided that the seismic action is correctly defined. This significant progress, which has been recently achieved, is also due to the collection of a great amount of information concerning the features of earthquakes and due to worldwide activity in research on the behavior of structures in seismic areas. However, the current code design methodologies fail in introducing some theoretical achievements in design practice, due to the following aspects (Gioncu and Mazzolani, 2002, 2006, Gioncu, 2006):

(i)   The design philosophy for earthquake loads is totally different from the design methodology for other actions, giving to structural designers many problems of assimilation of this unconventional design philosophy. While for dead and live loads, wind and snow, the structure must remain without damage after the maximum loads were reached, in case of seismic design, the admittance of plastic deformations during severe earthquakes implicitly anticipates the occurrence of structural damage. The philosophy of structural seismic design establishes the target level of safety that structures should resist for minor quakes without damage, moderate earthquakes with moderate structural damage and for major earthquakes, important damage without structure collapse. So, the most relevant performance criterion for a building structure surviving a strong earthquake is the total cost of damage. This damage control is very difficult to be quantified in a simple manner to be introduced as provision in a design code. *"What level of risk to public is acceptable? "* is a question very difficult to be considered by code provisions.

(ii)   The basic concepts of today's seismic codes were born almost 70 years ago, when the knowledge about seismic actions and structural response were rather poor. Today, the seismic design philosophy has grown within the new fields of Engineering Seismology and Earthquake Engineering, wherein many exciting developments are predicted in the near future. The challenge for a proper seismic structural design is to solve the balance between seismic demand and structure capacity. Seismic demand corresponds to the effect of earthquake on the structure and depends on the proper ground motions modeling. Structural capacity is the structural ability to resist these effects without failure. Looking to the developments in Engineering Seismology and Earthquake Engineering, it is clear that the major effort of researchers was directed towards the structural response analysis. Therefore, the structural response can be predicted fairly confidently, but these achievements remain without real effect if the accuracy in determining the seismic actions is doubtful.

(iii)   The prediction of the ground motions is still far from a satisfactory level, due to both the complexity of seismic phenomena and the communication lacks between seismologists and engineers. This remark can be confirmed after each important earthquake, when new and new in situ lessons regarding the characteristics of ground motions are learned, instead of providing in advance the missing data from the seismological studies. Therefore, the structural designer must be fully aware of the fact that *"... it useless to determine with great accuracy the structure response for a given seismic action if this one is established with a great level of incertitude".*

(iv)   The definition of a major earthquake action inevitably calls for a series of seismological knowledge, structural engineering judgments and safety policy. *The reduction (not the elimination) of uncertainties in ground motion modeling is now the main challenge in structural seismic design.* For this reason, the proper seismic design should be elaborated in close collaboration between seismologists, experts in faults characteristics, and structural engineers, who know very well the structural response to a given ground motion. In many cases, this collaboration is very difficult, due to the differences in education, being the differences between science and engineering. In order to fill this gap, engineers must especially work in the Engineering Seismology branch, in order to transform the qualitative knowledge of Seismology into quantitative values to be used by Earthquake Engineering and to partially eliminate the incertitude in determining the seismic actions as a function of the fault type.

(v)   A significant gap exists between the existing level of research results and the provisions of design codes. In many cases the research works are performed by professors and researchers, who are more interested in publishing their results for promotion among their colleagues, than in the transmission of new knowledge to those who will apply it (De Buen, 1996).

(vi)   Structural engineers are conservative professionals and not researchers in new directions of structural design. Their activities are driven by the need to deliver the design in a timely and cost-effective manner. They may also resist new concepts, unless these concepts are put into the context of their present mode of operation (Krawinkler, 1995). Therefore, during the elaboration of new codes it must be kept in mind that the design engineering community tends to be conservative. So, the new code provisions must be a compromise between new and old knowledge and procedures. Otherwise, the structural designers will reject the new methodologies.

(vii)   The implementation of new concepts in codes is constrained by the need to keep the design process simple and verifiable. Today, the progress of computer software has made it possible to predict the actual behavior of structures subjected to seismic loads using non-linear analysis. But designers are always pressed by the deadline to deliver the project, therefore, only for very special structures can they accept to use advanced design methodologies. For the majority of designed buildings, they request to use only simple procedures. This is the reason why, for instance, despite the progress in design methodologies for evaluating the structural ductility, the codes contain only simple constructional provisions, which do not always assure the required plastic rotation capacity in the members.

(viii)   Recognizing the need for code development based on transparent methodology, one must also recognize that it is necessary to underline some danger of this operation: over-simplification, over-generalization and immediate application in practice of the latest research results, without an adequate period of time during which these results can be verified.

10.2 PROGRESS IN SEISMIC DESIGN CODES

10.2.1 Design Codes

Initially there were no seismic codes and, only after the first earthquakes of the 20th Century, the engineering community decided to implement some level of lateral forces to avoid building collapse.

A design code is a document containing standardized requirements for the design of most types of buildings. It is a compendium of design regulations with the objective to guarantee that a building will be not damaged, or it will undergo reparable damage for low and moderate earthquakes and it will not collapse after a strong earthquake. Codes also regulate building construction and use in order to protect health, safety and welfare of occupants. Codes address all aspects of the construction, including integrity of structure and non-structural elements, fire resistance, seismic resistance, equipments, etc.

The first code was elaborated in 1927 in the USA for the Californian earthquakes. Following this code, the effort to elaborate codes was extended to all the world's seismic zones. For these codes, the current set of seismic design factors found in national standards is based on a measured combination of history of seismic events, state-of-the-art of research works and engineering judgments, very different in each country as a function of its experience of construction in seismic areas, coming from the nature and characteristics of ground motions, traditions and jurisdictions. Therefore, it is very important to analyze the evolution of seismic codes in the world's main seismic areas, in the context of the above-mentioned factors. This aspect is very important in order to know how concepts and provisions of one code could be used for the development of other codes, without making important errors. For this purpose, four main seismic areas are considered: the United States of America, Japan, China and Europe. The USA and Japan are selected, due to the very important activity in the improvement of code provisions, according to lessons learned after the Northridge and Kobe earthquakes. China is included as an example of special seismological aspects and high risk of earthquake, due to the massive scale of new development in populated centers. Europe is also selected due to the remarkable effort to unify the numerous and very different national codes in the frame of the Eurocodes, which are valuable, with some justified details, for all European countries (Gioncu, 2006).

Because it is not possible to present all details of the analyzed code provisions, only the development, the basic philosophy and the main characteristics will be reported.

10.2.2 United States of America

The presentation of the progress in seismic design codes starts with the USA, because the evolution of the world's codes was strongly influenced by the development of the USA code. The Uniform Building Code (1997) provisions have been recognized throughout the world as the leading reference for the design of earthquake resistant structures.

Seismological aspects. The territory of the USA is characterized by two very distinct seismic zones. So, the USA code must cover two different seismic demands for structural design. Intraplate (inland) crustal earthquakes shake the Eastern part, where the largest earthquake ever to hit the USA was the 1811-1812 New Madrid (Tennessee) one, with the magnitude M 8.6, due to the presence of an important rift system. The very high magnitude of these earthquakes, unusual for the intraplate faults, is due to very bad soil and liquefaction. There are also some intraplate crustal earthquakes produced in South Carolina, but the most active seismic regions are located along the Western shore of the country, where the Pacific and North American tectonic plates meet. Due to the almost complete oblique subduction of the Farallon plate under the North American plate (only two corners remain from this old tectonic plate, the Juan de Fuca plate at North and the Cocos plate at South), a system of strike fault lines has been developed. Some of these faults are known to be very active (such as the famous San Andreas fault), while others are presumed to be inactive, but they can give unexpected surprises (s.g. the Northridge earthquake, produced in an unknown fault belonging to the Los Angeles fault system).

Along the Alaska coast, the earthquake types are subduction-thrust, produced by crustal and intraslab faults. The majority of the sources are situated in shallow crust.

Considering all these aspects, the dominant earthquake types are situated in shallow crust, so the near-source effects are dominant for structural design.

Code evolution. It was strongly influenced by the last Californian earthquakes. The USA code system is a very special one, because each of the 41 states constitutionally has its jurisdiction for the regulation of construction. The evolution of codes is unique in the sense that the different organizations, agencies or institutions, on volunteer bases, are autonomously producing their codes, so that there are so many code models. Until 2000, each government usually adopted a model building code by ordinance, choosing one of the existing model codes, elaborated and published by private organizations (Bonneville and Bachman, 2002, Mahoney, 2002, Ghosh, 2002).

The development of codes for seismic-resistant constructions has been evolving over a number of years and it is applicable to all materials (Popov, 1991). Californian engineers and legislators were the initiators in the USA of the ever-improving seismic codes. San Francisco was rebuilt after the 1906 earthquake using the wind force provisions. It was only in 1927 that the Structural Engineers of California (SEAOC) developed the *Recommended Lateral Force Requirements* (known as the Blue Book) in the frame of the UBC. Here, the concept of lateral forces proportional to the mass was firmly introduced into practice. In addition to promulgating lateral force levels, seismic design regulations asserted that the design lateral forces must be applied in two orthogonal directions and distributed to all floor levels in proportional way (Diebold et al, 2008). Strong motion accelerogram data were recorded during the 1940 El Centro earthquake. The El Centro ground motion data were used as basis for seismic force levels introduced in the further codes for many years. In 1943 Los Angeles recognized the influence of the flexibility of a structure on the amount of the earthquake design forces. San Francisco engineers developed a relationship, stating that seismic forces are inversely proportional to the structure period. In 1971 SEAOC created the Applied Technology Council (ATC) as an independent organization for improving design practice and codes. The first results of this new organization were the 1978 UBC, the 1988 UBC and the 1988 Blue Book. Later on, the ATC was incorporated into the national model building codes as the National Earthquake Hazard Reduction Program (NEHRP), which is now the prompter of development of seismic regulations for new buildings. This code is elaborated under the patronage of an agency created in 1979, the Federal Emergency Management Agency (FEMA). The primary purpose of this agency is to coordinate the response to all disasters occurring in the USA. The American Institute of Steel Code was published in 1990 for steel structures and similar codes were elaborated for other structural materials. All the codes were published by the Building Officials and Code Administration (BOCA).

After 1990, especially after the 1989 Loma Prieta and the 1994 Northridge earthquakes, the improvement of the code provisions continued. In 1994, three model codes were issued by the International Code Council (ICC). In 2000, the ICC issued the first set of comprehensive and coordinated codes for the built environment, known as the International Building Codes (IBC) series. In addition to these code series, there are other professional societies or institutions, which elaborated their general or specialized codes. The situation of the existing codes is the following:

- *Uniform Building Code* (UBC), elaborated by SEAOC, which, historically speaking, was the earliest seismic design provision in USA. During the last period, the editions of 1991, 1994 and 1997 must be mentioned. Traditionally, the UBC code is used in the Western States. In California, this code is used as Californian Building Code (CBC).
- *National Earthquake Hazard Reduction Program* (NEHRP) provisions, elaborated by FEMA, with the editions of 1994, 1997, 2000, 2003 and 2009.
- *International Building Code* (IBC), elaborated by ICC on the basis of the 2000 NEHRP provisions, with the editions of 2000, 2003 and 2006.

The evolving process in the USA codification is now oriented to have only one seismic standard implemented everywhere in all the states, without any exception (Bonneville and Bachman, 2002, Gioncu, 2006).

Main code specifications. Regarding the design earthquakes, it was recognized that the design rules are constantly changing via new ideas, methods and research. The SEAOC Seismology Committee continues to review the UBC code with the intent to improving the recommendations. Each new important earthquake produces new provisions in UBC editions (Diebold et al, 2008):

Earthquake	UBC Edition	Code improvements
1971 San Fernando	1973-1976	Seismic zone 4 with increased base shear requirements
1985 Mexico City	1988	Separation of building to avoid pounding Design of steel columns for maximum axial forces Restriction for irregular structures Ductile detailing of perimeter frames
1987 Whittier Narrows	1991	Revision of site coefficients Revision to spectral shape Increased wall anchorage forces for flexible diaphragm buildings
1989 Loma Prieta	1991	Increased restrictions on chevron-braced frames Limitation on b/t ratios for braced frames Ductile detailing of piles
1994 Northridge	1997	Restrictions on use of battered piles Requirements to consider liquefaction Near-fault zones and corresponding base shear requirements Revised base shear equations using $1/T$ Spectral shape Redundancy requirements Design for overstrength More realistic evaluation of design drift Steel moment connection verification by test.

One can see that the Northridge earthquake, due to the severe damage produced generated the most important changes in the UBC provisions. Among them, the effects of near-source earthquakes, especially for strike earthquake types, as emphasized by the Northridge earthquake, which generated a crucial changing in seismic design philosophy. The 1997 UBC code is the only code which directly considers these effects, by increasing the seismic forces in function of the distance from the fault.

NEHRP was created in 1990, with the mission:

"*... to develop and promote knowledge, tools and practice for earthquake risk reduction, through coordinated multidisciplinary, interagency partnerships among* *the NEHRP agencies and their stakeholders, that improve the Nation's earthquake resilience in public safety, economic strength and national security"* (NEHRP, 2008).

The NEHRP agencies have established three long-term strategic goals:

(i)   Improve the understanding of earthquake processes and impacts.

(ii)   Develop cost-effective measures to reduce earthquake impacts on individuals, the built environment and society-at-large.

(iii)   Improve the earthquake resilience of communities nationwide.

Therefore NEHRP, after its creation, became to be the most active agency in mitigation of earthquake effects in the USA. The NEHRP 2003 recommended provisions are based mainly on the UBC provisions, but with some important improvements. So, due to the difficulties to determine the distance of site from the fault, the NEHRP 2003 code eliminates the UBC provisions to increase the near-source accelerations. In exchange, a seismic hazard map of the USA was elaborated. For structural design, national maps of earthquake shaking hazard provide essential information for determining the maximum accelerations, even for near-source events. The main problem of the USA codes is the definition of the maximum accelerations for Eastern and Western seismic areas with the two very different source types: interplate (strike-slip) and intraplate ones. For the second source, there is an important set of data, which is useful for the definition of the design acceleration without problems. Contrary, for the first source the data are scarce, due to infrequent occurrence of these earthquakes, so the definition of the design acceleration is very difficult and other criteria must be used. The intraplate faults give rise to low or moderate earthquakes, but in some exceptional cases (possible to be detected only for long return periods), the magnitude can be higher. These important differences give many problems in defining the design accelerations for the two seismic zones of the USA. In 1997 the NEHRP provisions stated that the design earthquake corresponds to one with a 2% probability of exceeding in 50 years (a return period of about 2500 years). But for Eastern USA, the design earthquake is defined as 2/3 of the maximum considered earthquake. The use of an earthquake with a return period of 2500 years as the basis for the design has created several criticisms, because the seismic hazard in various USA cities is different. It is well known that great uncertainty is involved in the estimation of ground motion with a long period (e.g. 2500 years). Thus, it is not appropriate to use such an earthquake as a basis for the design of buildings. Since seismologists have more confidence in the estimation of ground motion from earthquakes with shorter return periods, it should be more reasonable to use such an earthquake as the basis for defining the design earthquake. Therefore, it is proposed that the buildings situated in Eastern USA should be designed for serviceability against a design earthquake with a short return period, which can be estimated with less uncertainty. Then, the building should be provided with sufficient ductility to resist large infrequent earthquakes (Hwang, 2000). Due to these criticisms, one of the tasks for elaboration of the 2009 NEHRP provisions was the reassessment of hazard curves especially in low and moderate seismic areas (e.g. is a return period of 2500 years too long?) (Kircher et al, 2008).

10.2.3 Japan

The modern era of Japanese building construction can be considered to have its origin after 1868, during the Meiji restoration, a period when Japan was eager to embrace aspects of foreign culture and science, including the adoption of some foreign building construction practices. The AIJ (Architectural Institute of Japan) was formed in 1886 and the first city planning legislation was issued in 1888. The formation of the Earthquake Investigation Committee following the 1891 Nobi earthquake was an early step toward the development of modern earthquake resistant construction technology in Japan. Observations on the building performance under earthquakes during this period strongly influenced the modern earthquake engineering construction (Whittaker et al, 1997).

Seismological aspects. Japan is geographically located in a very complex and active seismic zone, composed by principal and secondary fault systems. The principal system consists in the zone where the Amur microplate (as the Eastern edge of the Eurasian plate) meets the Okhotsk microplate (as the Western edge of the North American plate), the Pacific plate and the Philippine plate. The movements of these boundaries are characterized by the Philippine, Pacific and North American plate's subduction under the Eurasian plate, resulting in important both crustal interplate and deep intraslab earthquakes. This principal fault system is associated to a secondary system produced by local faults, with more than 1500 active faults reported around Japan. The earthquakes are mainly concentrated in the Eastern and Northern parts of Japan and few events occur in the Western and Southern parts of Japan. Due to this very dense fault network, the Japanese earthquakes can be simultaneously produced by multiple sources. Therefore, in some cases, the secondary faults (e.g. the Kobe earthquake, where three sources were active) can be more damaging than the principal fault system. Because of this very complex fault system, all earthquake types can affect Japan.

Code evolution. The Japanese code was one of the first regulations in the world concerning the seismic actions. The Urban Building Law and Urban Planning Law were promulgated in 1920 to regulate buildings and city planning for six cities: Tokyo, Yokohama, Nagoya, Kyoto, Osaka and Kobe (Otani, 2004). The seismic building design code started in 1924; then the Urban Building Law was revised as a consequence of the 1923 Great Kanto earthquake disaster. A new version was published in 1950, when the Building Standard Law replaced the Urban Building Law with more elaborate provisions for structural design. Other details of structural design, such as structural analysis and proportioning of members, were specified in the Structural Standards issued by the Architectural Institute of Japan These standards, prepared separately for each structural material, supplement the law and can be revised more frequently to add new knowledge and provide rules for new materials as soon as they develop (Kuramoto, 2006). The Building Standard Law Order, revised in 1963, removed the building height limitation to 45 m and to 60 m in 1981, but with the approval of the Ministry of Construction.

The seismic design building code was radically changed in 1981 in the largest revision since 1924, after the 1968 Tokachi-oki and 1978 Miyagiken-oki earthquakes. The central feature of the revised seismic design code was the introduction of a two-phase earthquake design (Kuramoto, 2006):

(i)   The first phase earthquake design targeting the safety and serviceability of buildings during low and medium level earthquake activity.

(ii)   Second phase earthquake design was added to provide safety against severe earthquake motions.

Seismic provisions in the building code were significantly revised after the 1995 Kobe earthquake, which caused many fatalities and severe damage or collapse of buildings. Many lessons among scientists and engineers were learnt. The recognized need for a new generation seismic design led to the development of performance-based engineering for serviceability, reparability and life safety. The new 1998 Japanese regulations, elaborated by the Architectural Institute of Japan, were introduced in the revision of the Building Standard Law. The most significant revision towards the application of new structural elements and systems, with structural design performance-based requirements, was given in the 2000 Building Standard Law, Enforcement Order. The new code provisions encourage structural engineers to develop and apply new construction technology (Midorikawa et al, 2003).

Main code specifications. Two performance objectives were defined for the structural design under earthquake forces (Otani, 2004):

-    Damage limitation: protection of structural damage under frequent earthquake events, corresponding to a return period of approximately 50 years.

-    Life safety: protection of occupants' life under extraordinary earthquake events, corresponding to a return period of approximately 500 years.

In addition, structural specifications were prescribed for the method of structural calculation, the quality control of construction and materials, the durability of buildings and the performance of non-structural elements.

The intensity of design earthquake motions is established for two levels:

-    For rare earthquake events (Level 1).

-    For extremely rare earthquake events (Level 2), which correspond to five times that of level 1.

The intensity can be determined on the base of the recorded accelerations near the construction or of the generated artificial earthquake ground motions, compatible with the response acceleration spectrum. In case of generated ground motions, more than two accelerograms must be used. The duration of the artificially generated ground motions must be longer than 60 sec to excite long-period structure. In order to consider the effects of surface geology, the local horizontally layered deposits must be considered.

The performance requirement of building under Level 1 ground motions is that no damage is produced in structural members. Static linear elastic analysis of structures must be carried out for this level. The performance requirement of building under Level 2 ground motions is to not collapse, until a maximum interstory drift of 1/100. If this value is exceeded, special arguments must demonstrate that the cladding walls are capable of undergoing the corresponding deformation without failure. The new procedures allow the prediction of the maximum response against earthquake of structural members, non-structural elements and exterior finishing as well as curtain motions, without using time history analysis.

The provisions of the Japanese "Seismic Design Standard for Railway Facilities" (1999) give different spectra for near-source, interplate and intraplate earthquakes (Sato et al, 2001).

10.2.4 China

China is one of the countries which paid throughout the history a tremendous tribute in term of victims due to earthquakes. Now, due to the new economical development and urbanization, the seismic risk has increased dramatically. The new seismic Chinese code is one of the most up-to-date codes among the analyzed ones.

Seismological aspects. The continental China is a very active seismic zone, the majority of earthquakes being of crustal type. The density of epicenters is very high, especial in the zone of Tibet. The China's territory is characterized by the presence of some important blocks, jointed in different periods of world's history (Zhang et al, 2003, Gao, 2003). The South China block is influenced by the presence of the collision between the Indian and Eurasian plates, the largest continental collision in the World, producing the Himalayan Mountains and the Tibetan Plateau and regular earthquakes. In this seismic zone, very frequent earthquakes occur, having the characteristic of crustal collision earthquakes. The North China block is crossed by a series of very active intraplate earthquakes, which are the must murderous in the world. These earthquakes have sources along the ancient faults corresponding to the Sino-Korean and the Yangtze ancient blocks borders, the Eastern China territory being formed in a pre-Pangaea cycle by the collision between these blocks and later on by the collision with the Eurasia tectonic plate. During these collisions, the Japanese islands were also formed (Barnes, 2003).

Therefore, the majority of earthquakes occurring in China corresponds to crustal collision type.

Code evolution. Seismic design started in China very late in the 1950s. At the beginning, the codes of the former Soviet Union were adopted, the first draft being almost a copy of them. The 1964 first Chinese code was only the first draft, because the so-called Cultural Revolution stopped its official application, but the later codes TJ11-74 and TJ11-78 "Code for the Seismic Design of Civil and Industrial Buildings" followed its principles and main specifications (Li and Lai, 2000). After the lessons learnt as a result of the last important earthquakes during the 1970s, the new code was elaborated in 1989 by the China Academy of Building Research: "Code for Seismic Design of Buildings" GBJ11-89. A new version of this code, GB 50011, was elaborated in 2001, based on some new results from earthquake engineering investigations.

Main code specifications. The seismic intensity zonation map of China published in 1990 for the code GBJ11-89 is based on a probability method. It provides the intensity with 10% probability of exceedance in 50 years (Shen and Shi, 2004). At the same time, three levels of intensities are also defined: low earthquakes for probability of exceedance of 2%, moderate earthquakes for 10% and large earthquakes. Two design steps are considered in the code: the first refers to low earthquakes and the second to the large ones, corresponding to no collapse level (Shen and Shi, 2004).

The new code 2001 contains some very important improvements in comparison with the 1989 version (Wang, 2001). These modifications refer to the earthquake force levels, the seismic design parameters, the seismic design response spectra, the site classification, the input earthquake for time history analysis of structures, the minimum required shear force of story, the two phases of seismic design, the elastic analysis for design of sections, the design for minor earthquakes, the elasto-plastic analysis for the displacement response of structures for major earthquakes. The seismic intensity for the four degree is replaced by the peak accelerations: zone I, peak acceleration 0.05g, zone IIa with 0.10g, zone IIb with 0.15g, zone IIIa with 0.20g, zone IIIb with 0.30g, and zone IV with 0.40g. It is interesting to note that the corner periods T_g are given for near-source and far-source (considering both earthquake intensity and epicenter distance), in function of site classes (in seconds): zone I, 0.20 and 0.25, zone II, 0.3 and 0.4, zone III, 0.40 and 0.55, zone IV, 0.65 and 0.85 (the first values correspond to near-source, the second ones to far-source cases). These values generally increase with the soil conditions, the percentage not exceeding 40 to 50%, smaller for low intensities and larger for high intensities. A very detailed verification of story drifts for minor and large earthquakes is required, in function of structure type (including reinforced concrete, steel and masonry buildings).

10.2.5 Europe

The codification activity in Europe is characterized by three distinct periods. During the first, each European country elaborated its national code, in function of experiences and traditions in design against earthquakes. During the second, a set of pan-European model codes was developed by the European Committee for Standardization (CEN), in the frame of activity for the elaboration of the Eurocodes, being characterized by the co-existence of national and European codes. In the last periods, the Eurocodes will replace the national codes in all countries belonging to European Union.

Seismological aspects. Europe seismicity is characterized by very various tectonic conditions expressed by different source mechanisms. There are three main regions with different source types, showing the very complex tectonic behavior:

- *Mediterranean seismic region* (Marcellini et al, 1998, Vannucci et al, 2004) results from two major processing. In South, the tectonic displacements caused by the subduction of the African plate underneath the Eurasian plate. Exceptions are the Sicily and the Hellenic arc, where the intermediate intraslab earthquakes are dominant. The South European fault belongs to the Alpine-Himalayan belt, the second in importance after the Circum-Pacific Ring. The North fault system is due to the collision of some blocks against Europe: the

Iberian block, forming the Pyrenees Mountains and Italian block, forming the Alps Mountains. The Apennines Mountains are the result of collision between North Algerian and Adria Blocks. The actual territory of Greece has formed by the collision of a lot of mini-blocks, causing a very dense fault network. In the zone of the Carpathian Mountains curvature, the Vrancea fault is a special one, very different from the surrounding faults, being characterized by a narrow strip of the Black Sea block, which subducts the Pannonian plateau. The Anatolian fault system is due to the different movements of the Arabian and Eurasian plates, causing the most important strike-slip fault in the Mediterranean zone, very similar to the Californian San Andreas fault system.

- *Central European seismic region* is a zone where, during the tectonic evolution, a series of faults, horst and grabens created a fault network whose formation was due to the crustal weakness. In comparison with the Mediterranean region, the central region can be classified as a relatively stable area, where the seismic hazard is low. The exceptions are the Southern zones in the vicinity of the Alps Mountains, where collision earthquakes can occur.

- *Northern European seismic region* (UK and Scandinavian Peninsula) is largely influenced by the divergent movements of the Mid-Atlantic Ridge. The important earthquakes (magnitudes 5.8 to 6.0) can occur along the Norwegian coast only.

Code evolution. The evolution of the European seismic code is framed in the general evolution to constitute the European Union, composed for the first time, by some developed countries and developing countries and, finally, by all European countries. The unified codification process is under the constraint of some realities, the main being the existence of national codes, where special conditions of seismic risk, experience and traditions, very different for each country, must be included. An attempt to impose a single code model in all countries will be very difficult to achieve. Therefore, for an intermediate period, a proposal for a code model, having the purpose to unify the existing codes, must leave the liberty to introduce some specific aspects of each national code. The Eurocode 8 (EC 8) "Design of Structures for Earthquake Resistance. Part 1: General Rules, Seismic Actions and Rules for Buildings", now completed, was published by the European Standard Organization (CEN) as EN1998. Based on this model, some national codes were revised. For instance, the new Italian seismic code was published under the Ordinanza 3274 (2003) and, with some modifications, published in Ordinanza 3431 (2005) (Landolfo, 2005, Perago, 2005). The Italian code respects the general rules of EC 8, but contains some complementary provisions, resulting from Italian experiences in earthquake engineering. It can be considered as a model of how to integrate the EC 8 provisions to a national code. Now, the seismic regulations are included in the general structural code, which include all actions, materials and technologies. A similar integration of EC 8 in a national code has been done in Romania, where the special seismic conditions given by the features of the Vrancea intraslab earthquakes have been introduced.

Main code specifications. The tendency of Eurocode 8 is to introduce the non-linear analysis, static (push-over) or dynamic, as the reference method for the direct design of new buildings. Two recommended normalized spectra types have been setup from the database of available European records. Type 1 is proposed for large magnitude earthquakes (M >5.5) and Type 2 for low to moderate magnitude earthquakes (M <5.5). One can mention that the first type corresponds to intraplate earthquakes and the second to interplate earthquakes. In case of large magnitude earthquakes, the amplification is moderate and the corner periods are large. Contrary, for low to moderate earthquakes, the amplification is very large for reduced periods, but the corner periods are reduced. These spectra are not representative of intermediate intraslab earthquakes, like in Vrancea (Romania).

The second main specification refers to the performance requirements. The Eurocode 8 is based on two-level seismic design, with the following explicit objectives (Fardis, 2004):

- Protection of life under rare large seismic actions, by prevention of collapse of structures or part of them and preservation of structural integrity and residual capacity. The seismic actions are determined for 10% exceeding probability in 50 years (mean return period of 475 years).
- Limited property loss under frequent low seismic actions, by limitation of structural and non-structural damage due to lateral displacements of structures. These seismic actions are determined for 10% exceeding probability in 10 years (main return period of 95 years).

10.2.6 International Codes

A new initiative to create an international system of codes has been proposed by the International Organization for Standardization (ISO), which is a World Federation of National Standards body (Ishiyama, 2000, Gioncu, 2006). The works of preparing an International Standard is carried out through ISO Technical Committees (TCs). Each TC has several Working Groups (WGs). The draft elaborated by TC98 is the "ISO 3010. Bases for Design of Structures. Seismic Actions on Structures". The first edition was published in 1988 and the second in 2001. The ISO Standards are subjected to periodic review, usually every five years, so the last one is the 2006 edition.

The main characteristic of the ISO 3010 standard is to include only principles for determination of seismic actions and for seismic design, together with the main factors to be considered. It does not give specific values of the design seismic forces, but it includes information how to evaluate them. Generally, all standards prepared in the frame of ISO serve as guidelines for the preparation of national standards. Therefore, frequently, they are called the *"Code for Code Writers"*.

10.3 CHALLENGES FOR NEW DESIGN APPROACHES

10.3.1 Insufficient Impact of New Knowledge in Design Practice

Having the experience of the two last very damaging earthquakes, the USA and Japan have developed very extensive research programs in order to improve the new code provisions. All the world's scientists and structural engineers, involved in seismic design development, follow the obtained results, with the intention to incorporate them in the national codes. After a period in which these results were adopted without restrictions, the researchers were fully aware of the great difference between Californian and Japanese fault types and other earthquake sources, for which the seismic actions are very different and therefore, the characteristics of sources must be considered in the design provisions. The best example is the evolution the European codes. The strike-slip faults, with the same characteristics of the USA earthquakes, which produce high damaging, in the Mediterranean area exist only in the Northern Anatolia. With the exception of the Vrancea (Romania) seismic zone, which is a very special case among the European sources, the faults are characterized by collision, rift and graben systems, very different from the characteristics of the USA and Japanese last earthquakes. So, the American and Asian experiences cannot be assimilated in Europe without a very careful analysis. Therefore, the Eurocode 8 provisions differ from the American codes in a great extent.

There are some general lessons, which must be learnt from the last world important earthquakes as well as from the development of research works. There were some spectacular progresses during the last half century in understanding the physical phenomena concerning earthquake generation and propagation. At the same time, the complex and powerful mathematical models and computational resources to evaluate the structural response were developed.

Unfortunately, there is an insufficient impact of these advances on the practical design applications (Esteva, 2005). There are some general lessons, that must be learnt from the last important developments in seismic design, which are challenges for the next code generations (Gioncu, 2006), but the most important are:

- It is a stringent necessity to have common research works with the seismologists and engineering seismologists, in order to establish the presence and type of faults, especially in highly urbanized zones, where near-source effects can be very disastrous, even for low magnitude earthquakes.
- The code provisions must present different design philosophies and computation procedures in function of fault type, producing interplate, intraplate or intraslab earthquakes. They include different spectra, influence of ground motion duration, number of important pulses, effects of high vibration modes, differences in ductility demands, etc.

10.3.2 Seismic Demands and Structural Capacity Balance

The basic concepts of today's seismic codes were born almost 70 years ago, when the knowledge about the seismic actions and structural response were rather poor. Today, the earthquake-resistance has grown within the new multi-disciplinary fields of Engineering Seismology and Earthquake Engineering, where many exciting developments are predicted in the near future.

The challenge for a proper seismic design is to solve the balance between seismic demand and structure capacity (Gioncu and Mazzolani, 2006). The seismic demand corresponds to the effect of the earthquake on the structure and depends on the ground motion modeling. The structural capacity is the ability of the structure to resist these effects without failure. Looking to the developments of Engineering Seismology and Earthquake Engineering, it is clear that the major efforts of researchers have been directed towards the structural response analysis. Therefore, the structural response can be predicted fairly confidently, but these achievements remain without real effect if the evaluation of the seismic actions is not accurate and, therefore, doubtful. In fact, the prediction of ground motion is still far from a satisfactory level, due to both the complexity of the seismic phenomena and the communication lacks between seismologists and engineers. This remark can be confirmed after each important earthquake, when new and new in situ lessons regarding the characteristics of ground motions are learnt, instead of providing the missing data from seismological studies in advance. So, the reduction of uncertainties in ground motion modeling is now the main challenge in structural seismic design. This target is possible only if the impressive progress in the Seismology will be transferred into structural design. Three important developments during the second half of the last century contributed to the rapid advances in Seismology. First, the development of Plate Tectonic Theory, which offers now a clear and conceivable framework for the generation of the majority of earthquakes. Second, the establishment of fault mechanism producing different earthquake types, having very different ground motion characteristics. Third, the computer technology, which opened new possibilities to analyze a large amount of data and to model the fault rupture process, in order to simulate the ground motions. Any progress in Earthquake Engineering is impossible without considering the new amount of knowledge, recently cumulated in Seismology.

Today, after these improvements in knowledge, it can be assumed that there are three different earthquake types, which should be considered in the structural design: *interplate, intraplate and intraslab* earthquakes. There are such big differences between the ground motions generated by these earthquakes that the ignorance of these aspects can be considered as a shortcoming of code provisions. In this context, the engineering seismologists are now paying more concern to establish the differences in the main characteristics among these earthquakes. At the same time, the task of earthquake engineers is to take more care about the structural response of these earthquake types.

10.3.3 Critics of Current Design Methodologies

Since its introduction to Earthquake Engineering by Housner in the 1950s, the response spectrum has become an essential tool in structural analysis and design. But today, after the impressive development of the Earthquake Engineering Science, some shortcomings in current practice have been identified (Gioncu and Mazzolani, 2006):

- The codes historically were developed based on the experience of few recorded ground motions not sufficiently close to the causative faults, due to the absence of a dense network of recording stations. In the last period this situation is changed in some very urbanized seismic zones and the recent earthquakes, such as Northridge (USA), Kobe (Japan), Chi-Chi (Taiwan) and

Kocaeli (Turkey), yielded more near-source records. But, unfortunately, there is still a large number of seismic zones without adequate information.

- Due to the unknown fault network (e.g. Kobe earthquake, produced along an unknown fault) or some active faults different to the known ones (see Northridge earthquake produced by Elysian Park fault far from the very well-known San Andreas fault), it is practically impossible to design recording network stations in a such a way as to have records on the sites where the phenomenon of polarization shows the way to the maximum ground motion. So, the recorded values have a great incertitude and, only by chance, some of them are situated near the fault (e.g. Northridge, where, in spite of a large amount of recorded data, a very reduced number of records is available for near-source ground motions). In addition, these records can be influenced by some local site stratifications, so that they are available only for the recording station (see the 1977 Vrancea ground motions, recorded in Bucharest on a very poor soil condition, which led to the wrong conclusion that this source is characterized by long natural periods). Therefore, the design spectra developed based on these recorded motions are not, generally, available for other seismic sites.

- During recent decades, one can see a tremendous growth of urban areas in seismically active regions, which increases the risk that an earthquake occurs where there is a large concentration of population. The damage produced in these cases is well illustrated by the above-mentioned earthquakes. So, the so-called near-source earthquake must be taken into account in structural design, but these aspects are not considered by the response spectra used in practice. Only USA-UBC 97 and Chinese GBJ11-2001consider some aspects of the near-source earthquakes. The UBC proposes an increasing of acceleration values in Californian near-source zones, in function of the distance from the fault (but who knows exactly this distance?). The GBJ11-2001 proposes different corner periods for near- and far-source earthquakes. However, the increasing of acceleration or reduced corner periods are not sufficient measures to consider all the very damaging effects of near-source earthquakes.

- Peak ground acceleration, in the frame of the used design spectra, is the parameter associated with the severity of an earthquake. However, it has been generally recognized that it is a poor parameter for evaluating the earthquake damaging potential. For instance, one of the more significant shortcomings of the current design spectra is the fact they do not account for the duration of the input ground motions. The dissipation of seismic energy, as it is considered in the design philosophy today, is strongly affected by this duration. The near-source impulse type of ground motions (in case of crustal earthquakes) results in a sudden insertion of energy into the structure, which must be dissipated immediately by very few large yield excursions. Contrary, for long duration earthquakes (in case of far-source or deep earthquakes), the cyclical type ground motions with numerous yield reversals require a more steady dissipation of energy over a long period of time. The actual design spectra do not take into account this difference. So, the question is whether the structures are able to face these different situations.

-    In case of regions with low to moderate levels of seismicity, due to the lack of earthquake records, the required data are very scarce. It is a real mistake to use the data from other seismic zones, characterized by another source type, without checking if they are available for the zones under consideration.

The improvement of these aspects gives rise to new approaches in seismic design of structures, for which a deep knowledge of each earthquake type characteristics is required. In the last period, a very intensive activity in analytical and numerical modeling of the earthquake generation has been developed, giving a great contribution in the process of understanding the intricate phenomena produced during an earthquake and allowing the different earthquake types to be identified. The main task of specialists is to transfer this new knowledge to the structural practice.

10.3.4 Needs and Challenges for the Next Design Practice

After these critical points of view concerning the current design methodologies, it may be concluded that the process to setup a reliable design philosophy is very complex, due to the reason that so many factors, very difficult to be accurately accounted, are involved. Considering the conservative attitude of the design community, which does not like to accept some radical changes in design practice, a justified question arises: What must be the solution of this matter of fact, considering that the use of a methodology based on design spectra must remain available for current design in to the future? The answer is not simple and satisfactory, due to the complexity of the structural response during an earthquake. Considering current and future development, the proposed solution is based on three points (Gioncu and Mazzolani, 2002, 2006):

-    Verification at three levels of performances: serviceability, damageability and ultimate limit states.
-    Diversification of design spectra and behavior factor in function of the earthquake types for serviceability and damageability limit states.
-    Introduction of explicit ductility and fracture demands for ultimate limit state in function of the earthquake type, duration and number of large yield cycles, ground motion velocities, effects of strain-rate, etc.

After a long period when only the ultimate limit state is considered in the design practice, now all codes consider two levels, the serviceability and ultimate limit states. It is a reasonable approach, which has been accepted by the design practitioners. However, for the next code generation, it is an imperative issue to pass to a further step, considering three performance levels: (i) the serviceability for low and frequent earthquakes to protect against damage both structural and non-structural elements; (ii) the damageability for moderate and rare earthquakes, in order to limit the damage at a repairable level; (iii) the ultimate limit state for large and very rare earthquakes to protect the structure against collapse.

The second approach to have different spectra in function of the earthquake type is already given in some codes in different forms, but the application rules are not always clear. The above-mentioned UBC 97 code introduced, for the first time, the possibility to consider the effects of near-source ground motions for crustal earthquakes, by increasing the accelerations, but this aspect is not sufficient to consider all the effects of this earthquake type. Eurocode 8 proposed two types of design spectra, in function of the magnitude range. This specification can be considered as corresponding to two earthquake types: crustal interplate and intraplate earthquakes. Therefore, for the next generation codes, it is required to setup some clear criteria for establishing the pattern of design spectra in function of the earthquake type.

It is a matter of fact that the code provisions are very poor for the ductility demands. Instead of having methodologies to evaluate the ductility of structural elements, the code provisions just refer to some doubtful constructional requirements, in order to obtain a satisfactory ductility. Based on the SAC connection tests performed after the Northridge joint failure (SAC, 1996), it is accepted that members and connections should be capable to develop a minimum plastic rotation capacity of the order of 0.025 to 0.30 radian (FEMA 267, 1997). These values were adopted by many other codes. But some important questions arise concerning these tests, which were performed using quasi-static procedures with increasing displacement history and with a considerable number of cycles until the connection failure. Can these tests interpret the dynamic actions of ground motions with the same characteristics as the ones developed in near-source zones? It is very well known that the pulse velocities can be so high that they cannot be modeled in laboratory. Another question refers to the generality of the limit values of rotation capacities, considering the great differences in demand for different earthquake types, which must be reduced in case of intraplate earthquakes. Therefore, the next generation codes must introduce proper provisions for checking the ductility and fracture methodologies like the ones used for displacements and strength (Gioncu and Mazzolani, 2002).

10.4 CHARACTERISTICS OF EARTHQUAKES IN FUNCTION OF SOURCE TYPES

10.4.1 Main Factors To Be Considered in Seismic Design

In the frame of Earthquake Engineering, today the unique big challenge of seismic design is to provide reliable structures resisting the ground motions resulting from earthquakes. Earthquakes produce large forces of short duration which must be resisted by the structure without causing collapse and preferably without significant damage to the structural elements. The main factors, which must be considered in design process, in function of earthquake type (Table 10.1):

-   Interplate (subduction or strike-slip).
-   Intraplate (convergence or crust fracture)
-   Intraslab

These factors are:

- *Earthquake magnitude* is one of the fundamental parameters for characterizing the ground motion. It must be determined using the recorded data for the site, in function of the recurrence relationship. The magnitude must be specified for each performance level. Referring to the maximum expected magnitude, earthquakes can be classified as very strong (M > 8), strong (6.5 < M < 8), moderate (4.5 < M < 6.5) and low (M < 4.5).
- *Near-source effects* are very great in case of crustal earthquakes, considering all the characteristics of this special situation, in which the velocity of pulse, period of pulse, number of pulses, vertical components, etc. play a very important role in design provisions. In particular:
- *Pulse type,* in the sense that velocity pulse or acceleration pulse must be considered;
- *Number of pulses* is very important to know if the ground motions are characterized by much reduced number of pulses (only one or two pulses), reduced (for 3 to 4 pulses) or by a large number of pulses (5 to 10 pulses or more, in some special cases);
- *Period of pulses,* which can by very long (over 3 to 4 sec), long (1 to 3 sec) and short (less than 1 sec).
- *Spectrum type,* characterized by corner period and amplification for short periods, knowing that the differences in earthquake type are marked also by this aspect, due to the importance of high vibration modes. Two normalized spectra types are recommended to be used (available also in EUROCODE 8 (Fig. 10.1): Type 1 proposed for large magnitude earthquakes and Type 2 for moderate magnitude earthquakes.
- *Effects of soil conditions.* In function on source type, the effects are different. The effects can be very great, great and moderate.
- *Duration of ground motions* has a great influence on structure damage and collapse type. It can be very long (over 90 sec), long (under 90 sec), moderate (under 60 sec) and short (under 30 sec).

Table 10.1 Main characteristics of earthquakes

Earthquake type	Magnitude M	Near-source effects	Pulse type	Number of pulses	Period of pulses	Spectrum type (Fig. 10.1)	Soil condition effects	Duration
Interplate								
strike-slip	strong	very great	velocity	reduced	very long	T1	great	long
subduction	very strong	great	velocity	very reduced	long	T1	moderate	moderate
Intraplate								
convergence	strong	great	velocity	very reduced	long	T1	moderate	modest
crust fractures	low to moderate	moderate	velocity	very reduced	short	T2	great	short
Interslab	strong	no	acceleration	large	long	T1	very great	very long

Amplification of peak ground acceleration

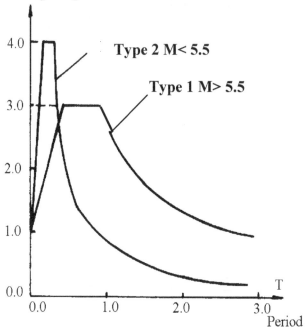

Figure 10.1 Elastic response spectra types (after Gioncu and Mazzolani, 2002)

10.4.2 Interplate Crustal Earthquakes

There are two interplate types: subduction and strike-slip earthquakes.

Subduction crustal earthquakes. These earthquakes are produced by faults situated in the Earth crust and characterized by the highest magnitudes recorded in the world: 1960 Chile M 9.5, 1964 Alaska M 9.2, 1957 Aleutian Islands M 9.1 and 2004 Sumatra M 9.0. Therefore, in function of recurrence studies, this earthquake type can be classified as a very strong one. The main effects are concentrated in the so-called near-source zones, assumed to be within a distance of about 20-30 km from the epicenters. Being very difficult to establish the exact distance between site and source, all areas where subductions exist must be considered potentially situated in near-source zones. Within this zone, the ground motions are significantly influenced by the rupture mechanism. The main effect is the velocity pulse, where the maximum values of velocity are very difficult to record due to the focusing effects and only by chance a recording station is located in that zone. Such a chance happened during the Chi-Chi (Taiwan) earthquake in an isolated station; the recorded velocity exceeded 350 cm/sec, without the certitude that there were not other sites with larger velocities. The fact that during the Northridge earthquake the recorded velocities did not exceed 200 cm/sec, does not mean that this value was not overcome in another points of epicentral area. The velocity spectra for Northridge earthquake shows velocities exceeding 350cm/sec (Krawinkler and Alavi, 1998). Therefore, the incertitude in this problem is very high and, for design purposes, a very high velocity must be assumed. Considering the number of important pulses, generally, this earthquake type is characterized by one velocity pulse. The period of pulses is very long, the maximum being recorded during the Chi-Chi earthquake from 6 to 11 seconds. Regarding the spectrum type, it corresponds to the interplate type (type T1), with moderate amplification and long corner period (0.7 to 1.2 sec). This earthquake type has, generally, short duration, with the exception in case of bad soil conditions, when the duration increases significantly.

Strike-slip crustal earthquakes. In comparison with the subduction faults, the number of important strike-slip faults is reduced: San Andreas (USA), North Anatolia (Turkey) and Alpine (New Zealand). But, due to the great earthquakes produced along the San Andreas and the North Anatolian faults in very urbanized areas, this earthquake type has been widely studied. The recorded magnitudes were not as strong as for subduction earthquakes, but the spectra show that the velocities can exceed the ones of subduction earthquakes. For instance, the spectra for for Loma Prieta, can exceed 500cm/sec, and for Kobe, even over 650cm/sec (Krawinkler and Alavi, 1998). The near-source effects are very important, due to the special characteristics of ground motions: large normal components, the possibility of two or three rupture sources (as in the Kobe earthquake), forward directivity effects, very long velocity pulse period, long duration and important effects of site soil conditions. The spectrum corresponds to the T1 type. Due to the particularities of this earthquake type, it cannot be used as background for the elaboration of codes in zones having other earthquake types.

10.4.3 Intraplate Crustal Earthquakes

There are two intraplate earthquakes:

Collision crustal earthquakes, produced by the suture of two ancient tectonic plates, such as the ones occurring along the Pyrenees, Alps, Apennines and Himalayan Mountains. The characteristics of ground motions are the same as those for subduction, with the exception of the magnitude, which the maximum recorded values not exceeding M 8.0.

Crust fracture earthquakes. This earthquake type has the most large spreading in the world, occurring in the areas considered stable, but able at any time to produce low to moderate magnitude ground motions (the maximum values, in normal soil conditions, do not exceed M 6.0). Many countries located in these regions are still in process to elaborate provisions for the seismic protection of buildings, which need to be very different from the ones of high seismicity regions. Due to the crustal position of source, the near-source moderate effects occur, with moderate velocity pulse of one, maximum two, cycles. The periods of pulse are very short and the corresponding spectrum is T2 type. The duration in normal conditions is very short, but the influence of soil conditions is very high, increasing both magnitude and duration of ground motions.

10.4.4 Intraslab Earthquakes

There are three intraslab earthquakes: shallow, intermediate and deep. Unfortunately, there are not sufficient records and data to establish the main differences between these three cases. Magnitudes are high, the maximum recorded values being around M 8.0. Due to the depth of source, the area affected by the ground motion is very large. Consequently, the near-source effects are insignificant. Generally, the velocities are low and pulses can exist only in recorded accelerations. The seismic waves passing through an important number of layers, the number of cycles is very large with long periods, corresponding to T1 spectrum type. The influence of site soil conditions is very high, leading to long duration of ground motions.

10.5 STRUCTURAL RESPONSE TO INTERPLATE CRUSTAL EARTHQUAKES: NEAR-SOURCE GROUND MOTIONS

10.5.1 Response of Structures for Near-Source Ground Motions

Although the libraries of ground motion records are becoming richer as far as the recording station coverage is denser in near-source regions, the sporadic occurrence of large earthquakes in heavily instrumented regions makes it difficult to investigate the spatial distribution of the structural response and provide a complete picture of the distribution of damage occurring in the region. To address this issue, seismologists are increasingly utilizing large-computer simulations to compute synthetic ground motions based on source parameters and models of the geological structures (Aagaard et al, 2001) (see Sections 7.5.4 and 7.5.5). Two scenarios, a strike-slip fault and a thrust (subduction) fault with magnitude of 6.0 and 5.8, respectively, were used to study the behavior of a 9-story moment resisting frame (structural period 2.3 sec), for a spatial position of the structure at each grid point in the analyzed region (Park et al, 2004). The impulse period is about 5-6 sec, greater than the structure period.

Figure 10.2 shows the story drifts distribution for the *strike-slip* ground motions, in case of fault-normal and fault-parallel components. For fault-normal component the distribution of story drifts shows that the maximum values occur along the fault, with systematic differences from the epicenter site to the end of the fault. In the first part, the drift at the top story is greater than at the other stories, but the drift at lower stories is also significant. Contrary, at the sites near the end of the fault in the forward directivity zone, the drift at the lower stories is greater than at the other floors. An important attenuation of story drifts can be noted in fault-normal direction, showing the importance of forward directivity. For fault-parallel component, the maximum story drift values occur near the epicenter at 2 km distance from the fault. The largest drift occurs in the top stories.

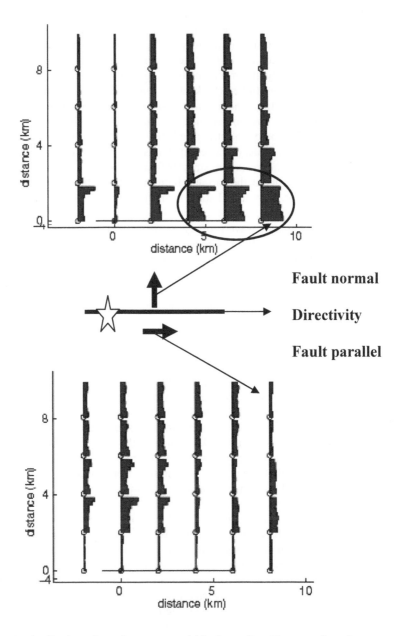

Figure 10.2 Distribution of maximum story drifts for strike-slip ground motions
(after Park et al, 2004)

Figure 10.3 shows the story drift distribution for *thrust (subduction)* ground
motions. The rectangular area represents the projection of the rupture surface and
the star is the position of the epicenter. The fault-normal direction of the largest

story drifts is located in the projection of the rupture surface, in the forward directivity zone. The maximum story drifts are concentrated at the higher floors, being more than twice the drift at the lower floors. For fault-parallel component, the story drift is fairly uniform along the height of the building

In general, these studies (Park et al, 2004) show the complexity of seismic design. At the same time, the research based on the numerical simulation of ground motions underlines the importance of the forward directivity zone and the dominant contribution on the structure behavior of the fault-normal components for both strike-slip and thrust (subduction) faults. The results of simulations show that the distribution of the maximum story drift varies with the location of the structure. The large story drifts at the top and bottom floors indicates that standard design procedures do not prevent soft-story formation for near-source ground motions in the forward directivity. Such localization of deformation in tall buildings can be interpreted in terms of wave propagation of pulse-type ground motions along the structure (Hall et al, 1995, Gioncu and Mazzolani, 2002). In case of modal response analysis, this means that it is compulsory to include the superior vibration modes in seismic design.

A main conclusion is that the design procedures must be different for strike-slip and thrust (subduction) faults, the most damaging ground motions being produced by the former.

The earthquake type, produced by strike-slip or subduction, can be established by means of seismological studies for each structure site. But the directivity of rupture, which can establish the normal or parallel components of seismic actions, is practically impossible to accurately predict in function of the structure position. Therefore, due to this incertitude, the structural design must consider the most severe situation, which is the normal to the fault actions.

The response of framed structures to near-source ground motions also shows special characteristics. Figures 10.4 compares the story ductility demand for a 20-story structure subjected to near-source Californian and Japanese ground motions, in comparison with the ductility determined from far-source earthquakes. Two cases are analyzed (considering also the earthquake type) (Alavi and Krawinkler, 2001):

(i)   Frame with moderate designed ductility $V = 0.4W$ ($q = 2.5$) (Fig. 10.4a) (V is the base shear force and W, the structure weigth). One can see that the maximum required ductility is concentrated at the top of the frame due to the effects of superior vibration modes. The larger ductility demands for Kobe earthquakes are the consequence of the multiple pulse, characteristic for this earthquake.

(ii)   Frame with high designed ductility $V = 0.15W$ ($q = 6.7$) (Fig. 10.4b). Contrary to case of moderate ductility, required ductility is concentrated at the base of the frame

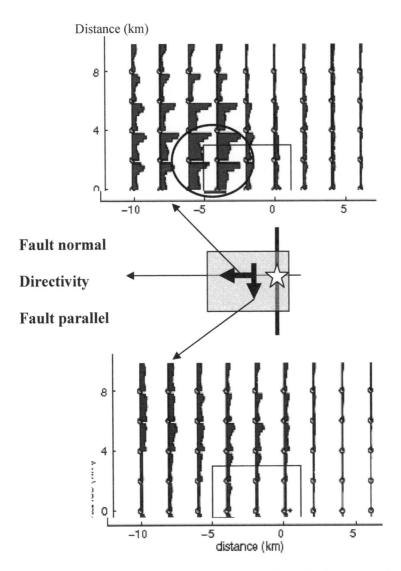

Figure 10. 3 Distribution of maximum story drifts for thrust ground motions: (after Park et al, 2004)

The particularity of the frame response to near-source ground motions is again prevalent. Contrary to the far-source earthquake (as are the requirements of common codes), the distribution of the demands over the height of the structure is highly non-uniform for the near-source records and depends on the designed ductility of the structure. For moderate designed ductility (rigid structure), the maximum ductility demands are concentrated at the top stories of the structure.

Contrary, for high designed ductility (flexible structures), the maximum ductility demands are concentrated in the first stories of the structure. Therefore, the severity of near-source ground motions leads to ductility demands, which are significantly larger and having less uniform distribution than those for the far-source records, which is represented in the common code provisions.

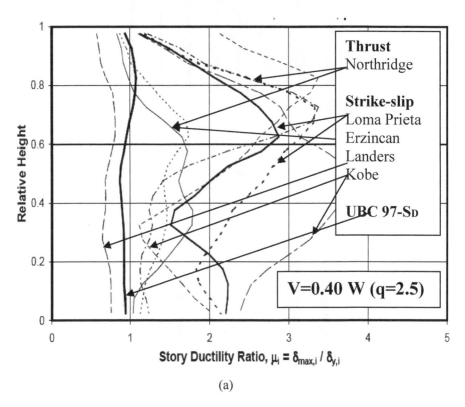

(a)

Figure 10. 4 (continues)

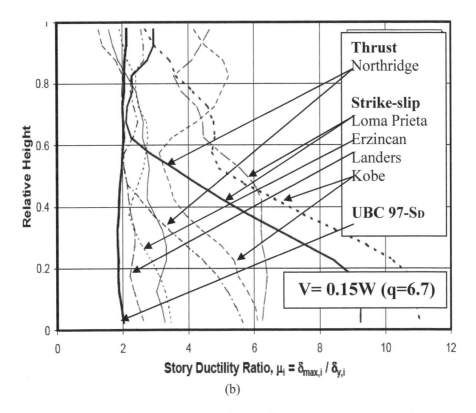

Figure 10.4 Ductility demands of steel frame for near-source ground motions:
(a) Moderate ductility frame; (b) High ductility frame (after Alavi and Krawinkler,
2001)

10.5.2 Response of Structures to Artificially Generated Pulse Input

The generation of artificially generated pulse input is presented in Section 7.5.3.
Various polygonal types of pulses are investigated by Alavi and Krawinkler
(2001), from the most basic ones being denoted as P1, half pulse, P2, full pulse and
P3, multiple pulse, in order to evaluate the response of structures to pulse loadings,
which represent the near-source ground motions. In all these cases, the pulse period
T_P is defined as the duration of a full velocity cycle. Other pulse types with
different configurations have also been studied (triangular and continuous curves),
but it was found that their response characteristics do not differ very much from
those of the three basic pulses.

In order to establish the artificially generated pulse input, three parameters need
to be evaluated: pulse type (P1, P2 or P3), pulse period T_P (seconds) and pulse
intensity (peak of velocity). Typical examples are presented in Figure 10.5 for the
Northridge and Kobe earthquakes. The obtained results are:

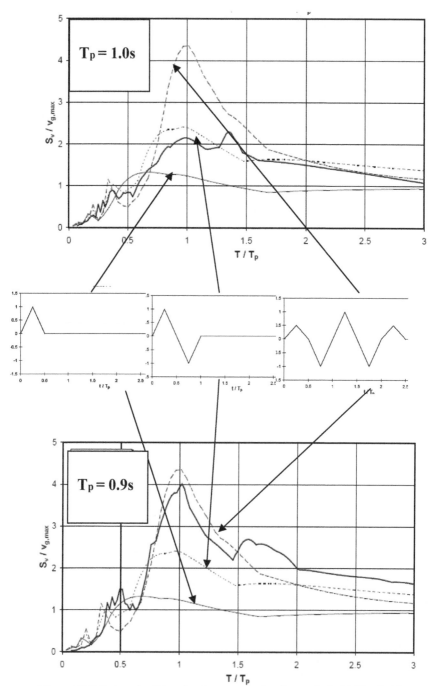

Figure10.5 Determination of pulse type and pulse period: (a) Northridge earthquake; (b) Kobe earthquake (after Alavi and Krawinkler, 2001)

- For the Northridge earthquake it is clear that Type 2 and Tp = 1.0 sec are the pulse parameters due to the fact that this earthquake is produced by a thrust fault characterized by a single pulse.
- Contrary, for the Kobe earthquake, produced by a strike-slip earthquake with multiple sources and two pulses, Type 3 and Tp = 0.9 sec are the pulse parameters.

Figure 10.6 compares the story shear distribution obtained from a standard design methodology, based on modal combination, and the one from a time-history analysis. The following observations can be made (Alavi and Krawinkler, 2001):

- For structures with T/Tp > 1.0, the dynamic shear force distributions are relatively smooth, but for structures with T/Tp < 1.0, these distributions show a clear effect of a wave traveling up the structure. Thus, for long period structures analyzed with standard design methodology, the early yielding has been expected in the upper stories (see the damaged buildings in Kobe).

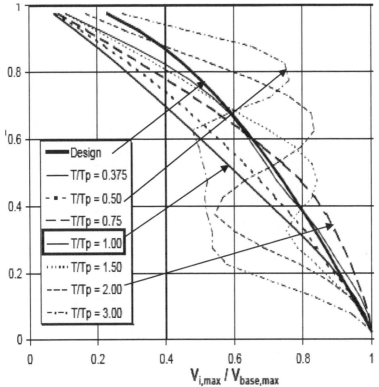

Figure 10.6 Normalized elastic story shear forces (after Alavi and Krawinker, 2001)

-   For structures with T/Tp < 1.0 the standard design methodology is not a good
    substitute for the dynamic time-history analysis, because the spectral analysis
    may not capture all important response characteristics of pulse-type ground
    motions, once the higher-mode effects become important.

As an example for checking the possibility to use artificial ground motions for
the evaluation of the structure behavior during a pulse earthquake, a one span-six
levels frame (Fig 10.7) has been designed by considering the first mode in
accordance with the Romanian seismic code, neglecting the effects of the superior
modes. The DRAIN 2D time-history program has been used for the elasto-plastic
analysis of the frame, by using a peak acceleration Ag = 0.35g. The study
parameter was the pulse period Tp, selected to cover the first three vibration modes
of the frame: T1 = 1.72 sec, T2 = 0.61sec, T3 = 0.5 sec (Gioncu and Mateescu,
1999, Gioncu et al, 2000). The results are presented in Figure 10.8. One can see
that the structure behaves elastically for pulse periods Tp smaller than 0.35 sec.
The first plastic hinge occurs at the top of the frame for 0.4 sec, near to the third
vibration mode. The increasing of the pulse period moves down the formation of
the plastic hinges. For Tp > 0.7 sec, corresponding to the second vibration mode, a
plastic mechanism occurs at the middle of the frame. For Tp = 1.8 sec,
corresponding to first vibration mode, a global mechanism is formed, but
associated with some story mechanism at the middle part of the frame.

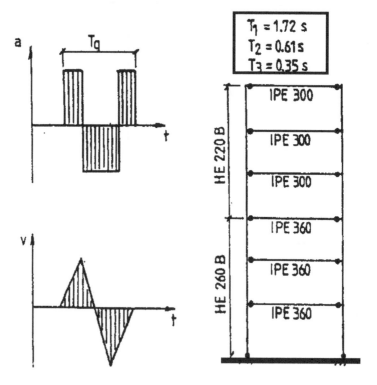

Figure 10.7 The analyzed frame (Gioncu et al, 2000)

Third vibration mode   **Second vibration mode**

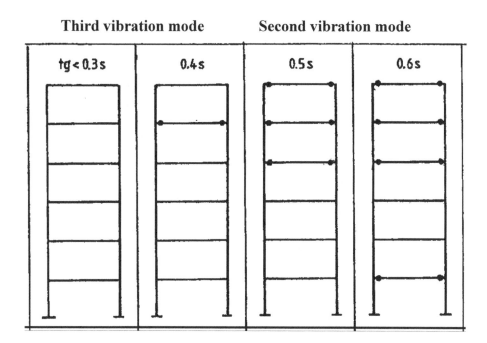

First vibration mode

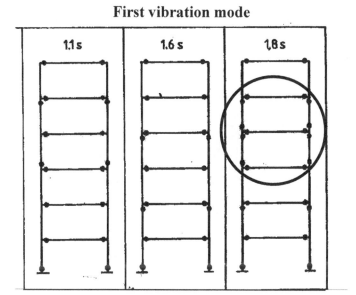

Figure 10.8 Formation of plastic hinges for ag = 0.35g and different pulse periods
(Gioncu et al, 2000)

Figure 10.9 presents the ductility demands for the analyzed frame, in function of the pulse periods. For periods near the superior modes, the maximum required ductility is concentrated at the top of the structure. For a period corresponding to the first vibration mode, the maximum required ductility occurs at the lowest levels, having very large values.

The study results show very clearly that the superior vibration modes give a very important contribution to the frame behavior for pulse periods smaller than the first vibration period, but the required ductility is smaller than the one corresponding to first mode. The frame behavior is dominated by the first vibration mode, when the pulse period is larger than the one corresponding to this mode, with a very large ductility demand. But at the same time, the ultimate limit state is influenced by the second vibration mode.

In the previous analyses it was considered that the ground motions are characterized by one important pulse only, which is generally the case of the thrust type earthquakes. But it is known that it is possible to have two or three important pulses for the strike-slip earthquakes, due to multiple sources. In case of $T_p = 0.4$ sec, Figure 10.10 shows the very large increasing of ductility demand at the top of the frame for 2,3 and 4 adjacent pulses caused by accumulation of plastic rotations.

The frame was resized to eliminate the formation of the story mechanism, strengthening columns by using the profiles HE450B and HE400B for the inferior and superior part of the frame, respectively (Fig. 10.11). This change in the frame configuration modifies the first three periods, which are now 1.35, 0.40 and 0.18 sec. Figure 10.11 shows that the formation of the middle level plastic mechanism is eliminated, the ultimate limit state being governed by the global mechanism.

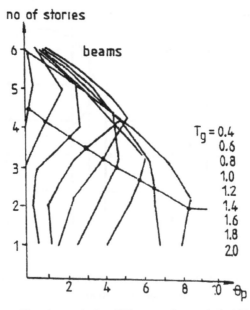

Figure 10.9 Ductility demands for different pulse periods (Gioncu et al, 2000)

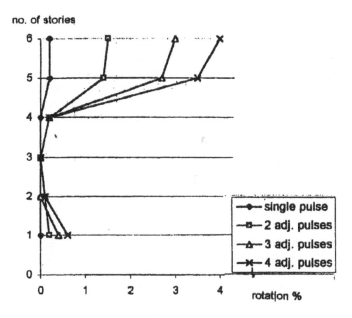

Figure 10.10 Influence of multiple pulses on ductility demands
(Gioncu et al, 2000)

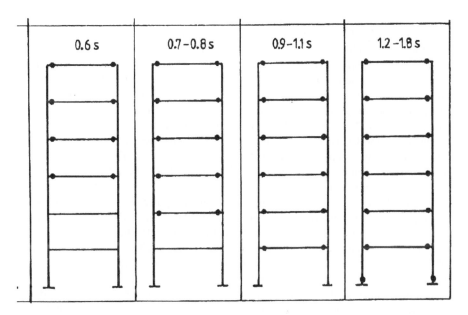

Figure 10.11 The behavior of the strengthened frame (Gioncu et al, 2000)

10.5.3 Response of Structures to Vertical Components

The effect of vertical components of ground motions on structures in near-source areas (mainly produced by P waves) focused the attention in the last decades, on the base of field observations concerning the damage which was produced by severe vertical vibrations (Elnashai and Papazoglou, 1997, Elnashai et al, 2004). The velocity in near-source field is very high. During the Northridge and Kobe earthquakes, values of 175 cm/sec were recorded at the soil level. The velocity of the vertical components dramatically increases near to the epicenter. So, in case of near-source fields, the velocity is the most important parameter in design, replacing the acceleration, which is the dominant parameter for far-source field.

Due to vertical components, the compressed RC columns are broken in an explosive manner and the steel columns undergo brittle tensile failure (see Northridge, Kobe and Greek earthquakes) (Carydis, 2005). As the vertical components are rarely considered in the structural analysis, it follows that there is a potential built-in deficiency in the majority of structures and their foundations to resist vertical earthquake induced vibrations. Elnashai and Papazoglou (1997) developed near-source spectra, with an important peak acceleration of 3.5 and corner periods 0.05 and 0.20 sec. The frame buildings are much stiffer in the axial than in the transverse direction and hence they possess very short periods in the vertical direction, framing in the range of 0.10 to 0.40 sec, corresponding to the maximum spectra amplification.

A relation between horizontal and vertical periods is presented in Figure 10.12. This amplification, combined with the severity of near-source vertical ground motions, suggests that large dynamic axial forces, acting both upwards and downwards (Fig. 10.13), should be expected in near-source field. In case of RC structures, the studies indicated that intermediate and top stories are more likely to undergo tensile forces. The axial forces due to vertical motions, being caused by accelerations whose magnitude is comparable with the one of horizontal motions, are larger than the corresponding forces due to horizontal motion. The contribution of vertical motion to axial forces in frames is ranging from 64 to 72% at the top stories and from 21 to 56% at the ground stories, in case of 4 to 20-story buildings (Papazoglou and Elnashai, 1996). Therefore, the contribution of vertical motions is always more significant in the upper floors than in the lower floors. These forces in the interior columns are even more significant, since the effect of horizontal motion is less important there.

The modeling of vertical component as a velocity pulse (Fig. 10.14a) with very short periods was performed by Gioncu and Mateescu (1998a). The artificial spectra were determined for various pulse periods $T_p = 0.05$; 0.10; 0.15; 0.20; 0.30 sec, and for symmetrical and asymmetrical pulse type (alpha = V_{max}/V_{min}). Figure 10.14b shows the spectra for $T_p = 0.10$ sec. It can be observed that the maximum amplification occurs when the first half pulse is maximum.

A time-history analysis was performed for a steel structure (Fig 10.15) for an asymmetric ground motion pulse (alpha = 1.6) and $T_p = 0.15$ sec. The evolution of plasticization in the columns was determined in function of the multiplier of the

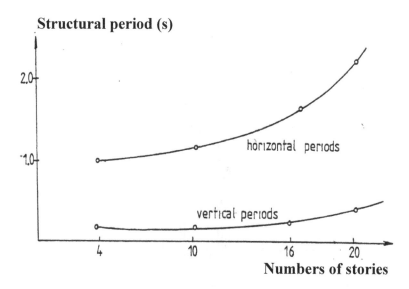

Figure 10.12 Horizontal and vertical periods (Gioncu and Mateescu, 1999)

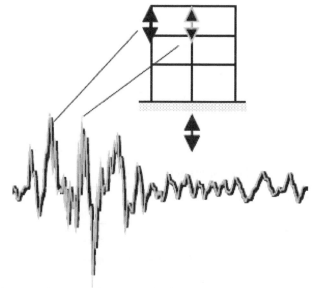

Figure 10.13 Development of axial forces in the upper columns
(after Chouw, 2002)

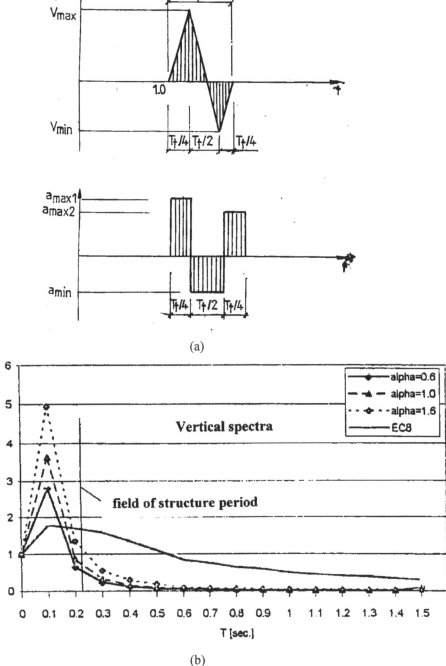

(a)

(b)
Figure 10.14 Vertical velocity pulse: (a) Puls model; (b) Spectra for $T_p = 0.10$ sec
(Gioncu and Mateescu , 1998a)

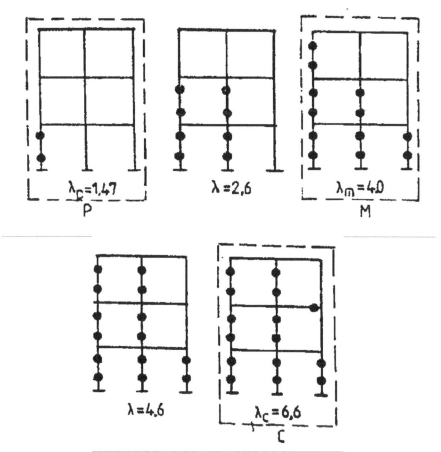

Figure 10.15 Time-history analysis for vertical components (Gioncu and Mateescu, 1998a)

gravitational acceleration. The mechanism was reached for 4g, meaning that the vertical components of the ground motion cannot produce the collapse of the structure, but can give important contribution to the structure behavior for horizontal seismic actions.

A very important problem is the evaluation of the effect produced by the asynchronism of vertical movements (Fig. 10.16), which can introduce very important internal forces, especially in beams and nodes. Generally, this effect is considered to be very important for long structures only (i.e. bridges), when the wave length corresponds to the dimension of the construction. It is true for the first vertical mode, but considering the superior vertical vibration modes, with shorter wave lengths, the building dimension can be attained. This problem must be solved in the future research works. There are some specialists who attributed part of damage produced during the Northridge and Kobe earthquakes to the effect of the vertical asynchrony of ground motions.

From the recorded ground motions in near-source areas, it is recognized that the vertical ground motions are much larger than the horizontal ones. Therefore, the vertical seismic component is a dominant parameter (in combination with the horizontal motions) in the various near-source field regions, mainly due to earthquakes caused by thrust (subduction) faults. As the vertical components of ground motions are often neglected in the seismic design of structures, most damage occurred in the near-source areas due to these vertical seismic components, especially in new buildings, which were designed respecting the modern code provisions for horizontal components.

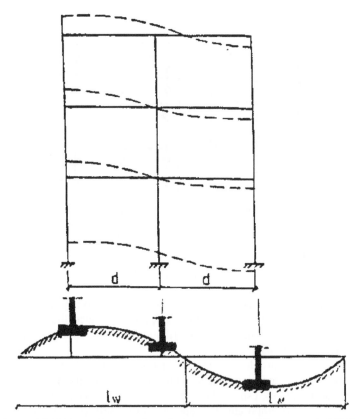

Figure 10.16 Asynchrony of vertical movements (Gioncu et al, 2000).

10.5.4 Problems of Structural Response Control

An effective design of structures subjected to seismic loads requires the use of plastic analysis where the ductility plays an important role. The structural response control depends on the ductility requirements, in function of the earthquake type and the available ductility of the individual members. The *ductility checking* should be quantified at the same level as in case of stiffness and strength (Gioncu and Mazzolani, 2002). The required ductility (global ductility) depends on the source and it is presented in the previous section for the interplate earthquakes. The main problems arise for evaluating the available ductility (local ductility), which must be determined in function of the main characteristics of ground motions. Due to the near-source effects characterized by high pulse velocities, the available ductility evaluation must consider the influence of strain-rate (Gioncu, 2000, Gioncu and Mazzolani, 2002)), which can shift the ductile plastic member response into brittle fracture (Fig. 10.17a), due to the increasing of the yield stress. At the same time, due to the strong pulse seismic loading and to the reduced number of pulses, the common ductile cyclic behavior is replaced by brittle fracture after the reduced number of cycles, due to the accumulation of large plastic deformations (Fig.10.17b) (Mateescu and Gioncu, 2000, Gioncu et al, 2000). The difference between the subduction (thrust) and strike-slip must be considered, because the strike-slip earthquakes are the most damaging events due to high velocities and large ground motions producing important plastic deformations.

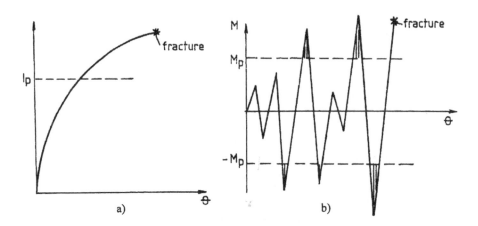

Figure 10.17 Fracture due to pulse with high velocity: (a) Fracture due to strain-rate; (b) Fracture due to accumulation of large plastic deformations (Gioncu et al, 2000)

Another important effect of near-source earthquakes is the *influence of the superior vibration modes*, especially for short pulse periods, introducing irregularities in the bending moment diagrams, especially at the middle high of the frame (Fig. 10.18a). The situation induces a strong reduction of the available ductility in the middle or top part of the frame. As the required ductility has a maximum just in these places, the local collapse of the building may occur due to insufficient ductility. This was a common phenomenon during the Kobe earthquake, where many buildings were damaged just in the middle stories. In case of large pulse periods, the reduction of available ductility occurs at the lower levels, producing a global collapse (Fig. 10.18b).

Due to the characteristics of the pulse seismic actions, developed with great velocity, and especially due to the lack of important restoring forces, the ductility demand could be very high, so the potential of the inelastic properties of structures for seismic energy dissipation has to be carefully examined. The *short duration* of the ground motions is a very important factor. A balance between the severity of the ductility demand, due to the pulse action, and the effects of the short duration must be seriously analyzed.

These aspects have an important consequence on the establishment of the reduction factor, because, due to short duration, the structure has not time enough to dissipate a corresponding input seismic energy. Therefore, these factors must have different values in case of near-source ground motions, in comparison with the most common far-source cases.

10.6 STRUCTURAL RESPONSE TO INTRAPLATE CRUSTAL EARTHQUAKES: LOW-TO-MODERATE GROUND MOTIONS FOR CRUSTAL FRACTURE

10.6.1 Main Characteristics of Intraplate Crustal Earthquakes

Much attention has been paid to seismic risk for interplate earthquakes and the results are consistent. In contrast, a relative interest has been shown for intraplate earthquakes, even if their consequences can be disastrous. As it was presented in 10.4.3, there are two intraplate types: (i) the *collision crustal earthquakes*, produced in the suture zones of the continental plates. These earthquakes have the characteristic of interplate subduction (thrust), presented in the previous section; (ii) contrary, the *crust fracture earthquakes* have very peculiar characteristics, with very important consequences on the structure responses. In the following, only the corresponding structure responses are presented

The earthquakes produced by crust fractures, due to their reduced rupture surfaces, have moderate magnitudes, generally under M 5.5. In very rare cases, a magnitude higher than M 7.0 has been recorded, but only in some special conditions (very bad soil or liquefaction phenomena), which can be identified by geotechnical studies. Due to this fact, the areas, where intraplate earthquakes occur by crustal fractures, are characterized by low to moderate ground motions.

.

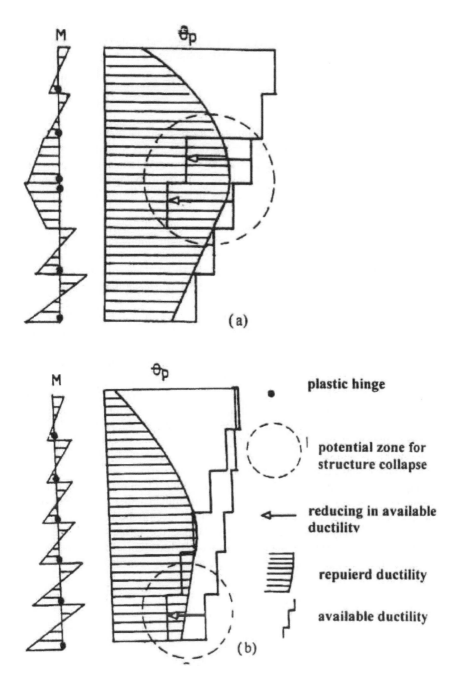

Figure 10. 18 Required and available ductility
(Gioncu and Mateescu, 1999 , Gioncu et al, 2000)

The depth of crustal fracture sources lays in the category of surface earthquakes. Therefore, the effects of near-source ground motions, presented in the previous section, are valid also here, but with reduced magnitude and some specific features, different from the interplate sources. In the following the main differences will be emphasized.

10.6.2 Elastic Spectra

In regions of low to moderate seismicity, local data are very scarce and, consequently, it is difficult to obtain reliable data necessary to establish proper elastic spectra. In this situation, the theory of multi-source data fusion, presented in Section 4.6, can be very useful. Therefore, the data from analogous regions with similar geological and seismo-tectonic features must be used for regions lacking in earthquake records (Chandler et al, 2001).

The spectra for crustal fracture earthquakes are very different in comparison with the spectra of interplate earthquakes. A comparison among the corresponding spectra in the US Eastern territory (intraplate earthquakes) and Western territory (interplate earthquakes) is presented in Figure 10.19. One can see that the difference consists in an important amplification of peak acceleration for very short periods (0.1 to 0.2 sec).

In Europe, a very significant situation is the case of Portugal, Italy and Greece, where the two source types, both interplate and intraplate, are present. In Portugal, the first type, located offshore, corresponds to the boundary between the African and Euroasian plates and it is capable of generating large magnitudes earthquakes (M 8.5), as well as effects at large distance from epicenter, long duration and long periods. The second type corresponds to seismic source located inland, capable of generating smaller, but still significant magnitudes (M 7.0), with effects at short distance, short duration and short periods (Azevedo and Guerreiro, 2007). In Italy and Greece, the interplate earthquakes are situated in Southern zones, due to the subduction of African plate under the European plate. The rest of the territories are subjected to intraplate earthquakes (mainly due to collision).

Earthquakes occurring in Australia are the result of intraplate movements (Chandler et al, 1992). The spectra for some Australian and Eastern Canadian earthquakes are plotted in Figure 10.44. All these spectra show very important amplifications for very short periods. Nevertheless, the influence of soft soil conditions is very important, reducing the peak amplification and increasing the pulse period (Fig. 10.45). This aspect was present in many recorded earthquakes in low and moderate seismic areas.

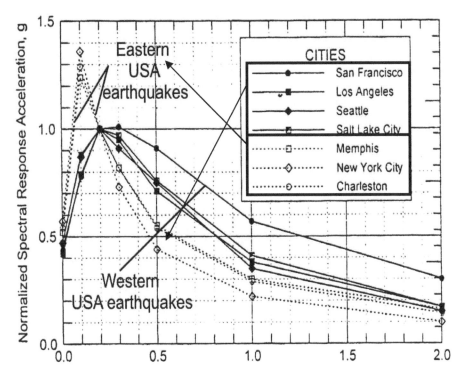

Figure 10.19 Comparison among elastic spectra in the USA
(Gioncu and Mazzolani, 2006, after Leyendecker et al, 2000)

All these spectra for American, European and Australian zones very clearly show the main characteristics of crustal fracture earthquakes: short periods, important acceleration amplification for these periods and important influence of site soil conditions.

Therefore, the intraplate crust fracture earthquakes have the characteristics of near-source ground motions, but they have very short pulse periods (0.1 to 0.4 sec), in contrast with the interplate earthquakes, which develop long pulse periods (larger than 1 sec). This fact involves the large participation of superior vibration modes in the structural design. The result is the removing of important drifts and ductility demands from the low levels of structures to the middle and top stories (Gioncu and Mazzolani, 2006).

10.6.3 Response of Structures for Low-to-Moderate Earthquakes

The primary objective of research in intraplate earthquake engineering is to develop methodologies for the structural design in case of low to moderate earthquakes. While important progress has been made in understanding intraplate earthquakes in a seismological context, far less is known in the earthquake engineering context, as far as the structural response is concerned (Hutchinson et al, 1996, Gioncu, 2008). The results of analyzed structures subjected to the low to moderate earthquakes are very scarce and not comparable with the abundance of data available for strong earthquakes. Due to this situation, a developed program of numerical tests was performed in Timisoara during 1996-2000, in order to underline the great differences between the structural responses in Vrancea (intraslab strong earthquakes) and in Banat (intraplate low to moderate earthquakes) seismic regions (Gioncu and Mateescu, 1997, 1998b). Some of the main results will be presented in the following.

The frame of Figure 10.20a was sized respecting the Romanian seismic code for $Ag = 0.16g$, according to the provisions mainly based on the Vrancea earthquake. As a consequence, the lateral distribution of seismic forces was selected on the base of the first vibration mode. The frame was analyzed using the incremental dynamic analysis, based on the time-history DRAIN 2D computer program, for an accelerogram recorded in Timisoara during the 1991 Banat earthquake, which is characteristic of the moderate earthquake type (M 5.6) (Fig. 10.20b). The time history of lateral displacements is presented in Figure 10.21 for the acceleration level producing the collapse mechanism, $Ag = 0.77g$.

One can see that the deformations are dominated by the superior modes of vibration. The formation of plastic hinges is presented in Figure 10.22. The formation of the first plastic hinge at the top level of frame, occurred at 0.43g, 2.7 times the design acceleration, meaning that the structure behaves elastically until this limit acceleration. The collapse mechanism is of global type, occurred at 0.77g, 4.8 times the design value of acceleration. The same conclusions were obtained for a six-story frame, its collapse mechanism occurring at 1.28g, 8 times the design acceleration. This fact very clearly shows that the methodology based on the load distribution corresponding to first mode do not correspond to the earthquakes where superior vibration modes are dominant.

An improvement the code provisions can be achieved by introducing two distributions of lateral seismic actions, which consider the effect of the first and the second vibration modes, are presented in Figure 10.23. The first distribution corresponds to the interplate ground motions (Vrancea earthquakes). The second distribution, corresponding to superior vibration modes, is recommended for intraplate ground motions (low to moderate earthquakes of Banat region).

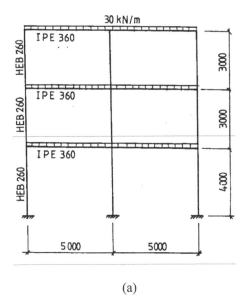

(a)

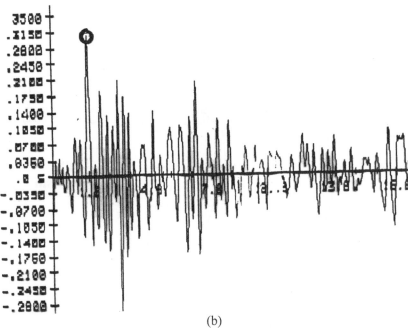

(b)

Figure 10.20 Analyzed structure for intraplate Banat earthquake:(a)Frame configuration; (b) Time history of seismic action (Gioncu and Mateescu, 1997)

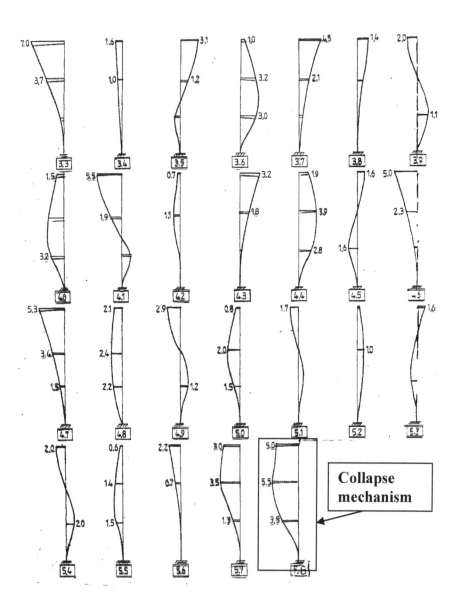

Figure 10.21 Time history of lateral displacements for Ag = 0.77 g, corresponding to collapse mechanism (Gioncu and Mateescu, 1997, Gioncu, 2008)

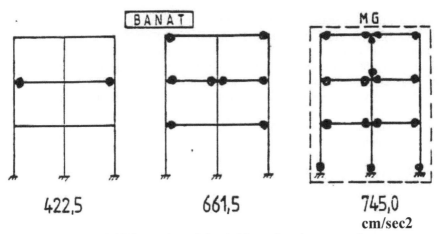

Figure 10.22 Formation of plastic hinges in a three-story frame (Gioncu and Mateescu, 1997, Gioncu, 2008)

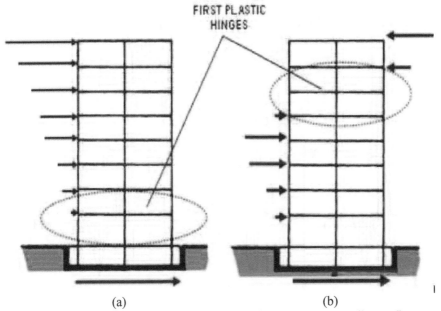

Figure 10.23 Lateral seismic action distribution: (a) Corresponding to first vibration mode; (b) Corresponding to second vibration mode (Gioncu, 2008)

10.6.4 Response of Structures to Artificially Generated Pulse Input

A set of artificial spectra was determined for interplate earthquakes, characterized by velocity pulse with short periods (Gioncu and Mateescu, 1998a, 2000). The characteristics of velocity pulse and the pulse types are presented in Figure 10.24. Considering the characteristics of the examined near-source ground motions for moderate earthquakes, the followings cases were considered:

- Symmetry or asymmetry of velocity pulse using different values of the parameter:
  alpha = vmax/vmin = 0.6; 1.0 (symmetry); 1.6
- Pulse periods
  Tp = 0.1; 0.2; 0.3; 0.4; 0.5 and 1.0 sec
- Number of pulses: one pulse, two adjacent pulses, two separate pulses.

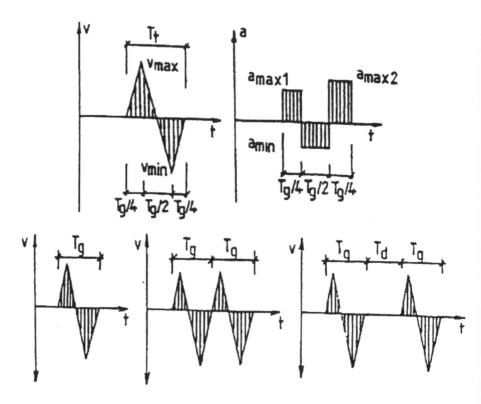

Figure 10.24 Characteristics and pulse types (Gioncu et al, 2000)

Figure 10.25 shows the spectra for a pulse period T_p = 0.2 sec and asymmetry ratio. One must notice that the asymmetric pulse with maximum velocity for the first half cycle (alpha = 1.6) gives the maximum amplification in the field of very short structure periods

A comparison among the amplifications for single pulse, two adjacent pulses and two separate pulses with different intervals in-between is presented in Figure 10.26. One can see that the highest amplification happens for two adjacent pulses (case of strike-slip earthquake). Two distanced pulses do not produce significant modification of the amplification corresponding to a single pulse. The amplification in the short structure periods is very high, exceeding the code values.

Figure 10.27 presents the spectra for velocity pulses, which are considered significant for low to moderate earthquakes. One can observe a very high amplification for short structure periods and that the spectra for T_p = 1.0 has multiple peak values, in contrast with the other velocity pulses having only a single maximum value.

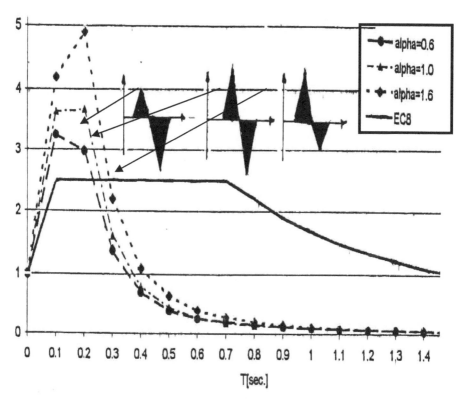

Figure 10.25 Influence of pulse asymmetry (Gioncu et al, 2000)

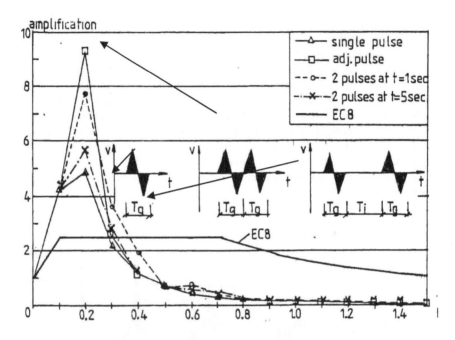

Figure 10.26 Influence of number and succession of velocity pulses
(Gioncu et al, 2000)

The frame of Figure 10.20a (with periods $T_1 = 0.90$sec, $T_2 = 0.26$ sec and $T_3 = 0.13$ sec), designed for acceleration of code 0.16g, was analyzed using different velocity pulses (Fig. 10.28) One can see that for very short pulse period (0.20 sec), a medium value between second and third structure periods, the plasticization began at the top of frame and the global collapses mechanism occurs at 1.6g, ten times of design acceleration. For higher velocity pulse period (0.50 sec), a medium value between first and second vibration modes, the collapse acceleration is 0.45g, 2.8 times the design acceleration. Contrary, for the highest pulse period, corresponding to the first period, the collapse acceleration is 0.22g, only 1.38 times the design spectra. This analysis clearly shows the importance of the highest vibration modes in determining the behavior of structures under low to moderate earthquakes.

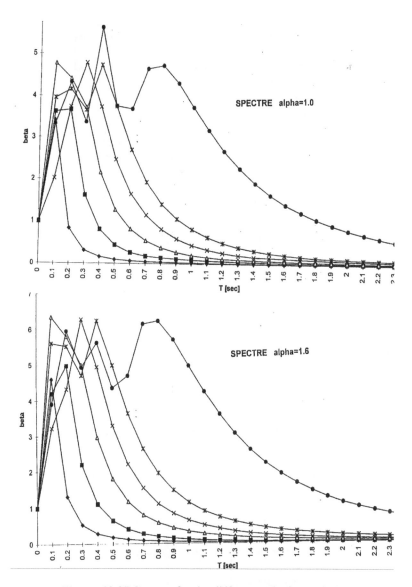

Figure 10.27 Spectra for the different velocity periods
(Gioncu and Mateescu, 1998c)

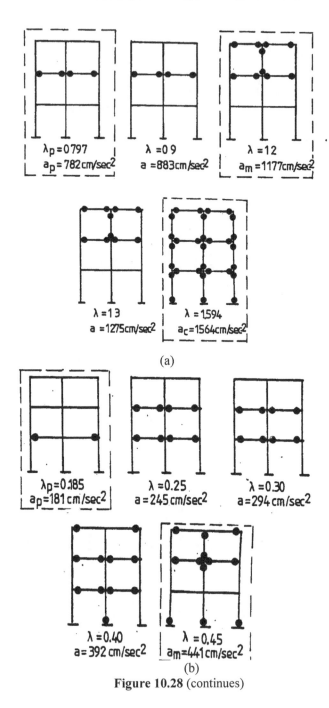

(a)

(b)

Figure 10.28 (continues)

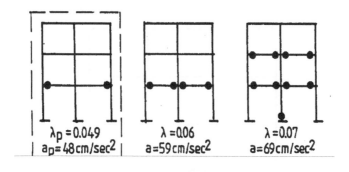

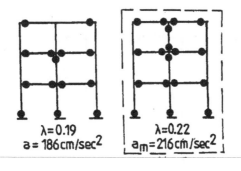

(c)

Figure 10.28 Behavior of three-story steel frame for different periods of velocity pulses: (a) $T_P = 0.20$s;(b) $T_P = 0.50$s; (c) $T_P = 1.0$
(Gioncu and Mateescu, 1998c)

The influence of the highest vibration modes was studied by Goel and Chopra (2005). The considered three-story frame is presented in Figure 10.29a ($T_1 = 1.01$s, $T_2 = 0.328$s, $T_3 = 0.172$s). Figure 10.29b shows four lateral distributions, Mode 1, ELF, equivalent lateral force, RSA, response spectrum analysis and uniform distributions. Observing these distributions, one can notice that there are not significant differences between the distribution corresponding to the first mode in term of values and direction of seismic actions. All floors of the building are pushed in the same direction, in contrast with the deformations determined using a time-history analysis. Contrary, for distributions according to vibration modes (Fig. 10.29c), the forces pull some floors and push others, conforming to the above-mentioned deformations. For the second vibration mode, one reversal occurs, i.e. the roof displacement changes direction. For the third vibration mode there are two reversals.

$T_1 = 1.01s; \ T_2 = 0.328s; \ T_3 = 0.172s$

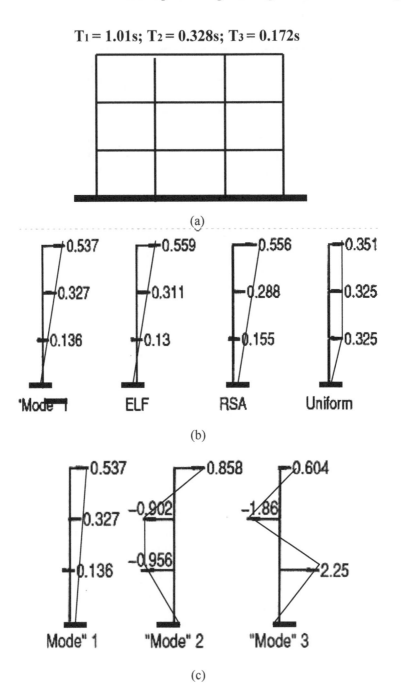

(a)

(b)

(c)

Figure 10.29 Analyzed three-story frame: (a) Typical moment-resistant frame :
(b) Code distributions; (c) Modal force distributions (after Goel and Chopra, 2005)

The frame was analyzed under the pulse type excitation using the time-history method, for velocities corresponding to moderate earthquakes. The first analysis considered the frame loaded by a velocity pulse of 50 cm/sec corresponding to the first vibration mode (Tp =T1= 1.01 sec) (Fig. 10.30a). Large lateral displacements (until 25 cm) occur and the global mechanism is formed. For the velocity pulse of 60 cm/sec and pulse period corresponding to second vibration mode (Tp = T2 = 0.328 sec), Figure 10.30b shows a maximum displacement of 4.5 cm, 5.6 times smaller than the displacement corresponding to the first vibration mode. The displacement profile corresponds to the second vibration mode. The collapse mechanism is formed at the top story. The response to the third ground motion (Fig. 10.30c) corresponds to a pulse with period equal to the third vibration mode (Tp = T3 = 0.172 sec) and velocity of 100 cm/sec. The maximum displacement is 3 cm, 8.3 times smaller than the one for the first vibration mode and the displacement profile corresponds to third vibration mode. The collapse mechanism involves the middle frame story.

These results demonstrate that the traditional analysis methods, mainly based on the predominance of the first vibration mode, cannot detect the effects of higher vibration mode, in case of ground motion characteristics which excite the building into higher vibration modes. The main effects are the formation of local mechanisms in the superior part of structure (Goel and Chopra, 2005). This is clearly the case of frames situated in low to moderate seismic regions, where the velocity pulse periods are always smaller than the first vibration periods of structures.

The force distributions corresponding to the formation of collapse mechanisms for the first three vibration modes are presented in Figure 10.31a. The first mode distribution produces a global plastic mechanism and the increasing force intensity can cause the building to rotate as a rigid body around its base. The second mode force distribution causes a local plastic mechanism in the third story and the increasing force intensity can cause the third story to rotate as a rigid body around the second floor. The third mode force distribution causes a collapse mechanism at the second story, which rotates as a rigid body over the first story .

The corresponding moments for the three distributions are presented in Figure 10.31b. One can see that the formation of a plastic mechanism requires that the applied lateral forces must be 2.8 and 3.68 times the ones of the second and third mode, respectively, in comparison with the first mode. This remark is a justification of the results of frame analyses, which show that a structure, designed for the first vibration mode, reaches the collapse of the top stories for accelerations higher than the design ones.

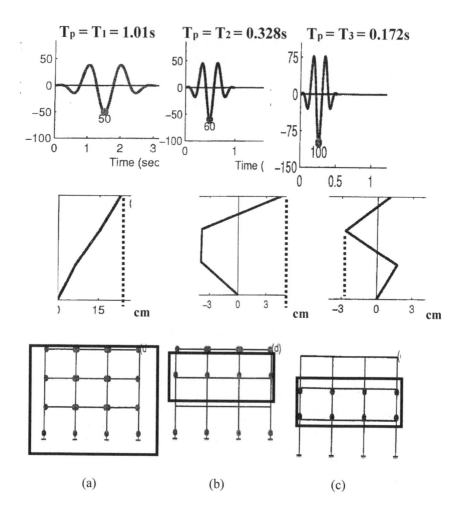

Figure 10.30 Structure response for a pulse corresponding to vibration modes.
Pulse characteristics, lateral displacement, collapse mechanism:
(a) First mode; (b) Second mode; (c)Third mode
(after Goel and Chopra, 2005)

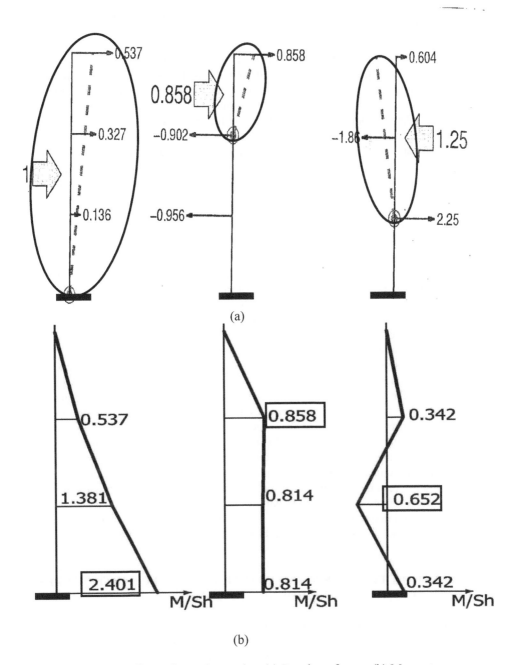

Figure 10.31   Effects of superior modes: (a) Resultant forces; (b) Moment
distribution (after Goel and Chopra, 2005)

10.6.5 Problems of Structural Response Control for Low-to-Moderate Earthquakes

There are many important problems in the structural response control for low-to-moderate earthquakes, which are different from the characteristics of the other earthquake types. First, being produced by surface sources, all the manifestation of *near-source effects* are present, but with some particularities: compared to the case of interplate earthquakes, all the characteristics are reduced, like magnitude, pulse periods, duration, etc. The numerical analyses presented in the previous section have clearly shown that the main characteristic for the structural response control is the pulse period, which is shorter than the period of the first vibration mode of the structure. This implies the influence of the *higher vibration modes* on the structure response, generally reducing the seismic demands, but moving them in different zones of the structure (i.e. at the building top). Since the demands are not as severe as in high seismicity regions, large inelastic deformations are not required even under the maximum credible earthquakes. The seismic design based on *limited ductility* design seems appropriate in moderate regions (Chang et al, 2001). Therefore, the use of the ductility demands, determined for the Californian earthquakes, also in case of moderate earthquakes is a mistake. Due to the short duration of moderate earthquakes, the structure has not time enough to dissipate a sufficient quantity of induced seismic energy. Therefore, the behavior factor, to be used in case of moderate earthquakes for considering this reduced capability of energy dissipation, has to be different from the one used in case of the other earthquakes. So, a *reduced behavior factor* is used in the design of structures under moderate earthquakes.

From the main results of the recent research works, the specialists are fully aware that the design philosophy to be used in regions of low-to-moderate earthquake must be different from the philosophy in regions of high seismicity. But, unfortunately, this philosophy is not developed as well. There are many countries belonging to low-to-moderate seismicity regions. Even in a country with strong earthquakes, there can be some zones of moderate seismic hazard level. Therefore, results and experiences on the new seismic design for low-to-moderate seismicity zones will be beneficial for many countries. In the last period, the increasing interest in this problem is documented by papers, reports, organized symposia and sessions in world conferences devoted to this subject.

10.7 STRUCTURE RESPONSE TO INTRASLAB DEEP EARTHQUAKES: LONG DURATION GROUND MOTIONS

10.7.1 Elastic Spectra and Duration

The intraslab earthquakes are capable of generating large magnitude between 7.5 to 9.5 and long duration earthquakes until 240 seconds. For these earthquakes, the number of the important cycles is very large. In addition, the influence of site soil conditions produced very important amplifications in peak accelerations. The seismic risk of the engineering structures in case of deep intraslab earthquakes may by tied to the duration of strong shaking (March and Gianotti, 1994). The seismic design procedures currently used in practice do not consider these events. In comparison with the sources of interplate and intraplate earthquakes, the number of sources of intraslab earthquakes is much reduced. Unfortunately, the information obtained from recorded ground motions is very scarce.

The main aspect of the intraslab earthquake is the duration, which cannot be defined by the spectra. It is well known that, for this earthquake type, the maximum amplitude of ground acceleration and the spectra are poor indicators of the destructive potential of seismic motions. The duration of strong shaking, which is related to the number of cycles of the earthquake motions, is a more significant parameter in determining the response of structure. It measures the damaging effect of strong motions, by the accumulation of plastic rotations as the effect of a number of the excursions in plastic field (Bommer and Martinez-Pereira, 1996).

There is little definitive information describing the duration of large earthquakes, but it is sure that this duration depends on magnitude and site soil condition (Marsh and Gianotti, 1994), as shown by the data below (duration in seconds):

Magnitude	Rock (sec)	Soft soil (sec)
7.0	16	32
7.5	22	45
8.0	31	62
8.5	43	86

It is important to know the difference between the duration of strong shaking and the total duration. The acceleration envelope of an earthquake with a total duration of 60 sec is presented in Figure 10.32. For this example the envelope is

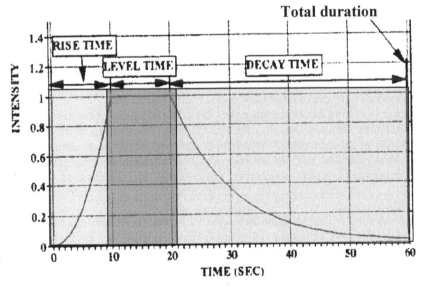

Figure 10.32 Acceleration envelopes (after Marsh and Gianotti, 1994)

composed by three fields, the rise time (10 sec), the level time (10 sec) and the decay time (40 sec). One of the maximum durations of 240 sec was composed by the rise portion of 42 sec, the level time of 48 sec, and the decay time of 150 sec.

It is understandable that only a part of this envelope, classified as *significant duration*, is important for the structure behavior, but there is not a very clear definition of the duration from the structural point of view (Bommer and Martinez-Pereira, 1996).The simplest definition is the bracketed duration which is the total time elapsed between the first and last excursion of a given level of acceleration (0.3 to 0.5g). A more advanced definition considers the accumulation of seismic energy.

10.7.2 Response of Structures to Intraslab Deep Earthquakes

The frame of Figure 10.33a has been analyzed, by using the incremental dynamic analysis, based on the time-history DRAIN 2D (1975) computer program, for the accelerogram recorded at Bucharest-INCERC in 1977 (Fig. 10.33b) (a_g = 0.24g). The corner period is very large (T_p = 1.6 sec), much larger than the first vibration mode of the frame (T= 0.74 sec). The obtained lateral displacements are shown in Figure 10.34. One can see that the displacements are dominated by the first vibration mode due to the difference between acceleration pulse period and the first vibration mode of the structure. Figure 10.35 shows the formation of plastic hinges and global mechanism, formed at 032 g, corresponding to 1.33 times the design acceleration.

10.7.3 Problems of Structural Response Control for Long Duration Earthquakes

The structural design demands for ground motions produced by intraslab earthquakes are very special and different from the ones of other earthquake types. Due to the deep fault position, the near-source effects are reduced, but the affected area around the epicenter is very large. In addition, there are some special aspects, which must be considered in the design practice: great influence of traveling paths, site soil conditions and, especially, the long duration of ground motions, associated to important numbers of large yield cycles (minimum 15-20).

In case of long duration earthquakes, the effect of cyclic actions on the ductility of structural members becomes the main factor (Anastasiadis et al, 2000). The limitation of local ductility of members is given by the plastic buckling of flanges (for steel structures, see Fig. 10.36a), or by the crushing of compressed concrete (for reinforced concrete structures). The result is the reduction of the rotation capacity, after some cycles, due to the accumulation of plastic deformations (Fig. 10.36b).

Examining the acceleration envelopes of Figure 10.37, one can see that there are two different situations: (i) Local plastic buckling occurs in the field of increasing of accelerations (curve 1), when the amplitude continues to increase after the local buckling (Figure 10.37a); (ii) Local plastic buckling occurs at the maximum peak acceleration (curve 2), when the amplitude remains practically

$$T_1 = 0.74s$$

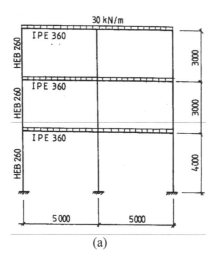

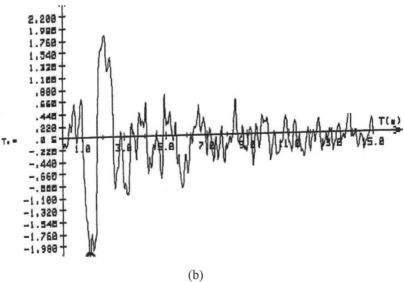

(b)

Figure 10.33 1977 Vrancea-Bucharest intraslab earthquake: (a) Analyzed frame;
(b) Recorded accelerogram (Gioncu and Mateescu, 1997)

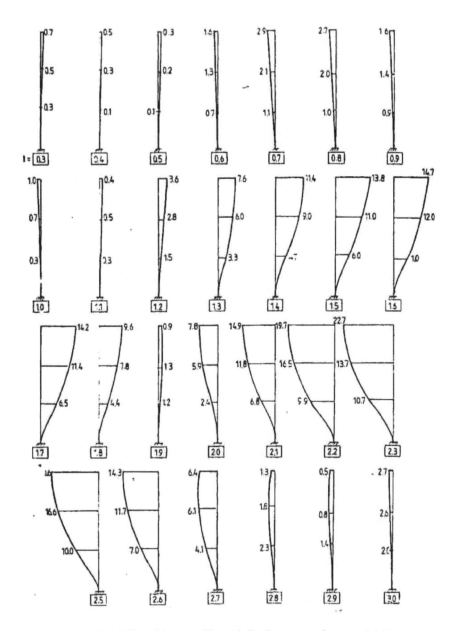

Figure 10.34 Time history of lateral displacements for ag = 0.216g
(Gioncu and Mateescu, 1997)

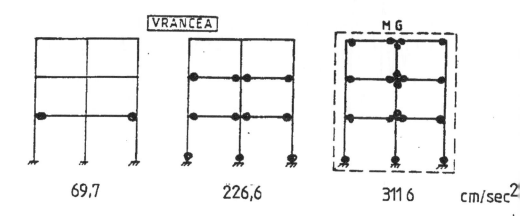

69,7   226,6   311 6   cm/sec²

Figure 10.35 Formation of plastic hinges in the three-story frame
(Gioncu and Mateescu, 1997)

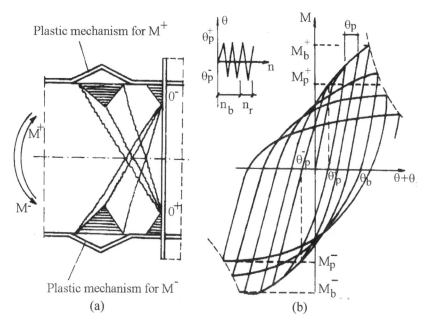

Figure 10.36 Effect of accumulation of plastic rotations
(Gioncu and Mazzolani, 2002)

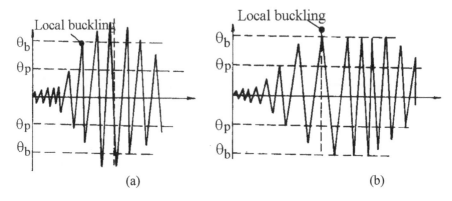

Figure 10.37 Two types of cycles: (a) With increasing of amplitude after the local
buckling; (b) With constant amplitude after the local buckling
(Gioncu and Mazzolani, 2002)

constant after the local buckling (Fig. 10.37b). The influence of the pulse number
produced after the local buckling is presented in Figure 10.38. The ductility
decreases as far as the acceleration amplitude increases (n_r is the number of pulses
occurred in the decay field of moment-rotation curve).

Another very important problem is the collapse by fracture due to the
accumulation of plastic strain after a large number of cycles (Fig. 10.39). This
number of pulses until the fracture is presented in Figure 10.39 in function of the
yield ratio (yield stress over tensile strength). One can see that the number of
pulses for high ratio (which can be the result of the influence of strain-rate) is
reduced, being in the range of practical values (n_b is the number of cycles occurred
in the field of stable moment-rotation curve).

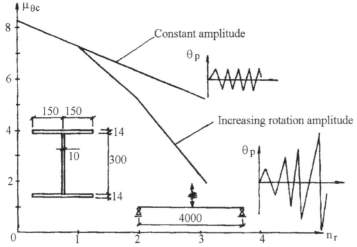

Figure 10.38 Influence of pulse number after the local buckling for constant and
increasing amplitudes (Gioncu and Mazzolani, 2002)

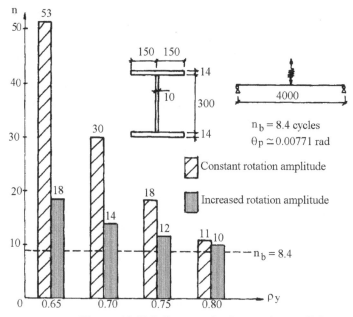

Figure 10.39 Influence of pulse number until fracture
(Gioncu and Mazzolani, 2002)

10.8 RECOMMENDATIONS FOR DEVELOPING SIMPLE BUT RELIABLE NEW CODE PROVISIONS

10.8.1 Two Topics for New Codes

Earthquakes produced in urban areas have demonstrated that the economic impact of damage, loss of function and business interruption is huge and that the damage control must become a more explicit design consideration (Fajfer and Krawinkler, 1997). The recent earthquakes have demonstrated that the behavior of structures depends on the earthquake type (Gioncu and Mazzolani, 2006). Therefore, the new codes, in order to improve the seismic design, must consider the following two topics:
- Performance-based design.
- Specific design and performance evaluation approaches for structures in function of the earthquake type.

As presented in Section 10.1.2, the implementation of these topics in codes is constrained by the need to be simple, but reliable, to be assimilated and not rejected, by the engineering community, which is generally conservative. For this reason, it must preserve the existing format of the present codes, the new aspects representing only some diversification of the principal design data.

10.8.2 Performance-based Design

In the recent years, a new design philosophy for building codes has been discussed within the engineering community. New building codes for earthquakes must adopt a performance-based design framework. The goal of a performance-based design procedure is to produce structures which have predictable seismic performance under multiple levels of earthquake intensity. Performance-based earthquake engineering implies design, evaluation and construction of engineering facilities, whose performance under common and extreme loads responds to the diverse needs and objectives of owners, users and society (Krawinkler, 1999, Leelataviwat et al, 1999).

In order to comply with these requirements, it is necessary to pass from the two levels of the present codes to the three- level approach (Gioncu and Mazzolani, 2002).

Limit state	Earthquake intensity	Avoided failure	Verification	Analysis method	Analsis object
Service-ability	low	non-structural elements	rigidity	elastic	story drift
Damage-ability	moderate	local collapse	strength	elasto-plastic	section capacity
Ultimate	severe	global collapse	ductility	kinematics	rotation capacity

- *Serviceability limit state,* under frequent low earthquakes, for which the structure must work in elastic field and the non-structural elements remain undamaged. To guarantee their integrity, the interaction of the structure and non-structural elements must be evaluated by means of an elastic analysis (Fig. 10.40), whose result is the reduction of the interstory drifts. The basic verification refers to the structure rigidity by controlling the story drifts, which must be framed within the prescribed limits.
- *Damageability limit state,* under occasional moderate earthquakes, where the non-structural elements are damaged (so the interaction structure-non-structural elements is neglected or just partially considered) and a very limited structural damage occurs. In this case, the analysis follows the elasto-plastic methodology (Fig. 10.40). The moderate damage to structural elements can be repaired without great technical difficulties. The basic verification refers to the structure members, by analyzing their section capacities.

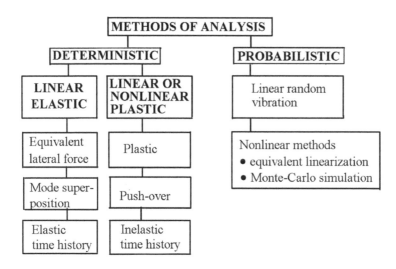

Figure 10.40 Classification of analysis methods (Gioncu and Mazzolani, 2002)

- *Ultimate limit state (survivability),* under severe earthquake, where all non-structural elements are completely damaged and a global mechanism is formed. The analysis follows kinematics procedures. The basic verification considers the available ductility, which is expressed by the rotation capacity of plastic hinges.

The performance-based design itself will not provide an improved or more predictable structural performance. The design only gives a set of drawings and instructions for the builder. The quality of the built product will depend on the development of methodologies incorporating new achievements in demand and response capacity of the structures (Krawinkler, 1999). As has been shown in this chapter, the demand cannot ignore the earthquake type and the response capacity directly depends on this demand.

10.8.3 Influence of Earthquake Type

Despite using very advanced design methodologies, most of the seismic design work cannot describe the actual structure behavior, as very significant differences exist between the predicted and the actual response. The reason for these differences is due to the neglecting of the influence of ground motion types. The code provisions classify the earthquakes only in function of their magnitudes and specific spectra. Therefore, a structure designed for a given magnitude and soil conditions, behaves in the same manner, regardless of whether it is situated in a zone influenced by interplate, intraplate or intraslab sources. This is clearly not true. This is a great mistake of the present codes, the structure behavior being very different for these sites, even if the magnitudes are the same. Therefore, the seismic design methodology must take into account these differences.

The first requirement for the next generation of codes is to complete the map of the seismic zonification in function of ground motion accelerations, by indicating on the map the *earthquake source type*:
-   Interplate crustal source:
        subduction (thrust)
        strike-slip;
-   Intraplate crustal source:
        convergence
        crust fracture;
-   Intraslab sub-crustal source.
The second step for the next generation of codes is to consider in seismic design the main particularities of each earthquake type.

The main problem is to select a *set of spectra* corresponding to earthquake type, normalized to the acceleration (Fig. 10.41) (Gioncu and Mazzolani, 2002) for the three earthquake types and for the three limit states. The values in this figure, concerning amplifications and corner periods, are given just as an example, the exact values being established for each case. One can see that the spectra for the limit states are different, contrary to the pattern of the present code spectra. For serviceability limit states, the spectra are characterized by high amplification for short periods, with a steep reduction in amplification for vibration periods exceeding the corner period. For damageability limit states, the amplification is not so high and the reduction more gentle. For ultimate limit states, no amplification occurs, because the structure is transformed into a plastic mechanism, which works as the ground motions impose, without any interaction due to lack of any vibration mode.

The differences depending on earthquake types are very large. The interplate and intraslab earthquakes are in the first category (Fig. 10.41a), being characterized by longer corner periods (especially for strike-slip and intraslab earthquakes). The amplification is higher for strike-slip and especially for intraslab earthquakes in comparison to thrust earthquakes, due to presence of a higher number of cycles. In case of intraplate, especially for crust fractures, the patterns of spectra are very different (Fig. 10.41b).

A very important step in improving the next code generation is the topic dealing with *different design philosophy* in function of earthquake type. The seismic-resistant structures are usually designed relying on their ability to sustain high plastic deformations. The earthquake input energy is dissipated through the hysteretic behavior of plastic hinges during a high number of cycles of seismic loading. According to this design philosophy, the structure may be designed for lower forces than those it has actually to resist. This seismic design philosophy has proved to be valuable in many great earthquakes and it corresponds to the methodology commonly included in the modern codes. But the recent events occurred in near-source areas produced a widespread and unexpected brittle fracture of joints with little or no evidence of plasticization of members, as it is assumed in code provisions. As a consequence, the amount of seismic input

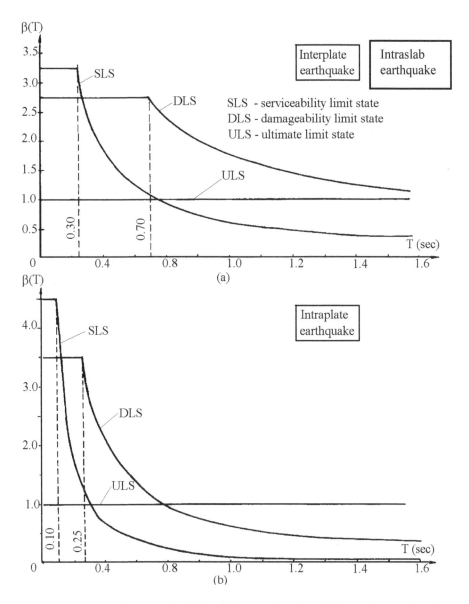

Figure 10.41 Design spectra: (a) Interplate earthquake; (b) Intraplate earthquake
(Gioncu and Mazzolani, 2002)

energy dissipated before the joint failure in the fractured members was not clearly apparent. From these circumstances, it has been questioned whether the current philosophy, based on the dissipation of seismic input energy by means of plastic rotations, is effective also in case of near-source earthquakes. Among the causes assumed to explain these brittle fractures, the effect of high velocity of the pulse seismic loading, producing important strain-rate effects in the structural material, is considered by the specialists to be the main influencing factor (Gioncu and Mazzolani, 2002). Therefore, there are two different types of member behavior influenced by the earthquake type:

-   In case of pulse seismic loading (Fig. 10.42a), characteristic of near-source thrust earthquakes, the great velocity induces very high strain-rate with the consequence of increasing the yield stress, reducing the ductility and producing the brittle fracture of members or joints at the first or second cycle. The reduction is brittle.

-   In case of cycle loadings (Fig. 10.42b), characteristic of near-source strike-slip, intraslab earthquakes and soft soils, an accumulation of plastic deformation occurs, producing a degradation of the structural behavior. The reduction is ductile.

Figure 10.43 presents a proposal for the values of the reduction factor q, in function of the earthquake type. This proposal refers to framed structures and normal soil conditions; for other structure types and bad soil conditions, the values can be modified accordingly. Only 1-2 pulses and short durations for thrust sources, 3-4 pulses and moderate durations for strike-slipe sources characterize the interplate crust near-source earthquakes. As a consequence, the seismic energy dissipation is very reduced for thrust sources and reduced for strike-slip sources.

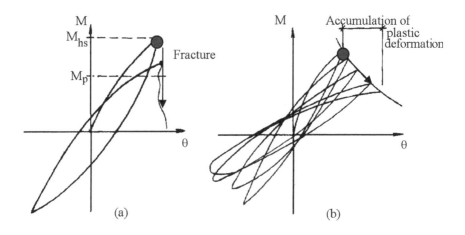

Figure 10.42 Member behavior under seismic actions: (a) Pulse seismic loading; (b) Cyclic seismic loading (Gioncu and Mazzolani, 2002)

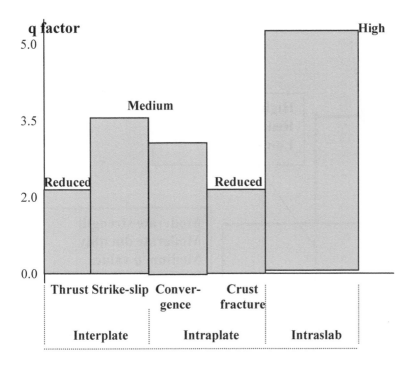

Figure 10.43 Proposed *q* factors in function of earthquake type

Consequently, reduced and medium values are proposed for the *q* factor. In case of crustal intraplate, for convergence sources, where both thrust and strike-slip earthquakes can occur, a medium reduction factor is proposed. For crust fracture with 1-2 pulses and very short duration, a reduced *q* factor is proposed. Contrary, a higher value for the reduction factor is proposed for intraslab earthquakes, characterized by a large number of cycles and long duration. Consequently, the design philosophy concerning strength and ductility must be different in function of the earthquake type (Fig. 10.44).

It is desirable to design structures in such a way that they behave in a known and predictable manner. Generally, the ultimate response can be as follows:

- Due to the reduced ductility and low *q* factor values (the case of thrust and crust fracture earthquakes), the structure must develop a large strength. The main design problem refers to the protection of structural members against brittle fracture produced by the first or second cycle due to the effect of strain-rate.

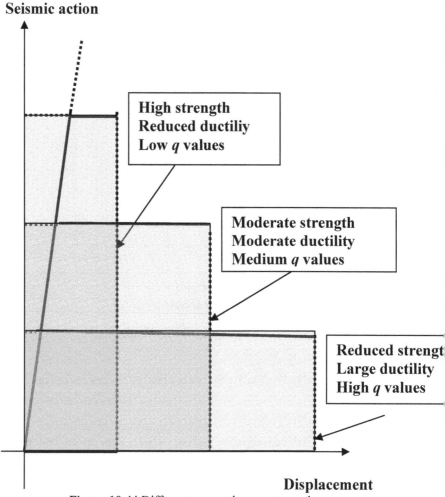

Figure 10.44 Different conceptions on strength,
ductility and q reduction factor values

- For moderate ductility and medium q factor values (the case of strike-slip or convergence earthquakes), the member fracture occurs after a reduced number of pulses. The design problem is related to the protection against this collapse type.
- For large ductility and high q factor values (the case of intraslab earthquakes), the main problem is the protection of structural members against the collapse due to the accumulation of plastic deformations.

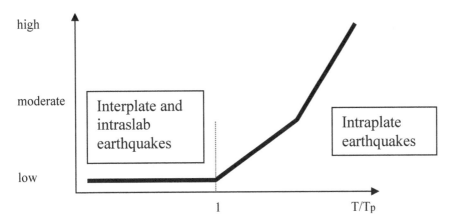

Figure 10.45 Effects of superior vibration modes

Another very important aspect in seismic design is the *effects of superior vibration mode,* producing a changing of seismic forces distribution in comparison with the distribution corresponding to the first vibration mode. The examples presented in Sections 10.5 and 10.6 have shown that these effects are very important only when the first natural period of structure, T, is higher than the pulse period of ground motions, Tp (Fig. 10.45). There are three different situations, in function of the earthquake types:

-   For *interplate crustal earthquakes*, the pulse periods are generally longer than the structure first period, so, the effects of superior vibration modes are reduced. But, if a high value of *q* factor is used (not recommended case), the structure results very flexible and the first natural period can exceed the pulse period, and consequently, the effect of superior vibration modes increases.
-   For *interplate crustal earthquakes,* the ground motion pulse periods are generally (especially in case of crust fractures) shorter than the first natural period of the structure. Therefore, practically in all cases, the effects of superior modes must be considered. The consideration of superior vibration modes reduces the plastic moments in the frame, having favorable effects. Taking account the reduced values of the involved magnitudes, it is possible to use a simplified design methodology, by considering the verification of serviceability limit state only, the other limit states being automatically satisfied.
-   For *intraslab earthquakes*, the influence of superior vibration modes can be neglected, the first mode being dominant. Only for very unusual flexible structures can these effects produce some changing in seismic forces distribution.

Regarding the *effects of vertical components* in near-source areas, they are very important only for the interplate crustal earthquakes, especially for normal-thrust sources.

10.9 REFERENCES

Aagaard, B., Hall, J., Heaton, T. (2001): Characterization of near-source motions with earthquake simulations. Earthquake Spectra, Vol. 17, No. 2, 177-207

Aktan, A.E., Ellingwood, B.R., Kehoe, B. (2007): Performance-based engineering of constructed systems. Journal of Structural Engineering, Vol. 133, No. 3, 311- 323

Alavi, B., Krawinkler, H. (2001): Effects of near-fault ground motions on frame structures. Blume Earthquake Engineering Center, Department of Civil and Environmental Engineering, Stanford University, Report No. 138

Anastasiadis, A., Gioncu, V., Mazzolani, F.M. (2000): Toward a consistent methodology for ductility checking. Behavior of Steel Structures in seismic Areas (eds. F.M. Mazzolani and R. Tremblay), Montreal, 21-24 August 2000, Balkema, Rotterdam, 443-453

Azevedo, J., Guerreiro, L. (2007): Current status of seismic isolation and energy dissipation R&D and applications for buildings, bridges and viaducts, in Portugal. 10th World Conference on Seismic Isolation, Energy Dissipation and Active Vibrations, Control of Structures, Istanbul, 28-31 May 2007

Barnes, G. (2003): Origins of the Japan islands: The new "Big Picture", Japan Review, Vol.15., 3-50

Bommer, J.J., Martinez-Fereira, A. (1996): The prediction of strong motion duration for engineering design. 11th World Conference on Earthquake Engineering, Acapulco, 23-28 June 1996, Paper No. 84

Bonneville, D.R., Bachman, R.E. (2002): The evolving process of US seismic code development. 7th National Conference on Earthquake Engineering, 7NCEE, 21-25 July 2002, Boston, USA

Carydis, P.G. (2005): The effects of the vertical earthquake motion in near field. Earthquake Engineering in the 21st Century, EE-21C, Skopje-Ohrid, 27 August-1 September 2005

Chandler, A.M., Hutchinson, G.L., Wilson, J.L. (1992): The use of interplate derived spectra in intraplate seismic regions. 10th World Conference on Earthquake Engineering, Madrid, 19-24 July 1992, 5823-5827

Chandler, A.M., Lam, N.T.K., Wilson, L., Hutchinson, G.L. (2001): Response spectrum modelling for regions lacking earthquake records. Electronic Journal of Structural Engineering, Vol. 1, No. 1, 2-14

Chang, S.P., Koh, H.M., Kim, J.K.(2001): Development of new seismic design concepts considering moderate seismicity in Korea. Proceedings of China-US Workshop, Beijing

Chouw, N. (2002): Reduction of the effects of strong vertical ground motions on structural responses. 7th National Conference on Earthquake Engineering, 7 NCEE, 21-25 July 2002, Boston

De Buen, O. (1996): Earthquake resistant design. A view from the practice. 11th World Conference on Earthquake Engineering, Acapulco, 23-28 June 1996, Paper No. 2002

Diebold, J., Moore, K., Hale, T., Mochizuki, G. (2008): SEAOC Blue Book: Seismic design recommendations 1959 to 2008. 14th World Conference on Earthquake Engineering, Beijing, 12-17 October 2008

DRAIN 2D (1975) A general purpose computing program for the dynamic analysis of inelastic plane structures with user's guide. Report 73-6 and 73-22, University of California

Elnashai, A.S., Papazoglou, A.J. (1997): Procedure and spectra for analysis of RC structures subjected to strong vertical earthquake loads. Journal of Earthquake Engineering, Vol. 1, No. 1, 121-155

Elnashai, A.S., He, L., Elgamal, A. (2004): Spectra for vertical earthquake ground motions. 13th World Conference on Earthquake Engineering, Vancouver, 1-6 August 2004, Paper No. 2309

Esteva, L. (2005): Earthquake engineering for seismic disaster mitigation in the 21st Century. World Conference on Disaster Reduction, Kobe, January, 2005

Fajfar, P., Krawinkler, H. (1997): Preface. Seismic Design Methodologies for the Next Generation of Codes. Bled, 24-27 June 1997, Balkema, Rotterdam, IX-X

Fardis, M.N. (2004): Current development and future prospects of the European code for seismic design and rehabilitation of buildings: Eurocode 8. 13th World Conference on Earthquake Engineering, Vancouver, 16 August 2004, Paper No. 2025

FEMA 267 (1997): Interim Guidelines Advisory. Report No. SAC-96-03

Gao, M. (2003): New national seismic zoning map of China. 1999-2002 China National Report on Seismology and Physics of Earth's Interior, Sapporo, 30 June-11 July 2003

Ghosh, S.K. (2002): Seismic design provision in US codes and standards. A look back and ahead. PCI Journal, January-February, 94-99

Gioncu, V. (2000): Influence of strain-rate on the behavior of steel members. Behavior of Steel Structures in Seismic Areas, STESSA 2000 (eds. F.M. Mazzolani and R. Tremblay), Montreal, 21-24 August 2000, 19-26

Gioncu, V. (2006): Advances in seismic codification for steel structures. Costruzioni Metalliche, No. 6, 69-87

Gioncu, V. (2008): Structural design problems in the low-to-moderate seismic regions. Part 1: Seismic risk and hazard; Part 2: Building's vulnerabiliy Seismic Risk. Earthquakes in North-Western Europe. Liege, 11-12 September 2008, 249-260, 261-272

Gioncu, V., Mateescu, G. (1997): Comparative parametrical studies concerning the structure response for frames subjected to Banat and Vrancea ground motions (in Romanian), INCERC Timisoara, Report A84/97

Gioncu, V., Mateescu, G. (1998a): Influence of vertical movements on behavior of steel structures (in Romanian) INCERC Timisoare Report A93/98

Gioncu, V., Mateescu, G. (1998b): Study concerning the displacement criterium for Banat ground motions (in Romanian), INCERC Report A92/98

Gioncu, V., Mateescu G. (1998c): Influence of velocity on ductility of steel structure (in Romanian). INCERC Report A93/98

Gioncu, V., Mateescu, G. (1999): Influence of the type of seismic ground motions. Inco Copernicus Recos Project, Final Report, INCERC Timisoara

Gioncu,, V., Mateescu, G. (2000): Methods to verity the strength and displacements of steel structures situated in zones with surface or deep earthquakes (in Romanian). INCERC Report, C6183/2000

Gioncu, V., Mateescu, G., Tirca, L., Anastasiadis, A. (2000): Influence of the type of seismic ground motions. Moment Resistant Connections of Steel Frames in Seismic Areas. Design and Reliability (ed. F.M. Mazzolani), E&FN Spon, London, 57-92

Gioncu, V., Mazzolani, F.M. (2002): Ductility of Seismic Resistant Steel Structures. Spon Press, London

Gioncu, V., Mazzolani, F.M. (2006): Influence of earthquake types on the design of seismic-resistant steel structures. Part 1: Challenges for new design approaches. Part 2: Structural response for different earthquake types. Behavior of Steel Structures in Seismic Areas. STESSA 2006 (eds. F.M. Mazzolani and A. Wada), Taylor & Francis, London, 113-120, 121-127

Gioncu, V., Mazzolani, F.M. (2009): Progress and challenge in the ductility control of steel structures. Behavior of Steel Structures in Seismic Areas, STESSA 09 (eds F.M. Mazzolani and J.M. Ricles), Philadelphia, 16-20 August 2009 (in press)

Goel, R.K., Chopra, A.K. (2005): Role of higher-"mode" pushover analyses in seismic analysis of buildings. Earthquake Spectra, Vol. 2, No. 4, 1027-1041

Hall, J., Heaton, T., Halling, M., Wald, D. (1995): Near-source ground motion and its effects on flexible buildings. Earthquake Spectra, Vol. 11, No. 4, 569-605

Hamburger, R.O. (2003): Building code provisions for seismic resistance. Earthquake Engineering Handbook (eds. W.F. Chen and C. Scawthorn), CRC Press, Boca Raton

Hutchinson, G.L., Wilson, J.L., Lam, N. (1996): Structural response under intraplate conditions. AEES Proceedings, Paper No. 5, 5.1-5.5

Hwang, H. (2000): Comments on design earthquake specified in the 1997 NEHRP provisions. 12th World Conference of Earthquake Engineering, Auckland, 30 January-4 February 2000, paper 0657

IBC (2006): International Building Code

Ifrim, M., Macavei, F., Demetriu, S., Vlad, I. (1986): Analysis of degradation process in structures during the earthquakes, 8th European Conference on Earthquake Engineering, Lisbon, 8/65-8/72

Ishiyama, Y. (2000): Revision of International Standards ISO 3010 basis for design of structures. Seismic actions on structures. 12th World Gonference of Earthquake Engineering, Auckland, 30 January-4 February 2000, Paper 1772

Kircher, C., Luco, N., Whithaker, A. (2008): Introduction to SDPRG and changes proposed for 2009 NHRP proposions. FEMA project'7, Seismic Design Procedures Reassessment Group

Krawinkler, H. (1995): New trends in seismic design methodology. 10th Eurpean Conference on Earthquake Engineering, Vienna, 28 August-2 September 1994, Balkema, Rotterdam, 821-830

Krawinkler, H. (1999): Challenges and progress in performance-based earthquake engineering. International Seminar on Seismic Engineering for Tomorrow (in honor of Professor Harashi Akiyama), Tokyo, 26 November.

Krawinkler, H., Alavi, B. (1998): Development of improved design procedures for near fault ground motions. SMIP98 Seminar on Utilization of Strong-Motion Data, Oakland, 15 September 1998, 21-42

Kuramoto, H (2006): Seismic design codes for buildings in Japan. Journal of Disaster Research, Vol. 1, No. 3, 341-356

Landolfo, R. (2005): L'evoluzione della normative sismica. Costruzioni Metalliche, No. 1, 54-66

Leelataviwat, S., Goel, S.C., Stojadinovic, B. (1999): Toward performance-based seismic design of structures. Earthquake Spectra, Vol. 15, No. 3, 435-461

Leyendecker, E.V., Hunt, R.J., Frankel, A.D., Rukstales, K.S. (2000): Development of maximum considered earthquake ground motion maps. Earthquake Spectra, Vol. 16, No. 1, 21-40

Li, Y., Lai, M. (2000): Several suggestions for the revision of code for seismic design of buildings, GBJ11-89 of China. 12th World Conference of Earthquake Engineering, Auckland, 30 January-4 February 2000, paper 1818

Mahoney, M. (2002): The world earthquake building codes, standards and provisions, A view according to FEMA. 7th National Conference on Earthquake Engineering, 7NCEE, 21-25 July 2002, Boston, US

Marcellini, A., Daminelli, A., Pagani, M., Riva, F. (1998): Seismic hazard of the Mediterranean area. 11th European Conference of Earthquake Engineering, Paris, 6-11 September 1998, TS2, 269-293

March, M.L., Gianotti, C.M. (1994): Structural response to long-duration earthquakes. Washington State Department of Transport, WA-RD 340.1 Final Report

Mateescu, G., Gioncu, V. (2000): member response to strong pulse seismic loading.. Behavior of Steel Structures in Seismic Areas, STESSA 2000 (eds. F.M. Mazzolani and R. Tremblay), Montreal, 21-24 August 2000, 55-62

Midorikawa, M., Okawa, I., Iiba, M., Teshigawara, M. (2003): Performance-based seismic design code for buildings in Japan. Earthquake Engineering and Engineering Seismology, Vol. 4, No. 1, 15-25

NEHRP (2003): Recommended provisions for seismic regulations for new buildings and other structures. Building Seismic Safety Council

NEHRP (2008): Strategic program for the national earthquake hazards reduction program (2009-2013). FEMA, NIST, USGS report 2008

Nishioka, T., Mualchin, L. (1997): Deterministic seismic hazard map of Japan from inland maximum credible earthquake for engineering. Structural Engineering /Earthquake Engineering, JSCE, Vol. 14, No. 2, 139-147

Oliveira, C.S., Campos-Costa, A., Sousa, M.L. (2000): Definition of seismic action in the context of EC-8, topics for discussion. 12th World Conference on Earthquake Engineering, Auckland, 30 January-4 February 2000, Paper No. 2552

Otani, S. (2004): Japanese seismic design of high-rise reinforced concrete buildings: An example of performance-based design code and state of practices. 13th World Conference on Earthquake Engineering, Vancouver, 1-6 August 2004, Paper 5010

Papazoglou, A.J., Elnashai, A.S. (1996): Analytical and field evidence of the damaging effect of vertical earthquake ground motion. Earthquake Engineering and Structural Dynamics, Vol. 23, 1109-1137

Park, J., Fenves, G.L., Stojadinovic, B. (2004): Spatial distributionof response of multi-story structures for simulated ground motions. 13th World Conference on Earthquake Engineering. Vancouver, 1-6 August 2004, Paper No. 1545

Perago, A. (2005): La Nuova Normativa Antisismica degli Edifici. Maggioli Editore

Petroski, H. (1985): To Engineer is Human. The Role of Failure in Successful Design. MacMillan, London

Popov, E.P. (1991): U.S. seismic steel codes. Engineering Journal, American Institute of Steel Construction, Third Quarter, 119-128

SAC (1996): Connection test summaries. Report No. SAC 96-02

Sato, T., Murono, Y., Wang, H.B., Nishimura, A. (2001): Design spectra and phase spectrum modeling to stimulate design earthquake motions: a case study through design standards of railway facilities in Japan. Journal of Natural Disaster Science, Vol. 23, No. 2, 89-100

Shen, J.W., Shi, S.Z. (2004): Large, moderate, small earthquakes and seismic fortification criterion. Acta Seismologica Sinica, Vol. 17, No. 5, 589-595

Somerville, P. (1998): Engineering characteristics of near fault ground motion. SMIP97, Utilization of Strong Motions Data, Seminar, Los Angeles, 8 May 1997, 9-28

Vannucci, G., Pondrelli, S., Argnani, A., Morelli, A., Gasperini, P., Boschi, E. (2004): An atlas of Mediterranean seismicity, Annals of Geophysics, Vol. 47, No. 1, 247-3006

Wang, Y. (2001): New code for seismic design of buildings in China. Institute of Earthquake Engineering of China, Academy of Building Research Report

Whittaker, A., Bertero, V.V., Wight, J., Higashino, M. (1997): Development of Japanese building seismic regulations. NISEE, University of California, Berkeley http://nisee.berkeley.edu/kobe/codes/html 2007

Zhang, D., Zhang, G., Zhang, P. (2003): Continental dynamics and continental earthquakes. 1999-2002 China National Report on Seismology and Physics of Earth's Interior, Sapporo, 30 June-11 July 2003

Appendix

Glossary

A.1 Engineering Seismology

A

Acceleration: Acceleration denotes the rate of change of the back and forth movement of the ground during an earthquake. It is commonly measured as a fraction of *g* (gravity constant on the Earth, 980 cm/sec/sec).

Acceleration time history: A paired dataset or series of acceleration values versus time.

Accelerogram: The record (or time history), showing acceleration of a point on the ground level or a point in a building as a function of time.

Active fault: A fault, which has moved during the recent geologic past and, thus, may move again in future.

Aftershocks: A series of smaller quakes following the larger shock of the earthquake. They can continue for a period of weeks, months or years.

Alluvium: The loose gravel, sand, silt, or clay deposited by streams.

Alpide (Alpine)-Himalayan belt: A mountain belt with frequent earthquakes, which extends from the Mediterranean region to the Pacific zone.

Amplification: Modification of the input bedrock ground motion by the overlying unconsolidated materials.

Amplitude: The size of the wiggles on an earthquake recording.

Ancient plate boundary (ancient fault): Boundaries and faults produced during the pre-Pangaea periods.

Artificial ground motion: The time history of a ground motion, calculated for engineering purposes by means of a deterministic or stochastic simulation.

Artificial seismogram: A computer-generated seismogram for models of seismic ground motion.

Asperity (barriers): Roughness on the fault surface subject to slip, being the area of a fault where the earthquake rupture usually begins.

Asthenosphere: The layer of the Earth below the lithosphere as a part of the upper mantle.

Attenuation: Decrease in amplitude and change in frequency content of the seismic wave with distance from the point where the fault rupture originated.

B

Basalt: Volcanic rock of basic composition typical of oceanic crust.

Basement: Harder and usually older rocks which underlie the main sedimentary softer and usually younger rocks.

Bedrock: Relatively hard, solid rock which commonly underlies softer rock, sediment, or soil.

Bifurcation: Bifurcation occurs in a dynamical system, when a small smooth change of a system parameter value causes a sudden qualitative change in its dynamical behavior.

Big one: The next great Californian devastating earthquake.

Blind fault: A fault which does not extend to the Earth's surface, so there is no evidence of it on the ground. It is buried under the uppermost layers of rock in the crust.

Block structure model: Dynamic theory which considers the equilibrium and bifurcation of the Earth's blocks (tectonic plates).

Body wave: A seismic wave which moves through the interior of the Earth, as opposed to the surface waves traveling near the Earth's surface. P and S waves are examples of it.

Butterfly effect: The great influence of small perturbation in the initial conditions for the behavior of nonlinear dynamic systems.

C

Chaos: A technical term referring to the irregular, unpredictable, and apparently random behavior of deterministic nonlinear dynamic systems, very sensible to small changes of initial parameters. The Theory of Chaos is widely accepted as being applicable to a variety of geophysical phenomena (i.e. tectonics and mantle convection).

Circum-Pacific ring (ring of fire): The zone surrounding the Pacific Ocean which is characterized by frequent and strong earthquakes and many volcanoes as well as high tsunami hazard.

Collision: A variation of subduction, where the subduction is destroyed and the continental margins enter into collision.

Collision zone: Zone where two continents collide as a result of continued subduction of oceanic plate. Continental collision is the cause of mountain building (orogeny).

Complexity theory: The analysis of large dynamical systems of elements, which interact in a complicated nonlinear fashion. Such systems sometimes organize themselves, often abruptly, resulting in a disaster (i.e. an earthquake).

Continental tectonics: A term used to include the large-scale motions, interactions and deformations of the continental lithosphere. It is often used in contrast to Plate tectonics.

Convergent plate boundary: The boundary between two plates which are moving toward each other.

Core: The interior part of the Earth, beginning at a depth of about 2900km. *Outer core* or upper zone of the Earth's core extends from a depth of 2900 km to

5100 km, presumed to be liquid. *Inner core* extends from a depth of about 5100 km to the center (6371 km) of the Earth, and is probably solid.

   Crust: The outmost layer of the Earth. The *oceanic crust* underlies the ocean basins, being 5 to 10 km thick and largely basaltic in composition. The *continental crust* underlies the continents, ranging in thickness from about 25 to 70 km and being granite in composition.

D

   Determinism: A philosophical proposition that every event, decision and action is causally determined by an unbroken chain of prior occurrences.

   Deterministic system: A dynamical system whose equations and initial conditions are fully specified, not stochastic or random.

   Diffuse zones: The seismic zones occurring along the plate boundaries or in areas between boundaries of tectonic plates, where the presence of faults is not very clearly defined.

   Digital recording: A series of discrete numerical digits for a recorded acceleration.

   Dip: The angle of a planar surface, such as a fault, in respect to the Earth's surface.

   Dip-slip: The up or down components of the slip parallel to the dip of the fault.

   Dip-slip fault: A fault in which the slip is predominantly in the direction of the dip of the fault.

   Directivity: An effect of a fault rupturing, due to an earthquake ground motion, which is more severe in the direction of the rupture propagation than in other directions from the earthquake source

   Displacement: The displacement of a point during an earthquake shaking, relative to the point at rest.

   Dissipative system: An open system which takes on and dissipates energy as it interacts with its environment

   Divergent plate boundary: The boundary between two plates which move apart.

   Duration: The time interval between the first and last peaks of strong ground motions. Also, it is a measure of time, when the ground motion exceeds a given threshold of shaking, such as 5 percent of the force of gravity.

E

   Earthquake: The sudden slip on a fault and the resulting ground shaking due to seismic waves and radiated seismic energy.

   Earthquake Architecture: An approach to the architectural design, which is based upon the Earthquake Engineering.

   Earthquake Engineering: A branch of Structural Engineering, defined as the engineering efforts to cope with the damaging effects of earthquakes, by developing specific methodologies for analyzing the effects of seismic actions on structures.

Earthquake prediction: The statement in advance of the event, predicting time, location and magnitude of a future earthquake.

Earth system science: A science which treats the entire Earth as a dynamic self-organized and synergistic physical system, which evolves as a result of positive and negative feedback among constituent systems (including the environment).

EASS: European Association for Earthquake Engineering.

ECEE: European Conference on Earthquake Engineering.

EMSC: European-Mediterranean Seismological Centre.

Engelados: The Giant who generates earthquakes, according to Greek mythology.

Engineering Seismology: A branch of Seismology, which uses seismological knowledge for the seismic design of buildings

Epicenter: The point on the Earth's surface directly above the focus (hypocenter) of an earthquake.

Epicentral distance: The distance in kilometers from a site to the epicenter of an earthquake.

Event: A general term used to represent an earthquake.

Evolutionary system: System that undergoes evolutionary processes.

F

Failure mode: The manner in which an earthquake produces a failure and the description of how the failure occurs.

Far-source (far-field): The field far from source.

Fault: A fracture or fracture plane in the Earth's crust along which the two sides are displaced relative to each other and parallel to the fracture.

Faulting mechanism: The type of faulting defined by the direction of the slip on the fault rupture plane; usually it is referred to by terms like strike-slip, reverse, thrust, normal or oblique.

Fault plane: A fracture surface along which blocks of rock on opposite sides have moved relative to each other.

Fault zone: The zone surrounding a major fault, consisting of numerous interlying small faults.

FEMA: Federal Emergency Management Agency.

Fling step: Occurs in the direction of the fault slip, normal to the fault plane.

Focal depth: The depth of the focus beneath the Earth's surface, commonly classed as shallow (0 to 70 km), intermediate (70 to 300 km) and deep (300 to 700 km).

Focus (hypocenter): The point at which the rupture occurs.

Focusing effect: A concentration of wave effects, due to the local surface profiles.

Footwall: The portion of the crust which lies below the fault or fault rupture plane.

Foreshocks: Relatively smaller earthquakes which precede the largest earthquake in a series, which end with the main shock.

Free-field: The ground area which is not influenced by the nearby buildings or other structures.

Frequency: The number of cycles or peaks of a periodic motion per second, or the inverse of the period. Usually measured in terms of hertz (1 Hz = 1 cycle per second).

G

g: A commonly used unit for acceleration due to the gravity of the Earth.

Geodesy (Geophysics): The study of the shape of the Earth as well as of the dynamical processes and forces leading to the observed shape

Geodynamics: The study of the dynamics of the Earth's interior, including heat transfer and convection current.

Geology: The science concerned with the study of Earth's materials (minerals and rocks), surface and internal processes as well as its history and life forms since its origin.

Geomorphology: The study of the character and origin of landforms, such as mountains, valleys, etc.

Geotechnical: The use of scientific methods and engineering principles to acquire, interpret, and apply knowledge of the Earth's materials to solve engineering problems. It portrays the specified ground characteristics in a given area.

GPS: Global Positioning System. This system allows extremely accurate position and time determinations anywhere on the Earth's surface.

Graben (rift valley): A down-dropped block of the Earth's crust resulting from extension or pulling of the crust.

Great (strong) earthquake: An earthquake with magnitude equal to or greater than M 8.0.

Ground failure: A permanent inelastic deformation of the ground resulting from an earthquake.

Ground motion: Movement of the Earth's surface due to waves which are generated by the sudden slip of a fault. It is measured by a seismograph which records acceleration, velocity or displacement.

Ground-motion parameter: A parameter characterizing a ground motion, such as peak acceleration, peak velocity, peak displacement, frequency contents, predominant period, magnitude and duration.

Gutenberg-Richter's relation: An empirical relation expressing the frequency distribution of magnitudes of earthquakes occurring in a given area and time interval.

H

Hanging wall: The portion of the crust which lies above the fault or fault rupture plane.

Horst: An upthrown block lying between two steep-angled fault blocks.

Hypocenter (focus): The point within the Earth where the rupture of the rocks initiates during an earthquake.

I

IAEE: International Association for Earthquake Engineering.

IBC: International Building Code.

Incoherence effect: Loss of coherence of seismic waves due to the scattering in heterogeneous medium of ground and superposition of waves arriving from different points of the source.

Intensity: A subjective measure of an earthquake at a particular place as determined by its effects on persons, structures and materials. It refers to the Modified Mercalli Intensity (with grades indicated by Roman numerals from I to XII).

Interplate earthquake: An earthquake which has its source in the Earth's crust at the interface between two tectonic plates.

Intraplate (inland) earthquake: An earthquake which has its source in the Earth's crust in the interior of a tectonic plate.

Intraslab earthquake: An earthquake which has its source in the slab, under the Earth's crust.

L

Large earthquake: An earthquake with magnitude ranges from M 6.6 to M 7.9.

Liquefaction: The sudden transformation of a granular material (soil) from a solid state into a liquefied state, as a consequence of the increased pore water pressure induced by vibration.

Lithosphere. The solid portion of the Earth in plate tectonics, including the oceanic and continental crusts and part of the upper mantle, being of the order of 100 km thick.

Local site conditions: A qualitative and quantitative description of the material properties of the soil and sedimentary rock layers above basement rock, which affects the ground motions during an earthquake.

Low earthquake: An earthquake with magnitude ranges from M 3.0 to M 4.5.

L wave: A wave trapped near the surface of the Earth and propagating along it, producing a sideways motion.

M

Magma: The molten rock which is found beneath the Earth's surface.

Magnitude: A measure of earthquake size which describes the amount of energy released, measured on a variety of scales: local magnitude (M_L), surface-wave magnitude (M_S), moment magnitude (M_w).

Mantle: The zone of the Earth below the crust and above the core.

Maximum probable earthquake: The maximum earthquake which could strike a given area with a significant probability of occurrence.

Mechanical processes: They produce earthquakes, due to loss of equilibrium based on the friction between two plates (in contrast with thermo-dynamical processes).

Micro earthquake: An earthquake with magnitude smaller than M 3.0.

Mitigation: Term used in Earthquake Engineering synonymous with reducing earthquake risk.

Moderate earthquake: An earthquake with magnitude ranges from M 4.6 to M 6.5

Multi-source data fusion: A new statistical analysis based on the fusion of data from different sources.

N

Namazu: A giant catfish who causes earthquakes, according to Japan mythology.

Near-source (near-field) site: The area around the epicenter, near the causative fault of an earthquake, where the dimensions refer to the source depth and the length of fault rupture.

NEHRP: National Earthquake Hazard Reduction Program.

Normal fault: A fault which exhibits dip-slip motion, where the two sides are in tension and move away from each other.

O

Oblique slip: A combination of strike-slip and normal or reverse slip.

Oceanic ridge: A line of ridges, formed by molten rock that reaches the ocean bottom, solidifies and topographically marks the fracture.

Oceanic trench: A narrow and elongated deep depression of the sea bottom caused by the subduction of one plate under another.

Orogeny: The process of forming mountains, especially by folding and thrust faulting in the collision zones: an episode of mountain building.

P

P wave: The primary and fastest waves traveling away from a seismic event through the crust and consisting of a train of compressions and dilatations of the material (push and pull). For this reason this wave is also called longitudinal wave.

Pangaea: A super continent, including most of the continental crust, which Alfred Wegener in 1912 postulated to have existed from about 300 to 200 million years ago. Pangaea is believed to have split into two proto-continents, Gondwana and Laurasia, which modern continents derived from, by means of continental drifts.

Path effect: The effect of the propagation path during seismic ground motion. It is implicitly assumed that source, path and site effects on ground motions are separable.

Peak ground acceleration (PGA): The maximum acceleration amplitude measured by a strong motion accelerogram during an earthquake.

Period: The length of time required to complete one cycle or a single oscillation of a periodic process. It is the inverse of the frequency.

Plate boundary zone: The zone of diffuse deformations, often marked by a distribution of seismicity and active faulting, where relative plate motions are accommodated.

Plate tectonics: A coherent theory and study of plate formation, movement, interaction and destruction, which explains seismicity, volcanism, and mountain building in terms of plate motions. There are major, minor and micro plates.

Precursor time: The time span between the onset of an anomalous phenomenon and the main shock.

Probability of exceedance: The probability that, in a given area or site, an earthquake ground motion will be greater than a given value during some previous period of time.

Pulse-type: A pulse of seismic action characterized by pulse amplitude, period and shape.

R

Rayleigh wave (R wave): A wave trapped near the surface of the Earth and propagating along it, producing forward and elliptic vertical surface waves.

Recurrence interval: The average time interval between consecutive events which occur repeatedly.

Return period: The average time between the exceedance of a specific level of ground motion in a specific location, equal to the inverse of the annual probability of exceedance.

Reverse fault: A fault which involves lateral shortening, where one block moves over another up the dip of the fault.

Richter magnitude: The magnitude of an earthquake based on the Richter magnitude scale.

Ridge: A geological form which features a continuous elevation crest.

Rift (rift valley): An extended feature marked by a fault-caused trough which is created in a crustal zone of divergent or other areas in tension.

Rupture velocity: The speed at which a rupture front moves across the surface of the fault during an earthquake.

S

S wave: Shear wave, essentially produced by shearing or tearing motions, perpendicular to the direction of wave propagation. Therefore, these waves are also called transverse waves.

Sedimentary rock: A rock formed from the consolidation of sediments.

Seismic design: Collects data from Engineering Seismology and uses the methodologies proposed by Earthquake Engineering to develop a complex examination of the structure behavior, including numerical analysis, structural conformation and solutions for constructional details.

Seismic gap: A section of a fault which has produced earthquakes in the past, but it is now quiet. For some seismic gaps, no earthquakes have been observed historically, but it is believed that the fault segment is capable of producing new earthquakes.

Seismic hazard: A probability of occurrence in a given area of a potentially damaging earthquake, at a severity level and at a specific period of time.

Seismic hazard map: A map showing the contours of a specified ground motion parameter for a given probabilistic seismic hazard or return period

Seismic network: A network of seismographs.

Seismic performance: Ability of structures to sustain their functions, such as its safety and serviceability, during and after a particular earthquake exposure.

Seismic risk: The probability that a specified loss of life or property will exceed some quantifiable level during a given exposure time.

Seismic station: A facility devoted to the measurement and recording of earthquake motions.

Seismic zoning map: A map used in building codes to identify specific areas of seismic design requirements.

Seismicity: The worldwide or local distribution of earthquakes in space and time; a
general term for identifying the number of earthquakes in a unit of time, or for the relative earthquake activity.

Seismograph: An instrument which detects and records ground motions along with timing information.

Seismology: A branch of Earth Science dealing with the study of earthquakes, the structure of the Earth and the mechanical vibration of the crust, caused by natural sources, such as earthquakes and volcanic eruptions.

Self-organized system: A system which changes its basic structure as a function of its experience and environment.

Significant pulse: Defined as the pulse having at least 50 percent of the maximum peak velocity.

Site category: The category of site geologic conditions affecting earthquake ground motions based on the geology, wave velocities, or other properties of the site soil.

Site classification: The process of assigning a site category to a site by means of geotechnical characterization of the soil profile.

Site effect: The effect of local geologic and topographic conditions at a recording site on ground motions. It is implicitly assumed that source, path, and site effects on ground motions are separable.

Site response: The modification of earthquake ground motion in the time or frequency domain, caused by local site conditions.

Slab: A subducted oceanic plate which under-thrusts the continental plate and is covered by the Earth's mantle; the intraslab deep earthquakes occur under the Earth's crust, in conditions of high pressure and temperature.

Slip: The relative displacement of formerly adjacent points on opposite sides of a fault, measured on the fault surface.

Statistical seismology: A new field which applies statistical methodologies to earthquake data in an attempt to predict the characteristics of the next earthquake.

Soft soil: A soil with a high degree of compressibility.

Soil profile: Vertical arrangement of the layers of soil.

Source depth: The depth of an earthquake hypocenter.

Source effect: The effect of earthquake source on ground motions, considering different source types.

Source-to-site distance: The shortest distance between an observation point and the source of an earthquake, which is represented as either a point or a rupture area.

Station: The site where seismic instruments (e.g. seismographs) have been installed for permanent observations.

Strain: Response of rock body to the applied stress

Stress: Action causing the earthquake.

Strike-slip fault: The slip parallel to the strike of the fault involving horizontal movements.

Subduction: The sinking of a plate under an overriding plate in a convergent zone.

Subduction thrust fault (thrust fault): The fault which accommodates the differential motion between the down-going oceanic crustal plate and the continental plate. This fault is the contact between the top of the oceanic plate and the bottom of the continental plate.

Subduction zone: A narrow zone of convergence of two plates, characterized by thrusting of one plate beneath the other.

Surface faulting: The faulting which reaches the Earth's surface.

T

Tectonic plate: A large and relatively rigid plate of the lithosphere which moves on the outer surface of the Earth.

Thermo-dynamical processes: Produce earthquakes at depth under conditions of high temperatures and pressures

Thrust fault: A reverse fault which dips with shallow angles.

Time history: An engineering term for a seismogram or time-dependent response.

Transform fault: A special strike-slip fault which accommodates relative horizontal slips between other crustal tectonic plates. It often extends from oceanic ridges.

Transform plate boundary: Plate boundary along which plates slide past each other and the crust is neither produced nor destroyed.

Travel time: The time taken for the seismic waves to propagate from one point to another along the ray.

Tsunami: A train of waves sent across the surface of the sea by a disturbance in the seabed, such as a submarine earthquake, landslide or volcanic explosion.

U

USGS : United States Geological Survey

V

Velocity: How fast a point on the ground is shaking as a result of an earthquake.

W

Wave length: The distance between successive similar points on two wave cycles (e.g., crest to crest or trough to trough).

Wave-passage effect: Differences in the arrival time of waves at different surface points.

WCEE: World Congress on Earthquake Engineering.

Weak crustal zone: Zones of low strength of the crust.

Wilson cycle: A successive recurrence of opening and closing ocean basins, due to plate motions on a timescale of about 100 million years.

A.2 Earthquake Engineering

A

Accidental torsion: Torsion induced in a structure by random unexpected irregularities: base rotation excitation and uncertainties in stiffness and masses

Active control system: Control of structural vibration using systems which produce forces in a prescribed manner.

Active mass damper (ADM): Active control which uses mass actuators, sensors and controllers, so that the combination of ADM and structure minimizes the vibration response.

AIJ: Architectural Institute of Japan.

B

Base isolation: Isolation of structures from ground shaking with special devices.

Base isolation system: A system in which the superstructure is isolated from the foundation by means of certain devices, which reduce the ground motions transmitted to the structure.

Base shear: The horizontal shearing force at the base of the structure. The maximum base shear, typically in the form of a fraction of the weight of the structure, is an important parameter in earthquake response studies and in earthquake-resistant design.

Behavior factor: Factor used for design purposes to reduce the forces obtained from a linear analysis, in order to account for the nonlinear response of the structure, associated with the structure capacity of seismic energy dissipation by the controlled local and global plastic mechanisms.

Braced frame: Essentially a vertical truss system of concentric or eccentric type provided to resist lateral forces.

Brittle: The inability to accommodate inelastic strain without loss of strength.

Buckling restrained braces: A brace conceived as a ductile brace core (also for compression forces) included in a filled tube and detailed so that the central brace can longitudinally deform independently of the material which restrains lateral and local buckling.

Building weight effect: Due to the building weight, the soil is overloaded, becoming stiff soil; this produces a change in the ground motions in this soil in comparison with a free soil.

C

Capacity curve: A relation between seismic base shear and peak relative displacement (typically the roof relative to the ground), transformed into the space of spectral displacement response and spectral acceleration response.

Capacity design method: Design method in which the elements of the structural system are chosen and suitably designed and detailed for energy dissipation under severe earthquakes, while all other elements are provided with sufficient strength so that the chosen energy dissipation can be maintained.

Capacity response spectrum: An acceleration spectrum format in which the period axis is converted in the displacement axis.

City-soil effect: In case of densely urbanized areas, a part of the vibrating energy of buildings is released into soil through the waves produced by the buildings themselves, significantly modifying the free-field ground motions, which are currently used in seismic design

Code of Hammurabi: The oldest known code containing specifications about the relation between owners and builders, including the responsibilities for safety by builders.

Concentrically braced system (CBF): A braced frame in which the members are primarily subjected to axial forces.

Confined concrete: Concrete is confined by transversal reinforcement.

Confined region: The portion of a reinforced concrete component where the concrete is confined by closely spaced special transverse reinforcements (spirals, closed stirrups or hoops), which restrain the concrete in direction perpendicular to the applied forces.

Conflagration fire: Large destructive fire, following an earthquake which destroys a complex of buildings.

Connections: Combination of joints used to transmit forces between two or more members (i.e. beams and columns).

Core system: Structural system consisting of flexible frames combined with a core (walls or braced frames) concentrated near the center of the building in plane.

CQC: The complete quadratic combination of modal response, based on random vibration theory.

D

Damage: Degradation of a building, due to earthquake effects. The degradation may affect the structural system, or the nonstructural components, such as partitions and exterior claddings, or the equipments, fixtures and furnishings.

Damper: Device added to a building to mitigate the response due to earthquake shaking.

Damping: Measure of the seismic energy dissipated by a structure.

Design code: A document containing standardized requirements for the design of most types of buildings.

Design earthquake ground motion: The earthquake effects which buildings and structures are specifically proportioned to resist.

Diaphragm: A horizontal system (roof, floor or other membrane or horizontal bracing) acting to transmit lateral forces to vertical-resisting elements.

Differential displacement: Displacement of adjacent support points.

Displacement response spectrum: The correlation between the displacement and structural period.

Dissipative structure: Structure which is able to dissipate energy by means of ductile hysteretic behavior and/or by other mechanisms.

Dissipative zone: Region of primary seismic elements, where plastic hinges may form.

Drift (ratio): The ratio of lateral inter-floor deflection to the height between the two floors involved. The allowable drift ratio under design loading is often prescribed in building codes.

Dual system: Structural system, in which the support to the vertical loads is mainly provided by a spatial frame and the resistance to lateral loads is contributed in part by the frame system and in part by structural walls, braced frames, infilled walls, etc.

Ductile: The ability to accommodate inelastic strain without loss of strength.

Ductile frames: Frames required to provide a satisfactory load-carrying performance under large deflections (i.e. ductility).

Ductility: The property of a structure or a structural component which allows it to continue to have significant strength after it has yielded or begun to fail. Typically, a well-designed ductile structure or component will show, up to a given point, an increasing strength as far as its deflection increases beyond yielding (for steel structures) or cracking (in case of reinforced concrete or masonry).

Ductility factor: The ratio of the maximum deflection or rotation (elastic plus inelastic) to the elastic (i.e. yield) one.

E

Eccentrically braced frame (EBF): A diagonal braced frame, in which at least one end of each bracing member connects the beam at a short distance from a beam-to-column connection or from another beam-to-brace connection.

ELF: Equivalent lateral force

Energy dissipating devices: Special devices, elements, or members installed in structures to provide energy dissipation.

Equipment: Mechanical, electrical or other components required for the system functionality.

Equivalent static analysis: A linear method available when one particular mode of vibration (mainly the first) is predominant as compared to the others; the structure is modeled by a SDOF system and an equivalent static lateral force can be used.

EUROCODES: A series of European standards providing a common approach for the design of buildings and other civil engineering works and construction products. Eurocode 8 deals with the design of structures for earthquake resistance.

F

Floor acceleration: Acceleration induced by an earthquake ground motion at the level of a building's floor.

Foundation: A substructure of the structural system, which transfers the actions from the building to the soil.

Fragility: The probability of a specific level of undesirable damage reaches a specified level of hazard.

Frame system: Structural system in which both vertical and lateral loads are mainly resisted by spatial frames.

Fully operational: The performance objective for which only very minor structural and nonstructural damage occurs.

G

Geometric irregularities: The set-back for the vertical configuration and the asymmetric plan-wise for horizontal configuration.

H

High strength steel: Steel used in structures for non-dissipative elements, which during an earthquake must remain in elastic range.

Hysteresis: A form of energy dissipation which is related to the inelastic deformation of a structure. A hysteretic loop is created by a force-deformation relationship to absorb and dissipate seismic energy.

I

IDA, Incremental dynamic analysis: Dynamic analysis of structures against collapse with incremental increase of the ground excitation intensity.

Inelastic response spectrum: A spectrum obtained by reducing the elastic spectrum, taking into account the structure capacity to dissipate seismic energy.

Infilled frame: A system in which the infilled elements (masonry or panel) play a remarkable contribution to the initial stiffness, reducing both drift and structural damage, especially for serviceability limit state.

Interstory drift: The displacement of the floor above a story relative to the floor below it.

ISO: International Standard Organization.

Isolation system: Collection of individual isolator units which transfer the ground motion forces from the foundation to the superstructure.

J

Joint: Area where two or more member ends, surfaces, or edges are attached by fasteners or welds.

L

Lateral force resistant system: A structural system for resisting horizontal forces, i.e. due to earthquake or wind (as opposed to the vertical force resisting system, which provides support just against gravity).

Life safe: The performance objective, for which significant structural and non-structural damage can occur, but the structure can be repaired. The risk of life threatening injury during the earthquake is low. The service interruption is less than three months.

Link: In EBF, it is the segment of a beam, which can be located between the end of a diagonal brace and a column or between the ends of two diagonals.

Low yield steel: Quality of steel suitable to be used in dissipative elements.

M

MDOF system: Multi-degree-of-freedom system.

Modified Mercalli Intensity (MMI): Qualitative description of earthquake intensity, with discrete levels from I to XII.

Moment frame: A building frame system, in which seismic shear forces are resisted by shear and flexure in members and joints of the frame.

Multi-level approach: A methodology developed to consider the different limit states for structural and nonstructural elements.

N

Natural mode: A characteristic dynamic property of a structure, when it freely oscillates.

Natural period: The time interval of each oscillation in a vibration system in free vibration.

Near collapse: The performance objective for which substantial damage, but not collapse occurs. The complete failure of nonstructural elements occurs. If experts decide the building can be repaired, the service interruption should be longer than three months. But in many cases, the repair is not practically convenient at this stage and the building must be demolished.

Nonlinear analysis: A calculation methodology, which includes material changes, such as yielding or cracking, and/or the nonlinear effects of large displacements, such as the increased base moment caused by lateral displacements or floor masses.

Nonlinear dynamic (time-history): Both seismic excitation and structural response variations are evaluated in time at many discrete points. The geometrical or material nonlinearities are accounted for. A set of carefully ground motion records or artificial generated time-history representations can give an accurate evaluation of the seismic performance of structures

Nonlinear geometry: Large displacement in the deformed structure configuration.

Nonlinear response: Non-linear relation between displacements, global degrees of freedom and the corresponding resisting forces.

Nonstructural elements: Architectural (partitions, exterior claddings, lighting fixtures, etc), mechanical or electrical elements, systems and components, which, whether due to lack of strength or to the way of connection to the structure, are not considered in the seismic design as load carrying elements.

O

Operational: The performance objective for which only minor damage occurs.

Overturning moment: The moment experienced at the base of the structure during an earthquake.

P

Participation factor: A relation between the maximum displacement of a MDOF system and a SDOF system.

Passive control: Control of the structural vibration by using hardware system alone, without sensors or control signal.

P-delta effect: Secondary effect of vertical forces on the laterally displaced structure, with important increasing of column axial loads, shear, moment and lateral deflection.

Performance-based design: The integrated effort of design, construction and maintenance needed to produce engineering facilities of predictable performance for multiple performance objectives.

Performance-based earthquake engineering: A seismic design or analysis approach, in which the behavior of a facility is characterized in the terms of functionality, safety, repair cost and restoration time, as alternative to the behavior of individual structural members or connections.

Performance level: A level of desired and quantifiable performance.

Performance objective: Expression of a desired performance level for a given level of earthquake ground motion.

PGA: Peak ground acceleration.

Pile foundation: This foundation type is used when the firm soil is at a considerable distance below the natural level of the site.

Plastic hinge: Location where a focused inelasticity occurs in a structural member.

Plastic rotation: The level of measured rotation of plastic hinge which occurs due to local yielding.

Pounding: The collision of adjacent buildings during an earthquake due to insufficient lateral seismic separation.

Progressive collapse: A structural failure which is initiated by localized damage and subsequently develops, as a chain reaction, into a failure involving a major portion of the structural system.

Push-over analysis (nonlinear static analysis): A step-by-step static nonlinear analysis (under constant gravity load), used to estimate the capacity of a structure to resist strong ground motions until collapse. Typically used in earthquake-resistant design studies, the analysis employs a selected lateral force profile which is increased in intensity until the analysis indicates that the structure is susceptible to collapse.

Push-over modal analysis: The analysis is based on the evaluation of the response for each significant mode (generally the first three modes) and the actual response is determined by using the SRSS procedure

Push-over curve: Relationship between seismic base shear and roof peak relative displacement

Push-over multi-modal pattern analysis: The analysis is performed using a pattern of lateral loads determined by a combination of the most important vibration modes

Push-over adaptive analysis: In comparison with modal analysis, the adaptive analysis considers the modification of both load pattern and structural rigidity at each step.

R

Randomness: Inherent variability of physical phenomenon.

Reduced beam section (RBS, dog-bone): Reduced beam section allows shifting the plastic hinge from the beam ends.

Redundancy: The capability of a structure to continue to carry loads after the damage or failure of one of its members, by developing new alternative load paths.

Residual deformation: The remaining deformation after the unloading of a structure.

Resonance: A tendency of a system to oscillate with large amplitude at some frequency than at others

Response spectrum: A set of maximum amplitudes (acceleration, velocity or displacement) of a single-degree-of-freedom oscillator (SDOF) subjected to a particular input motion, as the natural period of the SDOF is varied across a spectrum of engineering interest (typically, for natural periods from 0.03 to 10.0 seconds). The concept of response spectrum is used in building codes and in design methodologies.

Response spectrum analysis (multi-modal response method): Decomposition of linear dynamic response in eigenvector contributions, determining the maximum response of each mode with reference to the response spectrum and computing the final response by superposing these modes and using one combination mode (sum of absolute values, SRSS or CQC methods).

Robustness: The structural quality of being able to withstand unpredictable seismic loads with minimal damage.

S

SDOF system: Single-degree-of-freedom system.

SEAOC: Structural Engineers Association of California.

Seismic separation: The distance between two adjacent buildings, calculated to avoid the pounding of these buildings during an earthquake.

Shallow foundation: A type of foundation, which transfers the building loads to the soil near to surface, including spread shallow foundations and mat foundations.

Shape-memory alloy: An alloy, which remembers its initial shape after large deformations.

Soft story: A story of a building significantly less stiff than the adjacent stories (i.e. lateral stiffness is 70 percent or less than that in the storey above, or less than 80 percent of the average stiffness of the three stories above).

Soil dynamics: A branch of Soil Mechanics, which deals with the behavior of soil and foundation under dynamic loads.

Soil-foundation-structure interaction: A term applied to the consequences of the deformation and forces induced into the soil by the movement of foundation and structure. The common fixed-base assumption in the analysis of structure implies no soil-structure interaction.

Special moment-resisting frame: A frame structure consisting of beams and columns, which are rigidly connected. The bending moment can be transferred by connecting joints.

Spectral acceleration response: The maximum absolute acceleration experienced by an elastic SDOF system of specified period and damping, for a given earthquake.

Spectral displacement response: The maximum displacement relative to the ground of an elastic SDOF system of specified period and damping, for a given earthquake

Spectral velocity response: The maximum absolute velocity of an elastic SDOF system of specified period and damping, for a given earthquake.

Spectrum amplification factor: The ratio of a response spectral parameter to the ground motion parameter (acceleration, velocity or displacement).

SRSS: The square root of the sum the squares of forces and displacements.

Stiffness: Property of a structure to resist displacements.

Story drift: The difference of horizontal deflection at the top and bottom of the story.

Story drift ratio: The storey drift divided by the story height.

Strength: Property of a structure to resist forces.

Structural analysis: The calculation of structural deformations, internal member forces and structural displacements, for given external loading conditions.

Structural configuration: Related to dimension, form, geometric proportion and location of structural components, determined in such a manner to have a proper distribution of seismic loads.

Structural response: The set of deformations, displacements, internal member forces in a structure subjected to specified loads.

Superstructure: The portion of a structure above the surface of the ground.

T

Time-history analysis: The method is based on the direct numerical integration of the motion differential equations.

Toughness: The ability of material to absorb energy without losing significant strength.

U

UBC: Uniform Building Code.

Unreinforced masonry: The type of construction which employs brick, stone, clay tile, or similar materials, without steel bars or other strengthening elements.

V

V- braced frame: A CBF in which a pair of diagonal braces is connected to a single point within the clear beam span.

Vibration mode: Characteristic pattern of shape, in which the structure vibrates during an earthquake. There are many modes of vibrations and the actual vibration of a structure is always a combination of all vibration modes. But they need not all be excited at the same degree; there is just a reduced number, which is amplified due to the resonance effect.

Vulnerability: Expected damage corresponding to a specific value of a hazard parameter.

X

X- braced frame: A CBF system in which a pair of diagonal braces crosses near the mid-length of braces.

A.3 References

Aki, K., Lee, W.H.K. (2003): Glossary of interest to earthquake and engineering seismologists. International Handbook of Earthquake and Engineering Seismology (eds. W.H.K. Lee, H. Kanamori, P.C. Jennings, C. Kisslinger). Academic Press, Amsterdam, 1793-1856

Bozorgnia, Y., Bertero, V.V. (eds) (2004): Earthquake Engineering. From Engineering Seismology to Performance-Based Engineering. CRC Press, Boca Raton.

Chen, W., Scawthorn, C. (eds) (2003): Earthquake Engineering Handbook. CRC Press, Boca Raton

Index

Earthquake Engineering for Structural Design
Page numbers in *Italics* represent tables.
Page numbers in **Bold** represent figures.

Aagaard, B.T. 265; *et al* 266
Abruzzo region (Italy) 65
accelerations 281; rock *vs.* soil **240**
Added Damping And Stiffness Elements (ADAS) systems 356
Adria microplate 162, 194; Eastern zone 194
Africa and South America jigsaw 106
African seismic zones 211
Agricola 104-5
AISC **350**
Aktan, A.E.: *et al* 451-2
Alaskan coast 457
Alaskan earthquake (1964) 42-3, 45, 243; Cooper River bridge damage **46**; epicenter **45**; tsunami 45, **45**
Alavi, B.: and Krawinkler, H. 478-9, 483-4
Aleutian-Alaskan arc **186**
alluvial basins 242-3, **243**, 296-7
Alpide (Alpine)-Himalayan belt 173-5, **174**; Anatolian segment 174; Arabian segment 174; Azores-Maghrebian segment 174; Calabrian Arc 174; Hellenic Arc 174; Himalayan segment 174; Indonesian segment 175; New Zealand 189, **190**; Zagros-Makran segment 174
Alps **162**, 163
American continents: subduction 184
American Institute of Steel Code (1990) 458
amplification: building structure **310**; urban field 297

analysis methods 82, **521**
Anastasiadis, A.J.: and Klimis, N.S. 241-2
Anatolian segment (Alpine-Himalayan belt) 174
Andean Valley 45
animal behaviour: earthquake prediction 135
Annan, K. 6-7
Apennines Mountains 162-3, **163**
Applied Technology Council (ATC) 458
Arabian segment: (Alpine-Himalayan belt) 174
architect 21, 23
architectural demands 368
Architectural Institute of Japan (AIJ) 461; Japanese regulations (1998) 462; Structural Standards 461
Aristotle 104
artificially generated pulses 481, 484, 486
asthenosphere 143
astronomy 13
Australia: elastic response spectra 496; seismicity 204, **204**, 211
Azores-Maghrebian segment (Alpine-Himalayan belt) 174

Bacon, F. 105
Bam (Iran) earthquake (2003) 9; citadel 60, **61**
Banat (Romania) earthquake (1991) 278, **279**, **499**

Barcelona, Hotel de Las Artes (bracing system) **24**
Beijing, China 46
Belgium: seismicity   199; seismotectonic map **201**
Benfield Greig Hazard Research Centre (BGHRC) **202**
Bertero, B.B. 74
Bifurcation Theory 120, 125; evolution in cascade 120, **121**; fluctuations 120
Theory   of   Block   Dynamical Equilibrium 124
block   structure: equilibrium   122-3; lithosphere   dynamics   123;   models 120-4;   South   America   Plate   **122**; Vrancea   (Romania)   subduction   zone **123**
Blue Book (SEAOC) 458
Bokor, L.: continental rifting cycles 110;   future   world   **112**;   Wegener's Earth **108**
boundary element model (BEM) 296
boundary plate contacts 77
braced frames 348, **348**; buckling 348-9,   **349**;   cyclic   loading   348; dismountable bolted link concept **351**; dissipation systems 356; eccentrically 349, **350**; infilled 351-2, **352**; links 349, **350**; masonry infilled reinforced concrete frames 352; panel infilled steel frames 352, **353**; panel types **354**
bracing   systems   23-6;   ALCOA Building   (San   Francisco)   **23**; apartment building (California) **25**; Hearst Tower (New York) **24**; Hotel de Las Artes (Barcelona) **24**; John Handcock Building (Chicago) **23**; Tokyo   Institute   of   Technology (Japan) **24**; University Hall in Berkeley (California) **26**; Yokohama Institute of Technology (Japan) **25**
British Geological Survey (BGS): UK earthquakes **200**
Bucharest 136
Bucharest earthquake (1977) **237**
buckling 87

Building Code of Japan 238
Building   Officials   and   Code Administration (BOCA) 458
building   structures   367;   accidental torsion   374-5;   base   rotational excitation   375;   configurations   369, **369**;   conformations   **369-70**; geometric   irregularity   368;   mass irregularity   372-3,   **373**;   pilotis configuration   371;   pounding   of adjacent buildings 373, **374**; soft story 371-2, **372**; story stiffness 371; story strength 371; wave passage **376**; weak story 371, **371**
buildings 9; collapses 327; contents 88, **89**; cultural heritage 90; density 296-7;   design   21;   modern   88;   partition damage 88; strengthening 11; Turkey 57; well-structured 83
Burridge and Knopoff 128
Burridge-Knopoff model **128**

Calabrian Arc (Alpine-Himalayan belt) 174
California: apartment building (bracing system) **25**; earthquakes **188**; seismic zone 157, **159**; University Hall in Berkeley (bracing system) **26**
Californian segment (Eastern Pacific plate boundary) 172
Caltech Center 265
Canadian   segment   (Eastern   Pacific plate boundary) 171
Cardan 105
Carrubba, P.: *et al* 300
Central   American   segment   (Eastern Pacific plate boundary) 172-3
Central American seismic zones 209-10, **210**
Central Mississippi Valley **170**
Chaos Theory 124-32; application in Seismology 128-32; bifurcation 125; Burridge-Knopoff   model   **128**; butterfly effect 125; conclusions 132; deterministic   chaos   126;   linear equations of equilibrium 126-7; linear

vs. non-linear systems 125, **126**; non-linear behaviour (double pendulum) **127**

Theory of Chaotic Dynamics 125

Charleston zone 198

Chi-Chi (Taiwan) earthquake (1999) 57, 252, **252**, **258**, 474-5; collapse of RC buildings **58**; epicenter **58**

Chicago, John Handcock Building (bracing system) **23**

Chile earthquake (1960) 42, **44**, 184; destruction 42, **44**; tremors 42; tsunami 42, **44**

Chile earthquake (2010) xii

China: ancient seismoscope **104**; ancient times 102; emperor Kangxi 105; Haicheng town earthquake (1975) 133; Qinghai province earthquake (2010) xii-xiii; records of earthquakes 104; seismic code *see* seismic design codes (China); seismicity 196, **197**, 198; Shaanxsi earthquake (1556) 6, 35; Tangshan earthquake *see* Tangshan (China) earthquake (1976)

Chopra, A.K.: and Goel, R.K. 507, **508**, **510**, **511**

Choudhary, D.: and Subba Rao, K.S. 303

Circum-Pacific Ring 171, **172**

Code of Hammurabi 451

collapse of corner buildings **84**

collision zones 193

Complex Non-linear Systems ix

Complex Systems 114

computer program: DRAIN 2D 484, 498; DUCTROT M 324; SHAKE 238; SOFIA (SOil Frame InterAction) 364, 366

computer technologies: development 75, 97

Conception (Chile) xii

conformation: non-structural elements 88; structural element 84

construction: poor 83; vulnerability 81-8

Theory of Continental Drift 109

continental lithosphere **107**

continental plate boundaries 171

continental plates (collision) 160, **160**; subduction 192-8

convection currents **32**

convergent plate boundaries 145, 147, 149; continental-continental 151, **151-2**, 160, **160**; India-Eurasia plates collision **152**, **161**; ocean-continent 149, **150-1**; ocean-ocean 149, **150**

Croatian earthquakes **195**

crust 33; continental 143; fracture 164-5, **164**, 169; mechanical process 180; oceanic 143; sheared **169**; tensioned **166**; weak crustal zones 198-204

crustal interplate earthquakes *see* interplate crustal earthquakes

crustal intraplate earthquakes *see* intraplate crustal earthquakes

crustal subduction earthquakes 183-4

damage costs 1

De Buffon, C. 105

dehydration: intraslab earthquakes 205

Della Corte, G.: *et al* 377

Der Kiureghian, A.: and Keshishian, P. 247-8

Derleth: C. 73-4

Descartes 105

design codes *see* seismic design codes

design earthquake: characteristics 15, 20-2

deterministic approach 137

developed countries: GDP 8-9; seismic problems 9; urbanization 4, **5**

developing countries: GDP 9; seismic problems 9; urbanization 4, **5**

Diebold, J.: *et al* 459

Dietz and Hess 109

displacement **516**; velocity 509

divergent plate boundaries 145, 147, **148**; continental 147; oceanic 147

DRAIN 2D (computer program) 484, 498

Dubina, D.: and Stratan, A. 349

ductility 18, 471; bolts 346, **347**
DUCTROT M (computer program) 324
Dynamic Earth **206**
Dynamical System 114

Earth 32; cross-section **119**; lifetime 32; seismic movements 32; surface ix, 118
Earth Expansion Theory 109
Earth Science 114-15; Complex Systems 114; Dynamical System 114; System Theory 115; Theory of Complexity 115
Earthquake Architecture 21-6
Earthquake Engineering Research Institute (EERI) USA 29
Earthquake Engineering x-xi, 11, 13, 28-9, 72, 468; basic concepts 217; ductile collapse 327; earthquake type 19-20; methodologies 3, 20; residual resistance 327, **327**; robustness 327-8; structural response 20; tasks 18-20; *vs.* Structural Engineering 15-20
Earthquake Investigation Committee (Japan) 461
earthquake loads 454; design philosophy **19**; *vs.* seismic design 17; *vs.* wind loads 15-16, **17**
earthquake monitoring 2; regional seismic network 2, 12; seismic stations 2
earthquake prediction 133-6; animal behaviour 135; early warning 135-6; intermediate term 134; long-term 134; short-term 135; Time of Increased Probability (TIP) 134-5
Earthquake Research Institute (Tokyo University) 42
earthquake resisting structures 98-9
earthquakes: 20th Century 38-58, *39*; 21st Century 59-66; anatomy 179-82; Asia region **35**; characteristics *473*; deaths ix, **10**; demand x; depths 180; devastating 90; early period (1910-1950) 73; effects 3; exceptional *see* exceptional earthquakes; first record 102; frequency since Yr-1900 *34*; generation 12; hazard 75-81; historical 35-8, *36-7*; historical period (until 1910) 72-3; matured period (1950-1975) 73; moderate 34; modern period (after 1975) 73; most damaging (1990 to 2009) *36*; near-source 494; observations of after damage 217; occurring 117; phenomenon 179-80; recent and important 74; recur 75; reducing future losses 97; severe 32; small 224; strong 34, **224**; tectonic borders **33**; types **76**, 81, 468, 521-2; world's deadliest 35; worldwide 1, **34**; Yr-2000-2006 *59*
Eastern North America (seismicity) **166**, 198, **199**
Eastern Pacific plate boundary 171-3; Californian segment 172; Canadian segment 171; Central American segment 172-3; Mexican segment 172; South American segment 173
Eastern USA: design codes 460
economic losses 1, 6, **10**
El Centro earthquake (1940) 458; record 137
elastic response spectra **474**, 496; Australia 496; duration 512-13; Europe 496; US 496-7, **497**
Elnashai, A.S.: and Papazoglou, A.J. 488
Elsessner, E. 95-6
Engelados (Greek myth) 102, **103**
engineering aspects of earthquakes 80-1; ground motions 80; near-field sites 81; recurrence periods 81; rupture areas 80
engineering bedrock **238**
Engineering Seismology x-xi, 11, 13-15, 38; 3-D computer simulations 264-6; definition 14; scientific activities 28; tasks 14-15; *vs.* Seismology 12-15, 27-8
epicenters 33; North Antolian Fault **78**; source-site distance 231-2, **232**

Epicur 104
EPS-50: tensioned crust **166**
Erdik, M. 232
Esteva, L. ix
Eurocode-8 465-6, 471
Eurocodes 456, 464
European: earthquakes **276**, **277**, 282; elastic response spectra 496; intraplate crustal earthquakes 275; intraslab earthquakes **207**; seismic code *see* seismic design codes (Europe); seismic zones 182, **182**, 198-9, 201, **202-3**, 211
European Association for Earthquake Engineering (EASS) 28-9
European Committee for Standardization (CEN) 464
European Seismological Commission (ECS) 28
European-Mediterranean Seismological Centre (EMSC) 33
exceptional earthquakes 82, 97; design code 82

failure modes (shallow foundations) 302-3, **302**; horizontal sliding failure 302; overturning 303; shear failure 303
faults 157; area 224; asperity 221-2, **222**; barriers 222; lines 14; movements 76; networks 145, 218; normal 165, **167**; rift 165, **167**; thrust 164, **165**; types (diffuse zones) 155-70
Federal Emergency Management Agency (FEMA) 458; building contents (overturning) **89**; Iran Bam earthquake (2003) **61**; San Francisco's fault system **78**; seismic waves **229**; Shinh-kang Dam collapse (Taiwan earthquake) **91**; soil-structure system **361**
Fiber Reinforced Polimers (FRP) 328
financial losses 71, **72**
fires 90, 376-7; San Francisco earthquake (1906) **93**

First International Workshop on Rotational Seismology and Engineering Application 246
Foundation Engineering 298
foundations (structural) 298-306; bearing capacity method 302; displacements 300, **301**; impedance method 300; loading 299; mat 303-4, **304**; pile *see* pile foundations; seismic behaviour 299; slender buildings 303; spread shallow 300-3, **301**
France: earthquakes **200**; seismicity 198

Gandwanaland 108
Gazetas, G. 359
GBJ11-89 (Chinese design code) 463-4
GBJ11-2001 (Chinese design code) 469
geotechnical aspects of earthquakes 79-80; high density of built-space 79; soft soil conditions 79-80; travel path 79
geotechnical engineer 21
geotechnical local conditions 14-15
Germany: seismic zone 275
Geurrero gap 221
Ghersi, A.: *et al* 364
Gioncu, V. 116; Banat (Romania) earthquake (1991) **279**; building robustness 333; Californian earthquakes **188**; collision earthquakes **193**; design codes 454, 467; equilibrium **118**; European seismic zones **182**, **203**; evolution in cascade **121**; interplate crustal earthquakes **183**; and Mateescu, G. 488; and Mazzolani, F.M. 72; plastic hinges **501**; return periods (inter/intraplate earthquakes) **281**; Scandinavian zone **203**; soil model **238**
Global Positioning System (GPS) 2, 112-14, **113**
Greek philosophers 102, 104; Aristotle 104; Epicur 104; Pythagoras 104; Thales 104

Gross Domestic Product (GDP) 7-9; developing countries 9; developed countries 8-9; seismic losses **10**; transition countries 8-9

ground motions 15-16, 80, 179, 217-85; accelerations 246; attenuations 233, **234**; definition 217; earthquake duration 233, **234-5**; path effects *see* path effects; prediction 454; recorded 129; rotational component 246; six-degree-of-freedom 245-6; transational 246; travelled soil types 232-3; vertical *vs.* horizontal 492; world's largest 42

ground motions (crustal intraplate) 275-83; crustal zones 278; low-moderate 277-8; rifts 278

ground motions (factors influencing recordings) 219-20, **219**; site conditions 220; source 219; travelled path 220

ground motions (intraslab) 283-5; deep 284; intermediate 283; recorded 284; shallow 283

ground motions (near-source) 249-75; amplification 253, **253**; blind thrust faults 266-7, **267**; Earth block 266, 271; normal *vs.* parallel 269, **270**; Northridge earthquake 249; peak ground velocity 252-3, 255; pulses 260, **261**; rupture 267, **273**; rupture directivity 250-3; simulation 266; strike-slip faults 270, **272**; strong-motion recordings 254; structure response 476-80; velocity pulse period 259, **259**; vertical components 254, **254-5**

ground motions (spatial variation) 247-8, **248**; attenuation effect 248; ground velocity **258**; incoherence effect 248; site-response effect 248; wave-passage effect 248

ground motions (structures) 295-377; building's weight 296, **296**; city-soil effect 296; foundations *see* foundations (structural)

ground response (to earthquakes) 236-49; local horizontal layered deposits 236; local site conditions 236

Gujarat (India) earthquake (2001) 59

Haicheng-China earthquake (1975) 133

Haiti earthquake (2010) xii

Hellenic Arc (Alpine-Himalayan belt) 174, 207

Hess and Dietz 109

Himalayan Mountains 160, **161**, 193; Tibetan Plateau 160, **161**, 193, 463

Himalayan segment (Alpine-Himalayan belt) 174

Hispaniola island xii

Hotel de Carlo (collapse due to pounding) **85**

Housner, G.W. 40

Iceland Island: Mid-Atlantic Ridge 175, **176**

Idriss, I.M. **240**

Imperial Valley earthquake (1979) 255, **257**

INCERC Bucuresti **132**

India-Eurasia plates collision **152**

Indian Ocean earthquake (2004) 62; damage 622; deaths 62; epicenter 62, **62**; tsunami 62, **63**

Indonesian segment (Alpine-Himalayan belt) 175

inslab earthquakes *see* intraslab earthquakes

International Association for Earthquake Engineering (IAEE) 28

International Association of Seismology (IAS) 27

International Association of Seismology and Physics of the Earth's Interior (IASPEI) 27-8; scientific activities 27

International Building Codes (IBC) 458

International Code Council (ICC) 458

International Geoscience Programme (IGCP): Ural Mountains **196**

International Organization for Standardization (ISO) 466; ISO-3010 466; Technical Committees (TCs) 466

interplate crustal earthquakes 33, 76-7, 79, 81, 181, **181**, 183-4, **183**, 474; crustal subduction 183-4; near-source effects 475; strike-slip 187-9, **187**, 475; subduction 474

interplate *vs.* intraplate earthquakes 275, 277-8, 280-3, **280-1**

intraplate crustal earthquakes 33, 76-7, 79, **181**, 182, 190-204, 475, 494; collision 475, 494; crust fracture 475, 494; Europe 275; stable continental interiors 191; stable continental regions 275, 282; strongest in North America 38; types 190-1

intraplate *vs.* interplate earthquakes 275, 277-8, 280-3, **280-1**

intraslab earthquakes 76, **181**, 182, 205-9, **206**, 476; deep 205; dehydration 205; duration 513; Europe **207**; intermediate 205; seismic zones 207; shallow 205; structure response 514; temperature **206**

Iran-Bam earthquake (2003) 9, 60; Bam citadel 60, **61**; deaths 60; epicenter **60**; tourism 60

Irpina (Italy) earthquake (1980) 241-2, **241**

ISO-3010 466

Italian: collision earthquakes 162; design code 465; earthquake in Abruzzo region (2009) xi-xii; earthquakes **194-5**; old cities **93**; peninsula 162

Iwan, W.D. 312

Japan: seismicity 211; tectonic plates **43**; Tokyo Institute of Technology (bracing system) **24**; Yokohama Institute of Technology (bracing system) **25**

Japan Kanto (Tokyo) earthquake (1923) 40, 42, **43**; fire **43**; tsunami 42

Japanese earthquakes 461

Japanese segment (Western Pacific plate boundary) 173

Japanese seismic code *see* seismic design codes (Japan)

Joshua Tree record 251

Kashima (Japanese myth) 102, **103**

Kashmir earthquake (2005) 63

Keilis-Borok, V.I.: *et al* 134-5

Keshishian, P.: and Der Kiureghian, A. 247-8

kinematic theorem of plastic collapse 320

kinetic moments 305

Kirby's Theory 205

Klimis, N.S.: and Anastasiadis, A.J.. 241-2

Knopoff and Burridge 128

Kobe Hanshin Expressway (collapse) **94, 96**

Kobe (Japan) earthquake (1995) 1-2, 54, 96, 179; amplifications of ground motions 80; collapse of City Hall **55**; collapse of concrete building **55**; damage 54; design codes 462; epicenter 54, **54**; failure of connections **86**; forward directivity **225**; pulse types 226, **226**; pulses 483; records from ground motions 264-5; undamaged steel structure **88**

Kramer, S.L.: and Stewart, J.P. **363**

Krawinkler, H.: and Alavi, B. 478-9, 483-4

Kuramoto, H. 462

Lai, M.: and Li, Y. 463

Lake St. Francis 38

Landers (US) earthquake (1992) 251, **251**

Laplace 116

L'Aquila (Italy) earthquake 65; church damage 66; epicenter **66**; Sant Agostino church (collapse) **92**

L'Aquila (Italy) xi

Laurasia 108

Le Pichon, X. 109

learning from earthquakes 71-99
Leibnitz's aphorism 116
Lemery 105
Li, Y.: and Lai, M. 463
liquefaction 243-5, **244**, **245**; effect
  245; large blocks 244
Lisbon Portugal earthquake (1755) 37-
  8, **37**, 452; tsunami 37-8
lithosphere 143; block structure 123
Load-Unload Response Ratio (LURR)
  parameter 135
loads: static *vs.* dynamic 300
Logomen Shan Fault 63
Loma Prieta Cypress Viaduct (collapse)
  94
Loma Prieta earthquake (1989) 48, 255;
  deaths 48; epicenter 48, **51**; San
  Francisco-Oakland Bay Bridge
  damage **52**; velocity 255
Lorenz: E. 124-5
Los Angeles: San Andreas Fault 157
Lower Rhine Embayment 199
Lucerne record 251

magnitude 220; large 280; moderate
  179-80; radiated energy 223; scale
  223; strong 179
mantle convection 119
Mateescu, G.: and Gioncu, V. 488
Mazzolani, F.M. *34*; braced frames 348;
  building robustness 333; *et al* 349-50;
  exceptional earthquakes 97; and
  Gioncu, V. 72; Sant Agostino church
  (collapse) L'Aquila **92**; structural
  systems **341**
MCEER: seismic waves (building
  structure) **295**
Meditteranean seismic region 464-5
mega cities 4, 6, 9
Messina earthquake (1908) 40; tsunami
  40
Mexican segment (Eastern Pacific plate
  boundary) 172
Mexico City 136
Mexico City earthquake (1985) 47;
  amplifications of ground motions **80**;

building damage 47; epicenter 47, **50**;
  Pino Suarez building 84, **86-7**, 87; RC
  building collapse **50**; Steel Pino
  Suarez Building collapse **51**
Mid-Atlantic Ridge 147, 175, **175**;
  Iceland Island 175, **176**
Mitchell (English geologist and
  astronomer) 105
Miyatake, T. 266
modern buildings 88
moment resisting frames (MRF) 342-3,
  343; dog-bone concept 343-4, **345**;
  welded connection 343, **344**
Morgan, J. 109
Theory of Multi-Source Data Fusion
  137
mythological explanations 102

NASA: GPS **113**
National Earthquake Hazard Reduction
  Program (NEHRP) **338**, 458; mission
  459-60; recommended provisions
  (2003) 460
Nelson, S.A.: major tectonic plates **145**
New Madrid (Tennessee) earthquakes
  (1811-1812) 198, 457
New York, Hearst Tower (bracing
  system) 24
New Zealand: Alpide (Alpine) fault
  189, **190**; segment (Western Pacific
  plate boundary) 173
Newton: determinism 116
Newtonian Laws of Motion 115-16
NGDC: Alaskan earthquake (1964) **45-
  6**; Chi-Chi earthquake (1999) **58**;
  Chile earthquake (1960) **44**; Hotel de
  Carlo (collapse due to pounding) **85**;
  Lisbon Portugal earthquake (1755)
  37; Pino Suarez building (Mexico
  City) **86**
Nice (France): city-site effects 297,
  297-8
Niigata earthquake (1964) 243; soil
  liquefaction **79**
NISEE 367

NOAA: Indian Ocean earthquake (2004) **63**

North American Plate: subduction process **158**

North American seismic zones 209, **210**

North Antolian fault 54, **56**, 152, 187-8, **189**; epicenters **78**

Northridge (US) earthquake (1994) 1-2, 48; collapsed freeway 53; collapsed garage **53**; design codes 459; destruction 264; fault system **52**; forward directivity 225; ground motions (near-source) 249; pulse types 226, **226**; pulses 483; records **250**, 264-5; structural engineering 48; velocities 474-5

NYSED: plate boundaries **146**

ocean floor 152

Theory of Ocean Ground Expansion 109

Oceanography: continental plates (collision) **160**

Ohi, K.: *et al* 346

Oldham, R.D. 228

Orogeny 160, 192

Ortelius, A. 105

Otani, S. 462

Pacific Plate: subduction **186**

Pangaea 107, 110, 191

Panthalassa 107

Paolucci, R.: and Rimoldi, A. 242

Papazachos, B.C. 102

Papazoglou, A.J.: and Elnashai, A.S. 488

Pappin, J.W. 302-3

Parra-Montesinos, G.J. 337

path effects 227-36; crustal sources 230-3; deep sources 236

Pecker, A. 299

Pellegrini, A. 105

Peru Ancash earthquake (1970) 45

Peru Atico earthquake (2001) 59-60

Peru government 46

Petroski, H. 451-2

Philippine and Guinea segments (Western Pacific plate boundary) 173

physics 136

pile foundations 304-6, **306**; collapse mechanism 306, **307**; damage 305; kinetic moments 305; layer discontinuity effect 305; liquefaction effect 305; seismic waves 305; soil 305

Pino Suarez building collapse (Mexico City) 84, **86-7**, 87, 332

plastic buckling 514, 518, **518**

plastic hinges (structural design) 319-20, 322-4, **322**, **323**, **485**, **501**, **517**; cross-sectional classes 322; cyclic loads **325-6**; DUCTROT M 324; member behaviour classes 322; rotation **322**, **324-5**

plate boundaries 3, 144-5, **146**, 171, 180; continental earthquakes 171; convergent *see* convergent plate boundaries; divergent *see* divergent plate boundaries; Eastern Pacific *see* Eastern Pacific plate boundary; Northern Pacific 173; Southern Pacific 173; subduction 3; transform *see* transform plate boundaries; Western Pacific *see* Western Pacific plate boundary; zones 157

Plate Tectonic Theory 110, 118-19, 143, 191, 468; Super-Continents Cycle 110, *see also* tectonic plates

Port-au-Prince (Haiti) xii

post-earthquake investigations 72-3; damaged buildings 73; damaging earthquakes 73; multi-disciplinary 73

pounding damage 83; collapse of Hotel de Carlo **85**

practical point of view x

prediction of future earthquakes *see* earthquake prediction

Prigogine, I. 116-17

Principia (Newton) 115

probabilistic approach 136

professionals: code of conduct 26-7; seismic design 21, **22** pulse types (rupture) 226; pulse period 226, **227**

pulses **502**, **503**, **504**

Pythagoras 104

Quakes **204**

radiated energy: magnitude 223

Rassem, M.: *et al* 242

records: different sites 218; epicentral region 218; Joshua Tree 251; Lucerne 251; network station 218

regional seismic network 2

Reid 105-6

reinforced concrete (RC) buildings 48, 488; Chi-Chi earthquake (1999) **58**; Mexico City earthquake (1985) **50**; Romania Vrancea earthquake (1977) **49**; Turkey Izmit (Kokaeli) earthquake (1999) **56-7**

Renaissance (scientific development) 104-5; Agricola 104-5; Cardan 105

Rhine Valley rift system **168**, 199, 201, **202**

Rhodes earthquake (226 BC) 35

Richter scale 223; great earthquakes 224; large earthquakes 224; low earthquakes 223; micro earthquakes 223; moderate earthquakes 223

Rimoldi, A.: and Paolucci, R. 242

Ring of Fire *see* Circum-Pacific Ring

Roesset, J.M. 73-4

Romania: Timisoara City map **77**

Romania Vrancea earthquake (1977) 47, **49**; RC building collapse **49**

Romanian seismic code 498

Rome earthquake (1349) 35

rupture: duration 226; forward directivity 225; observations 225; pulse types *see* pulse types (rupture); space-time history 225

San Andreas Fault 77, 152; active faults **154**; aerial view **154**; Los Angeles 157

San Fernando veteran hospital (collapse) **92**

San Francisco: ALCOA Building (bracing system) **23**; fault system **78**; Oakland Bay Bridge damage **52**

San Francisco earthquake (1906) 40, 95; collapse of City Hall **41**; epicenter 40, **41**; fires **93**, 440; horses killed **95**; rebuiling 457-8; surface rupture **81**

Sant Agostino church (collapse) L'Aquila **92**

Saragoni, R.G.: *et al* 285

Sato, T.: *et al* 463

Scandinavian zone: earthquakes **203**; seismicity 201

Schettino, A. **169**

scientific point of view ix

SEAOC Seismology Committee 459

seismic: separations **370**; stations 2; structural design 359; vulnerability factors 366-77

seismic actions **524**

seismic area 6, 14, 456; Central Europe 465; Mediterranean 464; Northern European 465; urbanization **7-8**, 469, 519

seismic demand 454; structural capacity balance 467-8

seismic design 11, **11**, 20-6; architect 21, 23; bases of philosophy 16-20; bracing systems *see* bracing systems; design values 17-18; ductility 326-7; ethical dilemma 26-7; fire 377; free-field ground motions 360; human losses 74; improve structures 3; key milestones 95-6; lessons 95-9; modern 71; new aseismic systems 23; new challenges 1-29; professionals 21, **22**, 26-7; proper 82; rocking component 360; strategies 21-2; structural conception 23; translation 360; *vs.* earthquake loads 17-18

seismic design codes 3-4, 18, 22, 26; ductility 18, **19**; exceptional earthquakes 82; first code 456; mission 26-7; progress 456-7;

provisions 82, 94; reduction factor 18, **19**

seismic design codes (China) 463-4; code evolution 463; first code 463; GBJ11-89 463-4; GBJ11-2001 469; seismic intensity 464; South China block 463

seismic design codes (developments) 451-3; Code of Hammurabi 451; early prescriptive codes 452; international codes 453, 466; modern period 452-3; period of intuitive design 451; period of observations 452; period of personal computers 453

seismic design codes (Europe) 464-5; CEN 464; code evolution 465; EC-8 465-6, 471; Eurocodes 456, 464; Italian 465

seismic design codes (Japan) 461-3; AIJ 461; code evolution 461; Earthquake Investigation Committee 461; Urban Building/Planning Law 461

seismic design codes (next generation) 451-527; damageability limit state 520; implementation 455; needs and challenges 470-1; new codes 520; performance-based design 520-1; serviceability limit state 520; ultimate limit state 521

seismic design codes (USA) 457-61; American Institute of Steel Code (1990) 458; BOCA 458; Eastern USA 460; main problem 460; system 457; Uniform Building Code (UBC) 457, 458-9, 470-1

seismic design (factors to consider) 471-2; earthquake magnitude 472; ground motion duration 472; near-source effects 472; pulses 472; soil conditions 472; spectrum type 472

seismic design spectra **523**

seismic hazard 74-5

seismic movements: Earth 32

seismic region *see* seismic area

seismic risk: definition 74; losses 74-5; vulnerability 74-5

seismic risk mitigation 73-5, 89-95; building lessons 94-5; negative attitude of engineers 74; urban lessons 89-90

seismic waves 227-8; accelerogram **230**, **231**; amplification 241; body waves 228; building structure 295, **295**; crustal sources 230-1; L waves 228, 230; P waves 228, 231; R waves 228, 230; S waves 228, 231; surface waves 228; types 228, **229**

seismic zones 155, 182, **212**; African 211; Australia area 204, **204**, 211; Californian 157, **159**; Central America 209-10; Charleston zone 198; China territory 196, **197**, 198; diffuse boundary **156**; Eastern North America 198, **199**; European territory 182, **182**, 198-9, 201, **202-3**, 211; extension **158**; Germany 275; intraslab earthquakes 207; Japanese area 211; North American 209, **210**; Scandinavian zone 201, **203**; South America 210, **210**; subduction 184; Vrancea zone 207, **208**, 209, 211

seismic-resistant design 367

seismicity 220

Seismological Society of America 40

seismologists 2, 12, 21; *vs.* structural engineers 12-13

Seismology xi 12, 38, 112-20; advances 468; definition 13; revolution 112; source characterization 13, 14; tasks 13-14; *vs.* Engineering Seismology 12-15, 27-8

seismoscope (ancient Chinese) **104**

Semblat, J.F.: *et al* 296

SHAKE (computer program) 238

Shansi (China) earthquake (1556) 6

Shinh-kang Dam collapse (Taiwan earthquake) 89, **91**

Shock Dynamic Theory 109

Sichuan (China) earthquake (2008) 63; damaged buildings **64-5**; epicenter **64**

soil: conditions 79-80; model **238**; type 233, 236-41, 240-1, 299
soil classifications 239
Soil Dynamics 299
Somerville, P. 261
South America and Africa jigsaw **106**
South American Plate: block structure **122**; subduction **185**, 191, **192**
South American segment (Eastern Pacific plate boundary) 173
South American seismic zones 210, **210**
South China block 463
Statistical Seismology 136-7
steel structures 84, 87, 98
Stratan, A.: and Dubina, D. 349
strike-slip (crustal) earthquakes 187-9, **187**
strike-slip *vs.* thrust earthquakes 273-5, **274**
strong earthquakes: main lessons 71-5
structural analysis 74, 97
structural capacity balance: seismic demand 467-8
structural design 12, 90, 218; static loads 16
structural designers 298
Structural Engineering 12, 18; Northridge (US) earthquake (1994) 48; *vs.* Earthquake Engineering 15-20
structural engineers 12, 21, 368, 455; *vs.* seismologists 12-13
Structural Engineers of California (SEAOC) 458; ATC 458; Blue Book 458; SEAOC Seismology Committee 459
structural failure 71, 84
structural material 328-9, **328-9**; configuration 332-3; FRP 328; reinforced concrete 329, **330**; steel fracture joint **332**; steel members 331-3, **331**; unreinforced masonry 328
structural material (seismic-resistance) 334; concrete 336, **337**; concrete (fibre reinforced) 336; concrete (high-performance) 336; masonry 334; masonry (confined) 334, **335**;

masonry (reinforced grouted cavity) 334, **335**; pseudo-elasticity 339; shape-memory alloys 339-40; steel 339, **340**; steel (high strength) 339; steel (low yield) 339; tensile behaviour 337; well-confined columns 336, **338**
structural systems 341, **341**; active control systems 358, **358**; active mass damper system 359; base-isolated systems 355, **355**; bearings 356; bending moment diagrams **366**; bolted connections 346, **346-7**; direct approach 362, 364; dynamic analysis 364; energy dissipation systems 356, **357**; finite elements (FEM) 364; foundation deformation 361; foundation stiffness and damping 361; hybrid system 359; kinematic interaction 361; moment resisting frames (MRF) 342; pile foundations 364; response control systems 342, **342**; shallow foundations 362; soil 361; soil-foundation-structure interaction 359-64, **365**; standard procedures **360**; substructure approach 362, **363**; turned mass damper system 359
structural systems (steel) 341-59
structure: dual 83; well-conformed 83
structure behaviour (ground motions) 306-7, **308**; Alternative Path approach 333; capacity curve **315**; collapse mechanism 318-21, **319**; damping 307; ductility 314, 320-2; forced oscillations 306-7; free oscillations 307; interstory drift 316, **317**; linear elastic range 316; multiple resonance 309; non-structural elements **318**; redundancy 316, **320**; residual resistance **333-4**; resonance 309, **311**; robustness 315, 326; seismic loading 313; stiffness 314, 316, **317**; strength 314; time variability 306; velocity pulse 321; vibration modes 307, **309**

structure response: artificially generated pulses 502-13; ductility checking 493, **495**; ductility demands 480, **480-1**, **486**, **487**; framed structures 478-9; ground motions (near-source) 476-80; intraslab deep earthquakes 514; low-moderate earthquakes 498, 512; pulses **482**, **493**, **510**; stroy drift 477-8, **477**, **479**; vertical components 488, **490**, **492**

Subba Rao, K.S.: and Choudhary, D. 303

subduction 3, 157; American continents 184; continental collision 192-8; interplate crustal earthquakes 474; North American Plate **158**; Pacific Plate **186**; seismic zones 184; South American plate **185**, 191, **192**; zone 149, 184, **185**, 191

Sumatra earthquake (2004) 129; tsunami 129, **130**

super-mega cities 6

supercomputer simulations 264-5; processors of Caltech Center 265; wave propagation **265**

surface fault rupture 220-1, **221**; blind fault 220

Taiwanese segment (Western Pacific plate boundary) 173

Tangshan (China) 46

Tangshan (China) earthquake (1976) 1, 46-7, **47**, 77, 133; deaths 46-7; devastation **48**

tectonic borders: earthquakes **33**

tectonic plates 33, 109, 118-19, 124, 143-76, 191; ancient 144; boundaries *see* plate boundaries; Japan **43**; major 143, **145**; micro 144; minor 144; motions 144; slow-moving 144

thermo-dynamic process 180

thrust earthquakes: velocity 272

thrust *vs.* strike-slip earthquakes 273-5, **274**

Tianjin (China) 46

Tibetan Plateau 160, **161**, 193; Himalayan Mountains 160, **161**, 193, 463; seismicity **193**

Timisoara City (Romania) 77; numerical tests 498

transform plate boundaries 145, 152, **154**, 155; Mid-Atlantic **153**; oceanic **153**; zigzag **153**

transition countries: GDP 8-9

Trifunac, M.D.: and Brandy, A.G. **234-5**

tsunami 184; Chile earthquake (1960) 42; Great Alaskan earthquake (1964) 45; Indian Ocean earthquake (2004) 62, **63**; Japan Kanto (Tokyo) earthquake (1923) 42; Lisbon Portugal earthquake (1755) 37-8; Messina earthquake (1908) 40; Sumatra earthquake (2004) 129, **130**

Turkey: buildings 57

Turkey Izmit (Kocaeli) earthquake (1999) 54, 84, **85**; collapse of RC buildings **56-7**; epicenter **56**

Uniform Building Code (UBC) USA 372, 457-9, 470-1

United Kingdom (UK): earthquakes **200**; seismicity 198

United States of America (USA): El-Centro earthquake (1940) 42; elastic response spectra 496-7, **497**; Northridge earthquake (1994) 179; seismic codes *see* seismic design codes (USA)

Ural Mountains 196, **196**

Urban Building/Planning Law (Japan) 461

urbanization 2, 4-11, 71, 89-90; developed countries 4, **5**; developing countries 4, **5**; seismic areas **7-8**, 469, 519

US Geological Survey *see* USGS

USGS 33; Alpide (Alpine)-Himalayan belt **174**; Asia region **35**; Californian seismic zone **159**; Central Mississippi Valley **170**; Chi-Chi earthquake

(1999) **58**; China Tangshan earthquake (1976) **47-8**; Circum-Pacific Ring **172**; continental lithosphere **107**; convection currents **32**; convergent plate boundaries **150-2**; divergent plate boundaries **148**; GPS **113**; India-Eurasia plates collision **152**, **161**; intraslab earthquakes **206**; Iran Bam earthquake (2003) **60**; Japan Kanto (Tokyo) earthquake (1923) **43**; Kocaeli (Turkey) earthquake **85**; Loma Prieta Cypress Viaduct (collapse) **94**; Loma Prieta earthquake (1989) **51-2**; major tectonic plates **145**; Mid-Atlantic Ridge **175**, **176**; Niigata earthquake **79**; North Antolian Fault (epicenters) **78**; Northridge earthquake (1994) **52-3**; plate boundaries **146**; return periods (inter/intraplate earthquakes) **281**; San Andreas Fault **154**; San Fernando veteran hospital (collapse) **92**; San Francisco earthquake (1906) **41**, **81**; San Francisco earthquake (1906) fires **93**; San Francisco earthquake (1906) horses killed **95**; seismoscope (ancient Chinese) **104**; Sichuan (China) earthquake (2008) **64-5**; strike-slip (crustal) earthquakes **187**; subduction under North American Plate **158**; subduction zone **185**; Tibetan Plateau siesmicity **193**; transform plate boundaries **153**, **154**; Turkey Izmit (Kocaeli) earthquake (1999) **56-7**; Wegener's Earth **108**; worldwide earthquakes **34**, *36*
Uyeda, S.: Earth's cross-section **119**

Vasquez, J. **247**
velocity 253-9, **258**, **505**, **506-7**; different positions 271; dip angles 269, **271**; directivity effect 268; displacement 509; Earth block 269; moderate earthquakes 509; thrust earthquakes 272

velocity pulses: analytical models 262, **262**, 264; polygonial variation **263**; polynomial variation **263**; sinosoidal variation **263**
vibration modes 312, **527**
Vina del Mar (Chile) earthquake (1985) **247**
Vrancea (Romania) earthquake (1977) 82, 207, **208**, 209, 211, **515**; recorded accelerogram 310; subduction zone **123**, 131, **132**

water system 89
wave-propagation approach 312, **314**; single harmonically oscillating wave 312, **313**
Wegener, A. 106-10
Weisstein, E.W.: double pendulum behaviour **127**
Wenchuan earthquake *see* Sichuan (China) earthquake (2008)
Western Pacific plate boundary 173; Japanese segment 173; New Zealand segment 173; Philippine and Guinea segments 173; Taiwanese segment 173
Wilson, T. 110; Cycle Theory 110, **111**, 114
wind loads: *vs.* earthquake loads 15-16, **17**
World Conferences on Earthquake Engineering (WCEE) 28
world population growth 4, *5*, 6
www.nature.com **185**
www.wikipedia.com: earthquakes and tectonic borders 33; major tectonic plates **145**; Rhine's rift **168**

Yin, X.C.: *et al* 135
Youd, T.L. 245
Yushu-China (Qinghai province) earthquake (2010) xii-xiii

Zagros-Makran segment (Alpine-Himalayan belt) 174